CONTENTS

AN INTRODUCTION TO
ANIMAL BEHAVIOUR

Sixth Edition

Aubrey Manning
University of Edinburgh

Marian Stamp Dawkins
University of Oxford

CAMBRIDGE
UNIVERSITY PRESS

CAMBRIDGE
UNIVERSITY PRESS

University Printing House, Cambridge CB2 8BS, United Kingdom

Cambridge University Press is part of the University of Cambridge.

It furthers the University's mission by disseminating knowledge in the pursuit of education, learning and research at the highest international levels of excellence.

www.cambridge.org
Information on this title: www.cambridge.org/9781107000162

First published 2012
5th printing 2016

Printed in the United Kingdom by Bell and Bain Ltd, Glasgow

A catalogue record for this publication is available from the British Library

Library of Congress Cataloging-in-Publication Data

Manning, Aubrey.
  An introduction to animal behaviour / Aubrey Manning, Marian Stamp Dawkins. – 6th ed.
      p.   cm.
  Includes bibliographical references and index.
  ISBN 978-1-107-00016-2 (Hardback) – ISBN 978-0-521-16514-3 (Paperback)
  1. Animal behavior.   I. Dawkins, Marian Stamp.   II. Title.
  QL751.M22 2012
  591.5–dc23   2011027496

ISBN 978-1-107-00016-2 Hardback
ISBN 978-0-521-16514-3 Paperback

Additional resources for this publication at www.cambridge.org/9781107000162

PREFACE

It is 14 years since the last edition and the science or sciences of animal behaviour have progressed enormously. Topics which justified only a brief mention in an introductory text then, for example sperm competition as a factor in mating systems, have prospered to require textbooks of their own. How then to approach a new edition which cannot be allowed to become significantly larger.

It is our conviction that an introduction to the whole field, or at least a substantial part of it, remains as important as ever. Discussing new areas of research is best done from a firm basis of the basic concepts and for us these are still embodied in Niko Tinbergen's 1963 'Four Questions for Ethology' – function, evolution, causation and development. So the plan of our book remains essentially that of the other editions. There has been some extensive rewriting and we have tried to give good coverage to those areas where there have been important advances, notably in the evolution of behaviour and its development. We continue to give extensive references so that readers can easily get into the literature of areas that catch their interest. There is a great deal of new literature to be explored but we have never cited new work unless it really adds something. Often concepts are best illustrated by some of the now classical papers.

This edition benefits enormously from the encouragement we have had from Cambridge University Press to illustrate more widely and, in particular, to use colour photographs. Animals behaving provide stunning images and we are delighted with many of the new figures here. It remains to acknowledge the great help and support we have had from Martin Griffiths – it was he who first persuaded us to undertake a new edition. His team at CUP have helped us at every stage and it has been a pleasure to work with them.

Aubrey Manning
Marian Stamp Dawkins

Introduction

Wolves excitedly greet each other as members of the pack come together; a bumble bee uses its long tongue to reach the nectar at the base of a foxglove flower; a peregrine falcon stoops at high speed to strike a pigeon flying below; young cheetahs rest quietly together, very close to sleep (Fig. 1.1). The study of animal behaviour is about all these things and much more. It is about the chase of the hunter and the flight of the hunted. It is about the spinning of webs, the digging of burrows and the building of nests. It is about incubating eggs and suckling young. It is about the migration of a hundred thousand animals and the flick of a tail of one. It is about remaining motionless and concealed as well as about leaping and flying. Behaviour involves the static postures and active movements, all the noises and smells and the changes of colour and shape that characterize animal life.

Figure 1.1   Young cheetahs recline sleepily. Inactivity is as much a characteristic of behaviour as activity.

Animal behaviour is a very popular subject, not just with biologists but with the general public – so much so that it has even come to occupy a lot of prime time on television, the surest measure of real popularity! Since our earliest origins, human beings have always been fascinated by our fellow creatures. Apart from this intrinsic interest and the fact, which we hope to demonstrate, that the subject presents us with questions as challenging as any in science, the study of animal behaviour is also of great practical importance. The conservation of wild animals in their natural habitats and the welfare of those other species we have domesticated for our use are both topics which command a lot of public attention.

Experimental studies with animals are now controlled by law in many countries, and rightly so. In the past, we have gained valuable information about behaviour from experiments, for example deafening young birds or isolating a young monkey from its mother, which might now be deemed unacceptable. Nevertheless, we shall describe some such results as part of this book. Not to do so would be perverse, since they have advanced our understanding and by this understanding we are better able to design observations and experiments which are not invasive or cruel in any way. Animal behaviour workers are very conscious of their responsibilities in this regard and go to some lengths to minimize any kind of disturbance to the lives of their subjects. The Association for the Study of Animal Behaviour in the United Kingdom and the Animal Behavior Society of the USA collaborated to publish a collection of papers, *Ethics in Research on Animal Behaviour*, which considers such problems, not only for laboratory studies but also for fieldwork (ASAB/ABS, 2006). Even observing animals from a distance can disturb them and so needs to be done with great care. We are now able to learn a great deal about the behaviour of animals by tagging them for recognition or fitting them with tracking devices. Although the animals may then be released back into the wild, catching them may cause stress and the tracking or marking devices may even alter their behaviour. This means that the way in which we design any study of animal behaviour – experimental or observational – has moral, social and economic implications. The importance of animal welfare is being increasingly recognized around the world as more and more countries now have legislation, codes of practice or other guidance as to how animals should be treated. The study of animal behaviour has a particular contribution to make here, since understanding the behaviour of animals and how it has been evolved under natural selection can help us to improve welfare. Legislation will be effective only if it is based on good information about how animals live and how they respond to their environment.

For example, the captive breeding of cheetahs (*Acinonyx jubatus*) is likely to be one important tactic for the long-term survival of this wonderful animal. Zoos had kept them for many years but their breeding success was pitifully low until we obtained the results of fieldwork in Africa (see Caro, 1994). This showed that keeping males and females together – the standard zoo practice – was doomed to failure. Female cheetahs live and hunt alone, completely separately from males, except for the brief days of their oestrus. Once this aspect of the natural situation was reproduced in captivity, cheetahs proved quite easy to breed and several zoos – the cheetahs in Fig. 1.1 are at London's Whipsnade Wildlife Park – now have good numbers which could be the basis for re-introductions to the wild should this prove necessary.

We opened this chapter with a glimpse at the range of very diverse phenomena we are dealing with. Where in all this diversity do we start? What do people who study animal behaviour actually do and what do they want to find out?

There are two main approaches, the physiological and the 'whole animal'. Behavioural physiology is the study of how the body works, that is how the nerves, muscles and sense organs are coordinated to produce complex behaviour such as singing in a cricket or a bird. The 'whole animal' approach investigates the behaviour of the intact animal and the factors that affect it, for instance, what it is in the environment of the cricket or bird that prompts them to sing at a particular time or why they sing at all. 'Whole animal' questions of this latter type can be studied both by looking at wild animals in their natural environments and also by observing captive or domestic animals living under more controlled conditions: it depends on the exact question involved. Physiological investigations often require bringing animals into a laboratory environment because they will involve 'probing beneath the skin', as it were. For example, if we want to get at the mechanisms that give rise to the behaviour of singing, or those which organize an animal's responses to visual stimuli from predators.

In practice there is considerable overlap between the approaches. It is now possible to collect urine samples from animals in the wild – only tiny amounts are needed – and then back in the laboratory get an accurate measure of what hormones are circulating in the bloodstream. A few cells from the root of a recently shed hair or feather, or even those shed from the bowel with the droppings or shed from the inside of the mouth of a herbivore and left with its saliva on vegetation, can be cultured for DNA fingerprinting enabling relatedness within a social group to be worked out. The best understanding of behaviour nearly always comes when a combination of approaches is employed for, used well, physiological and 'whole animal' approaches complement each other – behavioural studies sparking off the search for physiological mechanisms and in turn putting physiological studies into a functional perspective. One of our aims in this book is to give some idea of the extent to which this is now possible.

Within the 'whole animal' approach, a distinction is often made between psychologists and ethologists, both of whom could be described as being interested in the behaviour of intact, functioning animals. Psychologists working with animals have traditionally been mainly interested in learning and have tended to work in laboratories on the learning abilities of a restricted range of species, often rats and pigeons. Ethologists have been more concerned with the naturally occurring, unlearnt behaviour of animals, often in their wild habitats. Although this distinction still exists to some extent, there is now a fruitful coming together of the two. Ethologists have become interested in the role of learning in the lives of wild animals and psychologists are beginning to ask evolutionary questions about the learning abilities of a much broader range of species and to study their responses to more natural stimuli. Another of our aims is to show how much psychologists and ethologists are increasingly learning from each other, to mutual advantage.

But nobody – physiologist, ethologist or psychologist – can rely solely on one source of information; they must approach problems at the level which is appropriate for them. Some physiologists like to emphasize that their methods are the more fundamental, and it is true that increasingly it becomes possible to explain certain aspects of behaviour in terms of the functioning of the basic units of the nervous system, the neurons. However, this is effectively a task without end and since the main function of the nervous system is to produce behaviour we must also investigate the end product in its own right. Even if we knew how every nerve cell operated in the performance of some pattern of behaviour, this would not remove the need for us to study it at a behavioural level also. Behaviour has its own organization and its own units which we must use for its study. Trying to describe the nest-building behaviour of a bird in terms of the actions of individual nerve cells would be like trying to read a page of a book with a high-powered microscope. Not only would it be incredibly laborious to discern the boundaries and make out the identity of each printed letter, we might miss out completely on the grouping of letters first into words, then sentences, then paragraphs and so on.

As we illustrated with our opening paragraph, behaviour includes all those processes by which an animal senses the external world and the internal state of its body and responds accordingly. Many such processes will take place 'inside' the nervous system and not be directly observable, although we are increasingly able to detect them indirectly through brain imaging or recording from single nerve cells. What we see from the outside may be an animal engaged in violent activity or one completely at rest, but both are behaving. The range of phenomena that are called behaviour immediately presents us with the problems of how to observe and how to measure it. In the physical sciences, and often elsewhere in biology, we have universally recognized units – molecules, milliamps, pH units, metres, etc. – for measuring and classifying our observations. When we watch animals we have no such framework. Behaviour is continuous for as long as life persists and so many different things may count. Put this way, the task would seem to be impossible and so to make it manageable, we have to abstract and simplify. In other words, we have to make decisions about what it is important to record and what can safely be ignored. Exactly how we do this may be different on different occasions or for different purposes.

For example, if we are studying the courtship behaviour of sticklebacks or pheasants we will probably decide that there is no need to record the number of times that the animals breathe. Breathing is part of that continuous activity which constitutes behaviour but it will often be judged irrelevant for behavioural studies of courtship. However, if we are watching the courtship of newts, the rules are changed. Male newts court females on the bottom of ponds or streams and their courtship is punctuated by trips to the surface to breathe. Halliday and Sweatman (1976) have found that males can postpone breathing up to a point if courtship is proceeding smoothly, but if there are delays or the female moves off they rush to catch up on their breathing. Thus records of breathing are an important part of any study of newt courtship patterns, unlike those of pheasants. The behavioural measures we use must be chosen to suit both the animal and the type of problem under investigation.

 ## Questions about animal behaviour

We can see, then, that animal behaviour presents us with problems of description and selection. There are other difficulties that arise from the very beauty and fascination of the subject which have made it so popular. Natural history TV programmes and books now offer films and photographs of such a high standard that we have become familiar with details of animal behaviour which used to be virtually impossible to observe. We can watch penguins fleeing from leopard seals by leaping up onto an Antarctic ice floe; we can watch in close-up the fantastic courtship displays of male birds of paradise; we can follow from start to frustrating finish the hunting chase of a cheetah who finally pulls down a gazelle only to be immediately driven off its prey by roving hyaenas. All this is splendid, but there can be a considerable temptation simply to gaze and wonder, rather than to think about analysing what remarkable things are happening.

We certainly must not lose our sense of wonder, but in order to use constructively the enthusiasm for further study which it gives us, we have to identify clearly what kinds of questions about behaviour need to be answered. In 1963, Niko Tinbergen, one of the founders of modern ethology, wrote a paper, as valuable today as when it was written, which he entitled, 'On the aims and methods of ethology'. There he suggested that there are four main types of question which we need to ask of an animal's behaviour. They are really four different ways of asking the question 'Why?' and we can get into a very unhelpful muddle unless we clearly understand the nature of the distinctions between them.

Firstly, Tinbergen asserts, we ask why the animal performs a piece of behaviour in terms of its **function**, i.e. how it helps to improve the survival of the animal or its success at reproducing itself. Secondly, we ask about the **evolution** of the behaviour, i.e. how it has changed over the course of evolutionary time, in just the same way that we might study how a whale's flipper or a bat's wing changed from the ancestral limb into what we see today. Thirdly, we have to deal with its **causation**, what factors, both internal and external, lead to the performance of that particular piece of behaviour at that particular moment. Lastly there is the question of the behaviour's **development**, how the behaviour of a young animal changes as it matures and what factors, again both internal and external, affect this process and its end point.

Function and evolution are obviously interconnected and it is by selection acting on the processes of development that behaviour becomes adapted to the environment in which it has to operate. **Adaptation** is a subject we shall examine more closely in Chapter 6; here we may note that it arises by selection between individuals over many generations following genetically based changes, or by modifying behaviour to achieve the best response during an individual's own lifetime. It will often be important to keep in mind the distinction between these two pathways to adaptiveness.

Tinbergen's ideas have been very influential amongst behaviour workers, so much so that 'The Four Questions' have entered into much of our thinking, certainly they have for the authors of this book. We want to integrate some of the diverse ways of studying behaviour, for full understanding can come only when we have answers to each of the questions.

Tinbergen directed his paper at ethologists but we should note that the same four questions can be asked no matter at what level of organization we are working. They are as relevant to studies of physiological mechanisms as to ecological ones.[1]

Of course, when one begins to examine particular cases it rapidly becomes obvious that function, evolution, causation and development are not isolated topics: they overlap extensively and studies directed at one topic will nearly always contribute to the others. Studies at different levels and on different animals will lead to emphasis

---

1 Confusingly, ethologists and physiologists use the words 'function' and 'adaptation' in rather different ways. For physiologists, 'function' often means the way an organ works – e.g. 'liver function is badly affected by too much alcohol'. They do not mean the contribution that one's liver makes to survival. Again, 'adaptation' to a physiologist often refers to sensory adaptation whereby a sense organ ceases responding to continuous or repeated stimuli, and has nothing to do with good fit to the environment.

being laid on this question or on that, but all will benefit from taking as broad an approach as possible. We can best illustrate this by taking a look at two very contrasted examples of well-studied behaviour patterns – the escape response of the cockroach and the courtship displays of male sage grouse.

The escaping cockroach

Anyone who has handled a cockroach will know that it behaves like greased lightning when an attempt is made to catch it. Toads, which are natural predators of cockroaches, seem to have almost as much difficulty as we do. Probably the first question that springs to mind is that of causation: we want to know how the cockroach manages to avoid the lightning strike of the toad's tongue. We want to explore both the mechanisms inside the cockroach and the stimuli impinging on it from the outside that enable it to detect that it is about to be attacked and to dash away so effectively.

The first clues about causal mechanisms come from watching the behaviour of freely moving cockroaches and toads, or rather analysing film of them because everything happens so quickly that the naked eye cannot follow it. Slowed-down film shows that when a toad is within striking distance then just before its tongue flips out of its mouth, the cockroach will turn rapidly and run. It seems, then, to have some way of detecting in advance that a toad is going to strike and from which direction before it actually does so. The cockroach always turns in the appropriate direction, away from the toad and about 16 milliseconds (ms) before its tongue becomes visible. At this point, the toad is apparently already committed to a particular direction and distance to its target. The cockroach's movement means that tongue may strike in the wrong place.

How does the cockroach predict that the toad is gearing up to strike? The most important cue seems to be tiny gusts of wind produced by the toad's preparatory movements. Camhi *et al.* (1978) showed that these slight air movements are picked up by the cockroach through many tiny wind-sensitive hairs on its cerci, paired appendages which protrude like little tails from the end of its abdomen (see Fig. 1.2). These cercal hairs are exquisitely sensitive and the cockroach will make a false run if a tiny gust of air is blown at a cercus. The smallest gust of air that generates escape behaviour is 12 mm/s with an acceleration of 600 mm/s^2. The hairs have to be physically moved, because if they are immobilized with glue, the cockroach is much less successful at escaping.

In behavioural experiments in which puffs of air were blown at cockroaches, it was shown that they started to show their escape behaviour just 44 ms after a puff started. Then, by measuring the wind generated by a toad when it strikes at prey, it was found that the critical (12 mm/s) gust of wind occurred on average 41 ms before the tongue emerged from the mouth – very close to the response latency of 44 ms shown by the cockroach to an artificial puff of wind. So it is the minute gust of wind generated by a

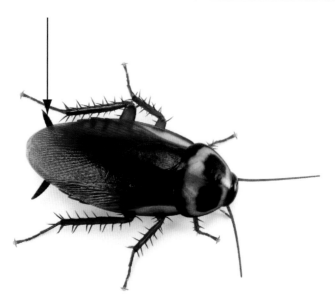

Figure 1.2   The cockroach *Periplaneta americana*. The cerci at the hind end are indicated by the arrow.

toad preparing to strike that sets off the escape turning behaviour. By waiting until the very last moment before turning, the cockroach evades the tongue that has already started to move and thus escapes because the toad cannot, at that late stage, change the direction of its strike.

So far, we can see that a great deal can be learned about the mechanism of the behaviour by studying the behaviour of whole animals – intact toads and intact cockroaches. Film records of strike and escape, experiments stimulating the sense organs of the cockroach and measurements of the wind stimulus produced by the toad give us a basic idea of the mechanism underlying escape behaviour at this level.

The next stage, and this is often the case with studies of causation, is to go 'inside the skin' and look at the same behaviour but at the physiological level. Close examination shows that there are about 220 hairs on each of a cockroach's cerci and that each hair is hinged so that it can be moved most easily in just two directions at 180° to one another. They move less easily in directions at 90°. Different hairs have different biases, which means that wind from certain directions moves some hairs and wind from other directions moves others. Thus the basis for the cockroach being able to discriminate wind direction so accurately appears to lie in the mechanical construction of the hairs. By recording from sensory nerve cells at the base of each wind-sensitive hair, Camhi *et al.* found that the nervous activity in these cells reflected the directionality of their hairs: the nerve cells responded much more to wind from some directions than from others.

Camhi *et al.* also found that there is a cluster of nerve cells, the terminal ganglion, at the hind end of the cockroach which contains a group of large cells known as Giant

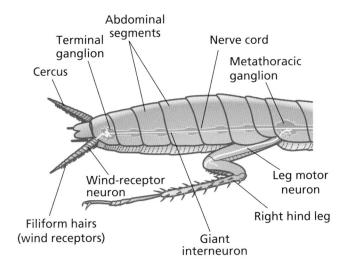

Figure 1.3   Hind end of a cockroach showing the cerci with filiform hairs, which are the wind receptors.

Interneurons (GIs). The GIs run up the nerve cord to the head, on the way passing through the thorax and linking to the motor nerves of the legs that do the rapid escape running. The GIs receive a pattern of information from the sensory nerves in the cerci which conveys the direction of the wind. They pass this on to the nerves which control the leg muscles: these nerves, in turn, command the legs to turn the cockroach and make it run in the opposite direction (Fig. 1.3).

When the insect is stimulated with a wind puff, the GIs become very active (as recorded by microelectrodes inserted into them) and so do the motor neurons to the muscles of the hind leg. It is possible to inactivate some of the GIs by injecting them with an enzyme, leaving other GIs intact. This has the effect of making the cockroach turn in the wrong direction in response to a puff of a wind. So it appears that some sort of comparison between activity in different GIs goes on in the normal cockroach, telling it which of its legs to move.

It is clear that for this piece of behaviour we have made very good progress, at both behavioural and physiological levels, towards understanding **causation**. For relatively few behaviours, particularly in vertebrates, do we understand the mechanisms in as much detail as this, right through from the detection of a stimulus to the command to the limbs. We can also give thoroughly satisfactory answers to the question of **function**, in the sense that we can readily understand how the escape behaviour aids the cockroach's survival. As we shall see, not all behaviour is so easy to interpret. However, note that the cockroach study does enable us to identify aspects of function in considerable detail. Whereas we might have assumed that it would be to the cockroach's advantage to run the instant it detected a toad – it might see its head turn for example – we can now understand the function of that 44 ms delay. It actually improves the insect's chances because it means that the toad's strike is now committed and if the cockroach can move a sufficient distance, the tongue is bound to miss – the toad cannot adjust. Many ethological studies of animals in their natural habitats have revealed just such detailed matching of behaviour to function.

What can we say about the related question, the **evolution** of this piece of behaviour? What were its origins and what sort of escape behaviour did ancestral cockroaches show? Since we have already seen that the function of the very rapid response is to evade predators, there must have been a kind of arms race between them and their prey, beginning with slower cockroaches and slower toads and other predators. Then, as now, both sides won some of the time, but some cockroaches would be a bit faster, perhaps because they had more sensitive cercal hairs which could detect smaller air currents. They would survive better and leave more offspring and sometimes their increased speed would have a genetic basis. When this was the case then their offspring would inherit their parents' extra speed. This would put selection pressure on the predators to become faster at grabbing cockroaches, which would in turn favour even faster cockroaches and so on.

This is a very general story suggesting how the skills of cockroaches became honed to a high degree over millions of generations. In this case it is difficult to be more precise because, for the most part, behaviour does not leave a fossil record and we do not have details of how the escape skills of modern cockroaches compare with those of their ancestors. However, we can deduce something of the course of escape behaviour's evolution by looking at the 'family tree' of such behaviour in living relatives.

Just as with bodily characteristics, so with behavioural ones we can get an idea of evolutionary history by comparing traits among groups of related species. If a wide range of species share a trait then we are fairly safe in assuming that it evolved before the different species diverged from a common ancestor. If a behavioural trait is shared by quite distant relatives then it pushes the time of its origin further back in time. On this sort of criterion we can be confident that the common ancestors of the whole ancient group of orthopteroid insects – the stick insects, crickets, grasshoppers and cockroaches – had escape responses organized very much along the lines just described. All of their descendants have cercus-like structures at their hind ends; all have giant fibres in their ventral nerve cords, often arranged in strikingly similar ways. Not all of them escape by running as does the cockroach; some jump, some may freeze into immobility to evade a predator's eyes, but the basic response and its mediation must have evolved very soon after the first insects appeared.

The last question we must address for the escape response concerns its **development**. As with questions about mechanism, questions about development can be asked at many different levels. Does a newly hatched cockroach escape as effectively from a toad as an adult or does it improve with practice? How do the connections between the sensory nerves and the GIs form so that escape behaviour can be accurate?

In a way, there is a simple and direct answer to all developmental questions: it is always true that an animal's genetic makeup and the environment in which it grows up both contribute to the final behaviour it exhibits. Neither genes nor environment on their own would produce a fully functional, behaving animal. But the way in which genetic factors and environmental ones interact may be very complex, and this contributes to the fascination of developmental studies. For example, a cockroach that has just hatched and has had no previous experience of wind turns away from a puff of air just as accurately as an adult cockroach (Dagan and Volman, 1982). From the very first puff, then, the escape behaviour is fully formed and very accurate even though the hatchling has only four sensory hairs on its cerci instead of an adult's 440. Nevertheless, this 'innate' behaviour can still be modified later in life. If the adult cockroach loses one of its cerci, its escape behaviour becomes at first very inaccurate but then gradually improves. Immediately after the loss, the cockroach erroneously turns towards a wind source instead of away from it. However, after 30 days, with or without practice, it improves markedly and turns consistently away from the wind source, despite still having only one cercus and no extra sensory hairs. Its nervous system has gradually changed its way of operating so that, by some mechanism not yet fully understood, the GIs on both sides of the body come to respond to sensory input from the one intact cercus and the cockroach is able to adapt to its changed sensory picture of the world and escape in the correct direction. So, even when a behaviour is fully functional at hatching, modification and improvement are still possible. The genes and the environment continue to interact throughout life.

The courtship of the sage grouse

Let us see how far we can get with the 'four questions' approach applied to a very different type of behaviour, the very remarkable and highly complex pattern of courtship displays performed by sage grouse *Centrocercus urophasianus*. These are large game birds of the American plains (Fig. 1.4 and front of chapter), which have been extensively studied by Wiley (1973a). For most of the breeding season the females are on their own. Once mated, it is they who nest, incubate eggs and rear the young, but during the early part of the spring, male grouse gather on communal display grounds or leks and the females visit these and select a male to mate with.

Obviously the courtship behaviour of the sage grouse takes us into realms far beyond the escape behaviour of the cockroach. To start with, we have to encompass the fact that several birds are interacting with each other and that what we call courtship is not a simple response over in a second or two – the cockroach turns and runs – sage grouse males may display for hours on end, even continuing through the night if there is enough moonlight! The females may not choose mates quickly, but visit the lek for several days in succession. Thus the four questions have really to be applied to the whole situation and it may be rather artificial and inadequate to apply them to individual elements of it.

Figure 1.4   Male sage grouse in full 'courtship strut' display on a lek or display ground. A female watches.

Nevertheless, we have to start somewhere and cannot be completely open-ended. We have to isolate something we identify as courtship and distinguish it from, say, feeding or parental behaviour. It helps to pick out behaviours which are to be seen only during courtship and not at other times. As far as the male sage grouse is concerned, the extraordinary display called the 'courtship strut', which is illustrated in Fig. 1.4, serves this purpose well for it is not seen outside the breeding season on the leks.

When we ask the four questions about the sage grouse's courtship, both the extent and the nature of the answers we can give are very different from those for the cockroach's escape response. In fact, they will take us far beyond the scope of this introductory chapter but they will give some idea of the problems and the intellectual challenge of behavioural studies.

Function provided few problems with the cockroach but is much more difficult here. What is the function of courtship behaviour? We can begin with some deceptively simple answers; it brings the sexes together, it stimulates the female and brings her into a receptive condition, for example. The trouble is that the sage grouse's performance, from both male and female, seems terribly wasteful. The males go on for hours, the females take ages to make a choice, some males are never chosen and all the time there may be predators around who may find courting birds are not as alert to danger as normal. Why don't the birds just pair off as they arrive at the lek; why are the females so choosy; why do males who are not being chosen go on courting? Again, we can ask why the males strut the way they do. What is there about this display which makes it functional, after all other game birds have different displays? Peacocks have huge flamboyant trains, cockerels (domestic fowl) and pheasants have neck ruffs which they spread, and so on. Why haven't they all converged on the same 'best design'? Each of these 'why' questions relates to function, but we cannot arrive at answers without considering the evolution of courtship as well.

Comparisons with related species, just as with the cockroach, enable us to suggest quite plausibly that the strut display evolved by the exaggeration and amplification of feather erection and postures which occur anyway when game birds are highly aroused. Perhaps the particular manner in which this occurred in grouse, as compared with peacocks, meant that natural selection favoured exaggerating air sacs in the former and tails in the latter. The protracted nature of the displays and the 'choosiness' of the females may help them to pick out the most dominant and persistent males as mates. We can be sure that this is absolutely crucial for them because they are going to get no further assistance from their mate; at least they must ensure that he is fit and likely to father strong progeny. Such explanations do not mean that the functions we ascribed earlier to courtship are wrong. It does bring the sexes together and so on, but evolutionary considerations certainly enable us to extend the concept of function and understand the remarkable and apparently maladaptive aspects of courtship.

We are unable to approach the question of causation of courtship in anything like the detail we could muster for the cockroach. No physiological work has been done on sage grouse, but we can be fairly confident that both males and females begin to congregate on the lek when their secretion of sex hormones rises in response to the lengthening days of early spring. Apart from this internal factor there will also be external factors that initiate courtship. Probably the sight and calling of other males is a powerful stimulus. Wiley's observation that males will continue courting all night if there is a full moon certainly suggests that the males need to be able to see each other. The arrival of females also arouses the males to great heights of new activity, as do the 'quacking' calls females make as they circle the lek before alighting.

The last question is development, which asks how courtship behaviour first emerges in young birds and if it changes as they age. The question is not a simple one and actually has several components. For example, do young males court as persistently as older ones? Is the form of their strut displays the same? Do females choose to mate with males of all age groups, and so on?

From Wiley's (1973a, b) studies it would seem that males begin to visit the leks during the first spring after they hatch. All birds have to mature rapidly in terms of body size and strength and, in addition to this, yearling sage grouse already have the full adult plumage. However, they appear to be incapable of displacing older males and setting up their territories for strutting close to the centre of the lek. This is the part that females head for and as a result there is no evidence that yearlings ever succeed in getting females to mate with them. Wiley believes that mortality is high among the males and young ones that survive gradually move their territories towards the centre of the lek as the old males die off. It may take them several years to do so and by choosing older

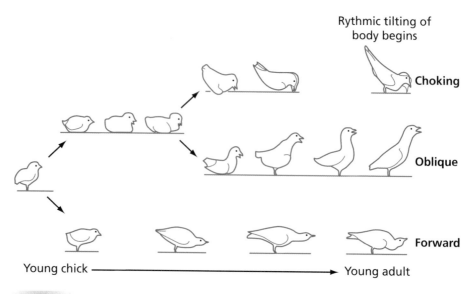

Figure 1.5   The manner in which postures and movements of young black-headed gull chicks (*Chroicocephalus ridibundus*) gradually become extended and stereotyped, developing into three very characteristic display movements of the adult gulls. We begin on the left with the more or less neutral posture of the young chick. As the chick matures, left to right, more discernible and regular postures begin to appear. These are (i) 'choking' (a display performed in the nest site; Figure 2.8, p. 45, shows the homologous display in the kittiwake); (ii) the 'oblique' threat posture and (iii) the 'forward' threat posture. Note how the same posture in the chicks – sitting with slightly extended wings – is a developmental stage common to both choking and oblique.

males the females, once again, will be ensuring that their mates are real survivors and accordingly fit. In Chapters 6 and 7 we look at the evidence for this and some of the other ideas that have been put forward to explain the sexual displays of males.

Going back to the details of the strutting behaviour itself, there is one as yet unexplained difference between the performance of yearling and adult males. Wiley (1973a) describes how young males strut with a quicker tempo, although they are just as stereotyped as mature males. For the moment, we know nothing more than this – it may or may not be a significant fact. It would be interesting to know how strutting begins in yearling birds performing the behaviour for the very first time; perhaps we could see early traces of the display, or elements of it, in much younger birds soon after fledging. The **causation** would not be the same, but the development of courtship might involve integrating various elements of other behaviour patterns to form a display which only comes fully together with sexual maturity. We have no evidence of this from sage grouse but not dissimilar developmental patterns are known for other birds, for example Groothuis' (1993) work with the threat and courtship displays of black-headed gulls (*Chroicocephalus ridibundus*) (Fig. 1.5).

So when we apply Tinbergen's four questions to the courtship of the sage grouse, not only are there many possible answers and few definite ones but, as we warned at the outset, we inevitably raise a whole host of important issues relating to the way behaviour is organized. We shall be tackling some of them in more detail as we go through this book. The big differences between our understanding of the cockroach and of the sage grouse result mainly from differences in complexity. It does not denigrate insects, which comprise the huge majority of all animals, to conclude that their behaviour is organized along simpler lines than that of birds. There are good adaptive reasons for this. Yet complex behaviour is bound to share many features in common with simple patterns. They are, after all, both products of nervous systems which, be they insect or avian, are constructed and operate along similar lines.

We mentioned earlier the problem of choosing appropriate **units** for behavioural studies and the sage grouse's courtship provides a good example of one approach to such choices. It is quite impossible to describe, even for a moment, every twitch of every muscle or every small movement of a part of the body of an animal. What we have to do is to look for patterns in what we observe. Instead of describing the behaviour of an animal in terms of every small movement we can see (head does x, left front foot does y, tail does z, then after 1.5 s, head does x1, left front foot y1 ..., etc.), we hope to detect a recognizable pattern. We might notice that the head, the left foot and the tail always move in a coordinated sequence. There would then be no need to describe over and over again the exact way in which they moved because, like a repeated motif on a wallpaper, there are sequences of movement sufficiently similar from one occasion to the next to be given names like 'head-up display', 'facing away' or 'licking'.

The one dramatic element of male sage grouse's courtship behaviour offers us the opportunity to isolate a clear behaviour pattern in this way. It is the display we have already referred to, called by Wiley (1973b) 'the courtship strut' (Fig. 1.4), in which the bird inflates a huge air sac under his neck and breast feathers and then suddenly deflates it with sharp snapping sounds that can be heard over a kilometre away.

A fairly full description (although not down to the muscle twitch level) is given by Wiley, thus:

The display begins from the strutting posture (tail fanned and cocked vertically, head raised, neck plumage erect) and consists in coordinated movements of the wings and the large oesophageal sac, together with associated movements of the legs, head and trunk. The oesophageal sac is inflated in the course of being twice lifted and dropped. Concurrently, the wings are extended forward and retracted twice. The culmination of the display follows immediately: a rapid compression and ballooning of the sac accompanied by a third excursion of the wings. Complex acoustic signals accompany this culminating action. These include two sharp snaps produced by the inflated air sac and an intervening whistle that rises and falls in pitch. The final 0.2 s second part sounds roughly like POINK. Preceding this sound are three low-pitched coos, probably produced by the syrinx, and two swishing sounds generated by the wings rubbing against the sides of the chest.

Now each male sage grouse will go through this whole astonishing sequence every 6–12 s when they are clustered together on their display grounds or leks and, as we have described, sometimes continue for hours at a time. The similarity of each sequence from male to male and from one performance to the next means that the strut display can easily be recognized as a 'behaviour pattern' or **unit** of behaviour characteristic of this species and different observers will know exactly what this term means. Using slowed-down films of sage grouse displaying, Wiley showed that for 45 consecutive struts by one male, the mean interval from the first wing swish to the peak of the first 'snap' sound was 1.55 s and that its variation was less than 1%.

Extreme stereotypy of this type is typical of, although certainly not confined to, postures and displays which ethologists have called 'fixed action patterns' (these are discussed in more detail in Chapters 2 and 3). Konrad Lorenz, who along with Niko Tinbergen is often seen as one of the 'founding fathers' of ethology, believed that fixed action patterns constitute part of the behavioural makeup of a species in just as fundamental a way as the shape of its bill or the colour of its tail feathers constitute its basic physical makeup. Today, we recognize that most behaviour patterns are not completely fixed and that there is often a degree of flexibility in their performance. We also know that extreme fixity or stereotypy of behaviour can sometimes be the result of captivity or environmental restriction, as we shall discuss in Chapter 6. The usefulness of continuing to use the term fixed action pattern in the present context is it draws attention to the remarkable stereotypy of the display and provides a handle for the quantification and further study of a male sage grouse's courtship. For example, we could compare the number of struts made by different males and see whether females were more likely to choose males that displayed the most.

We have already referred to some of the operations of the cockroach's nervous system and related them to the whole animal's behaviour as we can observe it. It is nowhere near as simple to do this for the sage grouse, but clearly having some reasonable and consistent units of behaviour for measurement is a start. We can begin by looking at ways in which units of behaviour may relate to the operation of the nervous system. Accordingly, we turn now to a brief description of the way in which the basic units of nervous systems work. Then we will try to make some comparisons between how the nervous system operates to produce very simple behaviour patterns, such as reflexes, and more complex behavioural units, such as the courtship strut. We will find that many of the principles that operate at the neuronal level are also to be found when we look at reflexes and more complex behaviour.

 ## Units of the nervous system

The basic structural and functional unit of the nervous system is the nerve cell or neuron. Neurons are connected to each other in many complicated ways (Fig. 1.6), some individual cells being in contact with hundreds or even thousands of

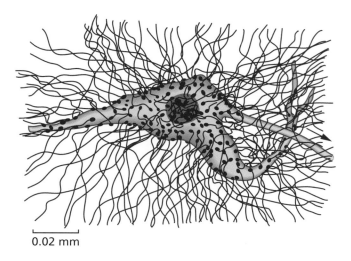

0.02 mm

Figure 1.6   Reconstruction of the cell body of a single motor neuron from the spinal cord of a cat. Parts of two dendrites can be seen (one branched) and the axon is indicated by the arrow. The cell and its dendrites are covered by synaptic knobs (boutons terminales) showing where the fine axons from hundreds of other neurons in the cord make contact with it.

others. These connections are quite vital to the working of the nervous system since it is through them that nerve cells transmit information to each other. Each nerve cell has branching processes called dendrites that receive information from other cells; it has a cell body or soma and, connected to this, a long tubular axon that may extend over a considerable distance and transmits information to other cells (Fig. 1.7).

For the most part, information is transmitted through the nervous system by means of electrical impulses. However, at the junctions between cells, called synapses, there is usually chemical transmission. The synaptic terminal of the stimulating cell secretes small packages of a highly specific chemical – known as neurotransmitter – across the tiny gap that separates it from the next cell. This neurotransmitter diffuses rapidly across the gap and when it arrives at the other side of the synapse, it gives rise to new electrical activity in the receiver cell; the excitation is once again transmitted electrically along the next axon until it reaches to the end of that. (Some synapses do not employ this chemical package method and are completely electrical – two cells being close enough together for direct electrical propagation.)

Each nerve cell is electrically active all the time and, when not transmitting information, maintains a potential difference across its cell membrane (known as the membrane or resting potential) which means that the inside of the cell is slightly electrically negative relative to the outside, i.e. it is polarized. 'Depolarization' is the term applied to changes in the membrane potential of a cell away from its normal value of 60–80 mV and these changes are the way cells transmit information.

(a)　　　　　　　　　　　　　(b)

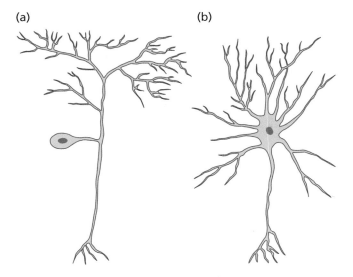

Figure 1.7 Simplified drawings of (a) an invertebrate (arthropod) and (b) a vertebrate (mammalian) neuron. Real neurons have much more extensive branching than shown here.

Being stimulated by another cell through its dendrites causes a change in the membrane potential away from its resting value. This may be a very brief depolarization, but sufficient to pass on information if it gets propagated along the cell axon. Now, all cells are poor conductors – an electrical impulse simply does not travel very far down the cell (certainly not as far as if the cell were made of copper wire). Sometimes, when one cell is stimulated by another, all that happens is that there is a tiny electrical signal in the receiver cell which travels a few millimetres down a dendrite and then fades out because of the poor conductivity of the cell. The small induced signals are postsynaptic potentials (PSPs) and, although they do not last very long or travel very far, they nevertheless have a very important role in determining the kinds of operations that neurons can perform.

However, if PSPs were the only way in which nerve cells responded, long-distance transmission of information would be impossible. Accordingly, many neurons supplement PSPs with another type of signal altogether – a massive electrical change known as an action potential or nerve impulse (Fig. 1.8a). The nerve cell is capable of responding to a tiny PSP by generating an explosive action potential that can travel along the entire length of a long axon without a decrease in its amplitude. Sometimes one PSP alone is not enough to trigger a nerve impulse, but several PSPs coming rapidly over a short period of time, or from several different sources, will give rise to a nerve impulse. In this way, the nerve cell shows **summation**: it adds together the effects of several sub-threshold PSPs to give rise to the all-or-nothing response of the nerve impulse (Fig. 1.8b). At other times, the presynaptic neuron will deliver a barrage of several

(a)

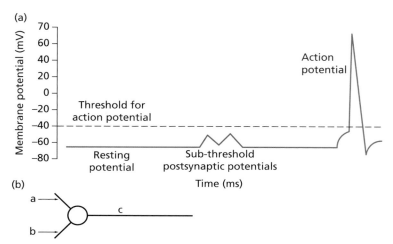

(b)

(i) A postsynaptic neuron, c, is shown diagrammatically with two presynaptic neurons, a and b.

(ii) Spatial summation: the responses of presynaptic neurons a and b are summed by the postsynaptic neuron c.

(iii) Temporal summation: the successive responses of one presynaptic neuron are summed.

(iv) Facilitation: c responds more strongly with each succesive input from a.

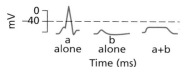

(v) Inhibition: c responds less strongly if a and b are active.

Figure 1.8 (a) Three examples of the electrical activity that can be recorded from neurons: the resting potential when the cell is at rest; postsynaptic potentials that do not lead to a 'nerve spike' or action potential; and the massive action potential itself. (b) Some operations performed by neurons.

large action potentials to the receiver cell and each action potential results in more transmitter release than did the one before. Thus each PSP is larger than the one before, a phenomenon referred to as synaptic facilitation.

Interestingly, cells do not just excite one another. Some neurons also inhibit others. Their transmitters hyperpolarize the postsynaptic membrane, making it less likely that receiver neurons will respond. As we will see throughout this book, the concept of inhibition is of very great importance for the understanding of all animal behaviour.

Having looked at some of the basic attributes of nerve cells – excitation, inhibition, sub-threshold responses, summation and facilitation – we can now turn to behaviour itself and see how these same phenomena are reflected in behavioural units which represent a more complex level of operation in the nervous system, and must involve the coordinated activity of several neurons at the very least.

Reflexes and more complex behaviour

Reflexes are often considered as among the simplest units of behaviour. The reflex which causes us to close our eyes when something flashes towards us, or to withdraw a foot when we step on a sharp object, is functionally very important. Yet it seems so very different in scale both from the action of individual nerve cells on the one hand, and on the other from the kinds of behaviour with which most of this book will be concerned: courtship strutting, building a nest, running through a maze to get food and so on. Reflexes are somewhere between the two – not too dissimilar from the escape response of the cockroach but much simpler than a courtship display.

We are clearly dealing with a continuum here, particularly as complex behaviour may itself incorporate many reflexes. The swallowing reflex is the culmination of elaborate food-seeking behaviour and the reflexes controlling balance and walking are involved in almost everything animals do. Nevertheless, there is much to be learnt from comparing the properties of nerve cells, reflexes and more complex behaviour and such comparison is a useful way of emphasizing the unity of the study of behaviour.

In 1906, Charles Sherrington published *The Integrative Action of the Nervous System*. Sherrington, more than any other single person, can be regarded as the founder of modern neurophysiology. In this book he considered the way in which reflexes operate and how the central nervous system integrates them into adaptive behaviour, combining information gathered from different sources, arranging sequences of action and allocating priorities.

In the first few chapters of his book, Sherrington discusses some of the properties of reflexes and contrasts them with those of the same movements when elicited by direct stimulation of the nerves to the muscles concerned. We shall, in turn, use part of his classification to compare the properties of reflexes and more complex behaviour.

Latency

Reflexes and complex behaviour both show latency in response – there is a delay between giving a stimulus and seeing its effect. The latency between a dog encountering a painful stimulus with its leg and showing the flexion reflex in which it withdraws its leg usually lies between 60 and 200 ms. Only a small fraction of this delay is due to the time taken for nerve impulses to be conducted along axons; most of it is due to the delay at the synapses (a term we owe to Sherrington) between one neuron and the next. It is hardly surprising to find delays between stimulus and response in complex behaviour, for in the chain between receptors and effectors there are often dozens of synapses to cross. We have already noted the crucial 44 ms latency between the arrival of the tiny air movement which signals the movement of the toad's tongue and the first turning away of the cockroach, a delay which is actually beneficial.

Although it is often difficult to measure latencies for complex behaviour (unlike the cockroach escape situation, it is often impossible to fix the time of stimulus onset precisely), nevertheless, results are sometimes vivid. Wells (1962) describes how, when

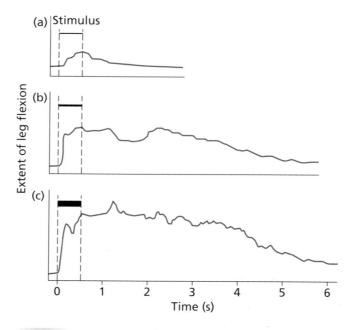

Figure 1.9   The extent and persistence of the dog's flexion reflex with the three strengths of stimulus, each of 0.5 s duration. The area enclosed between the line and the x-axis gives a measure of the 'amount' of the response. Even with a weak stimulus (a) the after-discharge represents 75% of the total 'amount'; with the strongest stimulus (c) it represents over 90%. Note that the latency of the response decreases as the stimulus increases in strength.

a tiny shrimp is presented to a newly hatched cuttlefish, there is no detectable response for perhaps as long as 2 minutes. Then the nearest eye of the cuttlefish turns to fixate on the shrimp. There is a further delay, but usually only a few seconds, before the cuttlefish turns towards the shrimp so that both of its eyes are brought to bear. Another brief delay follows and then it launches its attack and seizes the shrimp with its tentacles.

With reflexes, it is found that the stronger the stimulus, the shorter the latency (Fig. 1.9) and some evidence suggests that the same is true for certain complex behaviour. Hinde (1960) measured the latency between presenting various frightening stimuli to chaffinches and how soon they gave their first alarm calls. Just as with reflexes, the stimulus known on other grounds to be the strongest produced the shortest latency.

Summation

We have seen that individual neurons sometimes respond only after they have received several postsynaptic potentials. This shows that they are able to summate excitation coming at different times (temporal summation) or from different places (spatial summation). Sherrington gives several clear examples of summation at the level of reflexes. The scratch reflex of the dog is elicited by an irritating stimulus anywhere on a saddle-shaped area of its back. The hind leg on the same side is brought forward and rhythmically scratches at the spot. Weak stimuli – say, a series of 5 or 10 touches given in rapid succession – may not evoke any response, but after 20 or 30 scratching appears: the stimuli have been summed in time. Figure 1.10 shows the spatial summation of stimuli from two areas of skin 8 cm apart. Neither is strong enough alone to provoke scratching, but they are effective when given together.

Dethier (1953) studied the stimuli which cause blowflies (*Phormia regina*) to extend their proboscis before they drink. The flies can detect sugars and other food substances with sensory hairs on their fore-tarsi. They search for food by running over a surface and extending the proboscis when the front legs encounter anything suitable. As measured by this proboscis extension, the flies can detect sugars at very low concentrations. Dethier found that when only one leg is dipped in a solution, the lowest concentration of sucrose to which 50% of the flies respond is 0.0037 M. However, if both legs are stimulated together, twice the number of sensory fibres are sending signals into the brain; their effects summate and now 50% of flies respond to only 0.0018 M.

With more complex behaviour, summation frequently occurs between stimuli of quite different types perceived by different sense organs. We all know how the sight and smell of food summate when we are hungry. Beach (1942) showed that male rats respond sexually to a combination of olfactory, visual and tactile stimuli from a receptive female. Young males do not respond unless two such sources are available – it does not matter which two. Mature males, with previous sexual experience, will respond to one type of stimulus alone.

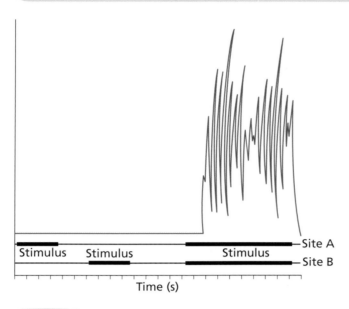

Figure 1.10 Spatial summation leading to the appearance of the scratch reflex in the dog. The tracing represents the movements of the dog's leg when scratching. Site A and Site B are two points on the shoulder skin. Weak stimuli, given singly first at A then at B do not evoke the reflex. When both points are stimulated simultaneously the reflex appears with a latency of about 1 s.

'Warm-up' or facilitation

Sherrington found that some reflexes do not appear at full strength at first but, even with no change to the stimulus, their intensity increases over a few seconds. Neurons, as we saw, show synaptic facilitation, each successive PSP being larger than the one before. At a behavioural level, Hinde (1954) found that chaffinches show a similar type of 'warm-up' effect when shown an owl. The birds' response to the owl is to give a mobbing call. Counting the number of calls given by a chaffinch in successive 10-second periods after the owl is shown to it indicates that it begins by calling at a relatively low rate and that the maximum calling rate is not reached for about 2.5 minutes, after which time it gradually declines (Fig. 1.11).

Sherrington was able to show that 'warm-up' in some reflexes is due to summation of stimuli which come to evoke a response from more and more nerve fibres, producing a stronger contraction. He called this phenomenon 'motor recruitment'. Some analogous process probably occurs with complex behaviour, but what we commonly see is not only a change in the intensity of response but in the nature of the behaviour as well. Sherrington (1917) provides an excellent example from what he calls the cat's 'pinna reflex'. Repeated tactile stimulation to the cat's ear first causes it to be laid back. If stimulation persists, the ear is fluttered; thirdly the cat shakes its head and when all else

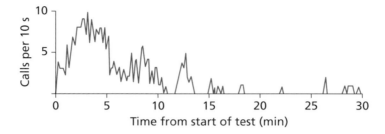

Figure 1.11   The 'warm-up' and subsequent 'fatigue' of alarm calling when a chaffinch is presented with a stuffed owl in its cage. The maximum rate of calling occurs after about 2.5 min.

fails to remove the irritation, it brings its hind leg up and scratches. Clearly there is more involved here than the recruitment of a few extra motor neurons. Mechanisms which control patterns of movement such as ear-fluttering and head-shaking must be recruited. Perhaps all these mechanisms are activated in some way by stimuli to the ear but their thresholds are different. That for laying back the ear will have the lowest threshold, with successively higher ones for the other three patterns. Workers studying complex behaviour frequently rank the patterns they observe on a similar intensity scale of increasing thresholds. A similar threshold idea has been used by Bastock and Manning (1955) to explain the fact that a male fruit fly (*Drosophila*) switches from one courtship pattern to another when courting a female whose behaviour remains constant.

Inhibition

Inhibition operates at every level within the nervous system. As we have seen, nerve cells may sometimes actively inhibit each other's transmission of information. Similarly, the prevention of one activity's occurrence while another is in progress constitutes inhibition at the behavioural level. In many ways, inhibition is just as important for the coordination of behaviour as excitation and to see why we can again turn to Sherrington's work on reflexes.

   Muscles are commonly arranged in antagonistic pairs, such that one flexes a portion of a limb and the other extends it. Clearly it would be impossible to both extend and flex the same limb at the same time: Sherrington showed that excitation of one member of a muscle pair is accompanied by inhibition of its antagonist. Such inhibition is not absolute, and an inhibited muscle does not simply go limp. Once it is stretched by its antagonist then its own 'stretch reflex' (p. 29 has a fuller description) will tend to make it contract. Although the antagonist muscle may override it, it will take up the slack, so to speak, in an active fashion. Much finer control of movement is possible if muscles can be made to work against one another in this way. Mutual inhibition allows them to take the lead in turn during limb movements and to alternate flexion and extension of the limbs. Sherrington found that it is not only antagonists on the same limb which

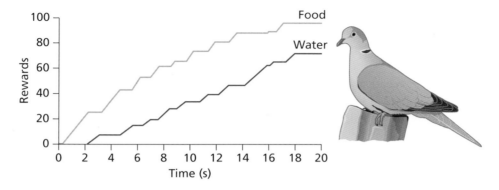

Figure 1.12 Cumulative number of food and water rewards obtained by a dove which was both hungry and thirsty. The dove could choose between pecking one key which yielded food or another which yielded water. Note how it starts by repeatedly pecking the food tray, then switches to the water key for several pecks before switching back for another bout of food pecking, and so on.

inhibit each other, but also muscles located on opposite limbs and which have antagonistic effects during locomotion. When the flexors of one limb are contracting, the flexors of the opposite limb are inhibited. Reciprocal inhibition of this type is one of the basic integrating mechanisms for walking and without it coordination of the different limbs would be impossible.

The role of inhibition in complex behaviour is superficially less obvious than that of excitation. We stimulate an animal and the conspicuous result is that it makes a response. But in so doing it has made a swift transition which requires the inhibition of its behaviour prior to the stimulus and other behaviour which it may be stimulated to perform at the same time. Sherrington saw reflexes as 'competing' for the final common pathway, i.e. the muscles whose action is common to several different reflexes. In an analogous way, we can see the different systems controlling patterns of complex behaviour like fighting, feeding and sleeping competing for the control of the animal's musculature. Such systems are obviously incompatible in the sense that only one behaviour can occur at a time. The 'integrative action' of Sherrington's title refers in part to the necessity for the nervous system to allocate priorities since there will often be conflicts. Which stimuli should an animal respond to; which can it safely ignore for the moment? Inhibition of action will be as crucial as excitation. For example, many animals need to take in food and water over the same time in order that digestion can proceed normally, but they do not try to do both at once! In a study on doves (*Streptopelia*), McFarland and Lloyd (1973) gave ad lib food and water to birds which had previously been deprived of both. The doves alternated their behaviour smoothly between bouts of feeding and drinking (Fig. 1.12). Thus, long before they had assuaged their hunger, they abruptly stopped feeding and switched to drinking, then after a short time they switched back again, continuing in this fashion until some time later they gradually ceased both activities. What is striking about their behaviour is that they did not 'dither' between feeding and drinking, spending only a few seconds on each. Rather, sustained bouts of each activity in turn inhibited the other. We do not know the exact mechanisms involved here but McFarland and Lloyd found that it was the time spent in feeding and drinking, not amount of food or water consumed, which seemed to determine the points at which switch-overs occurred. They called the process 'time-sharing' and it certainly suggests alternating inhibition of one activity by the other.

Sherrington found that when inhibition was removed from a reflex, it returned at a higher intensity than it had previously. Figure 1.13 shows this phenomenon, which Sherrington called 'reflex rebound', for the scratch reflex. We commonly observe that when a particular type of complex behaviour – for example courtship – has not been elicited for some time, it has a lowered threshold and is performed with high intensity when it is, at last, evoked. Vestergaard (1980) found that when laying domestic hens are kept on wire so that they have no substrate in which to dustbathe, they start dustbathing quickly and dustbathe in very much longer bouts when eventually given access to litter than do hens kept all the time on litter. It is possible that the system controlling dustbathing shows something akin to reflex rebound.

Kennedy (1965) interpreted some aspects of the behaviour of aphids (*Aphis fabae*) along these lines. The behaviour of winged aphids alternates between periods of flight and periods of settling and feeding on leaves. If an aphid settles on an 'unattractive' surface – an old leaf for example – it does not stay long and soon takes off, but flies relatively weakly and soon settles again. Conversely, if it has settled on an attractive young shoot, it stays for a long period but, when it takes off, flies vigorously and for a long time. By an elegant series of experiments, Kennedy was able to exclude any simple explanation for this relationship based on physical exhaustion during flight and recovery after resting and feeding on a young leaf. He suggested that there is mutual inhibition between the systems controlling flight behaviour and those controlling settling. As with reflexes, activation of the settling system may temporarily inhibit the expression of the flight system but, at the same time, gradually lower the threshold for flight. In Chapter 4, we shall discuss other evidence that the systems controlling complex behaviour do inhibit one another and consider the various other interactions that occur between different control systems.

Feedback control

Very commonly, reflex or complex behaviour consists of a steady output of some activity which has to be held at a given level. When we 'stand at ease', our body is evenly balanced over the pelvic girdle and easily corrects for any slight jostling we may receive. To do so, the muscles of the legs and back must be held at a constant level of tension and if shifted away from this level, they must correct to bring the body upright again. Analogously, animals under normal circumstances maintain a very constant

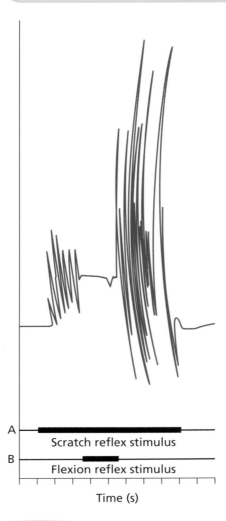

A

Scratch reflex stimulus

B

Flexion reflex stimulus

Time (s)

Figure 1.13 Inhibition of the scratch reflex by the flexion reflex. The stimulus denoted on line A evokes the scratch reflex, but this response is inhibited when the stimulus on line B evokes the flexion reflex. The moment B is removed the scratch reflex returns, and much more vigorously then before – an instance of 'reflex rebound'.

body weight and they eat and drink sufficient for their needs at regular intervals. If a surplus is available they do not overeat. In times of scarcity, they spend a higher proportion of their time in searching for food and consume more when the chance arises to replace any deficit.

Both these examples show us behaviour acting as a **homeostatic** system ('homeostasis' means literally 'same state') and serving to preserve the status quo. In the first case, this was achieved through reflex systems controlling the leg and trunk muscles; in

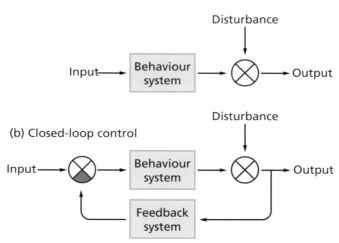

(a) Open-loop control

Disturbance

Input⟶ [Behaviour system] ⟶ ⊗ ⟶ Output

(b) Closed-loop control

Disturbance

Input⟶ ⊗ ⟶ [Behaviour system] ⟶ ⊗ ⟶ Output

[Feedback system]

Figure 1.14   Diagrams of simple open- and closed-loop control systems. The output from the behavioural system is affected by disturbance factors (the crossed circle represents interaction). In open-loop control (a) if the output is affected by disturbance no correction takes place. In the closed-loop arrangement (b) the results of the disturbance feed back to affect the input of the behavioural system. The dark segment of the crossed circle represents an interaction between the feedback mechanisms and the input, which tends to bring back the output to its original value. The feedback system produces its own output, which is proportional to the disturbance and changes the input to the behavioural system both to the right amount and in the right direction.

the second case, it was achieved by a series of more complex systems regulating the search for food, feeding and satiation. In both cases the operation requires that the end result (posture and balance whilst standing, state of nutrition) is monitored in some way. When it deviates from a set value a signal is sent to the control mechanisms to correct the imbalance and bring the end result back to the set value again. This idea is shown diagrammatically in Fig. 1.14 and its application to feeding and drinking behaviour is discussed in more detail in Chapter 4.

The homeostatic control of posture is understood quite thoroughly at a neurophysiological level. In some cases we know the paths of the neurons involved and can actually identify the structures which function as parts of the control system. One such is illustrated in Fig. 1.15, which represents a typical muscle in the limb of a mammal, such as would be involved in maintaining posture. Motor neurons which have their cell bodies in the ventral horn of the spinal cord run to the muscle; it is their activity which determines the tension developed by the muscle. In parallel with every skeletal muscle, and embedded within its fibres so that they contract and relax with it, are muscle spindles. These are specialized sense organs for recording the degree of tension in the muscle. Their sensory nerves run back to the spinal cord and, entering through the

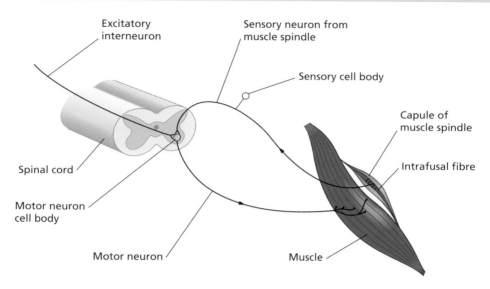

Figure 1.15 A simplified diagram of some of the neural pathways involved in the stretch reflex. Further explanation is given in the text.

dorsal root, synapse with the motor neurons to the muscle. Thus a loop is closed which forms the basis of the stretch reflex, already mentioned.

When a muscle is stretched by the contraction of its antagonists, the muscle spindles are stretched also and their sensory fibres increase their rate of firing, stimulating the motor neurons so that the muscle contracts. It is easy to equate the units of the closed-loop control system in Fig. 1.14 with those of this real muscle mechanism. The output (state of tension in the muscle) is affected by a disturbance (being stretched by other muscles), and a feedback mechanism (muscle spindle) records the change and feeds back to change the input (motor nerve) and restore the original input. This is a simplified picture of the real situation which, in fact, includes other regulatory mechanisms allowing for very fine graded control over the muscle contractions involved both in the maintenance of posture and in movements, but it serves to illustrate the reality of feedback control at a reflex level.

Not all behaviour involves feedback control. When a movement must be made very rapidly, there is simply not time to modify the movement while it is in progress. The flip of the toad's tongue towards the cockroach is just such a case. The strike of the mantis is very similar. It moves towards a fly and orientates its body very slowly and precisely (operations which certainly involve feedback control), but once aimed, the strike is an all-or-nothing movement. If the fly moves after the strike is initiated, this makes no difference to the movement and the mantis strikes in the wrong place. Such behaviour which occurs without feedback is said to be under 'open-loop' control (as opposed to the 'closed-loop' control of a homeostatic system). The difference between open and closed systems is shown in Fig. 1.14.

So, despite the many levels at which the behaviour of animals is studied, we can see that there are certain principles – excitation, inhibition, summation, facilitation and feedback control amongst others – that appear to be common to many different levels. Studying single neurons and studying the behaviour of whole animals may require very different techniques and, in many cases, very different concepts. Nevertheless, it is valuable to keep these common principles in mind. It is often possible to break down complex behaviour patterns into smaller units, some of which are immediately equatable with reflexes. However, we cannot always explain behavioural observations using reflex terminology, nor is there any point in trying to do so. There *are* differences in complexity and these often require different types of approach, as we have seen for the cockroach versus the sage grouse.

 ## Diversity and unity in the study of behaviour

It will be clear by now that the study of animal behaviour presents us with a great diversity of subject matter, levels of analysis and questions to be answered. At one end, neuroethology merges into cellular physiology and biochemistry. At the other, the study of animal groups moves into evolutionary theory and ecology. Psychologists, ethologists, physiologists and behavioural ecologists all contribute to this diversity, so much so that it may be tempting to think that there is no sense in which there is a single 'study of animal behaviour' at all.

There are, however, two very positive reasons for regarding it as a single subject of rich diversity rather than as a collection of isolated disciplines. The first is that the same questions, particularly those of mechanism, survival value and ontogeny, can be asked at all levels. It is as important to ask 'why' (in the survival value sense) an animal has a nervous system with sensory, motor and interneurons connected in particular ways as it is to ask why male sage grouse display. It is as important to ask about causation (how does it work?) when looking at the way in which a bird assesses the amount of food in its territory as it is when recording from a single nerve with a microelectrode. Studying adaptation without looking at mechanism tends to become sterile, since it is essential to understand the constraints imposed by what the animal's body can do before we can understand the evolutionary significance of what it actually does. Studying mechanism without asking about adaptation tends to give lists of facts with no context. Evolution through natural selection is the unifying theme of all biology and the study of animal behaviour, at all levels, is at its best when questions about mechanism and questions about evolutionary significance are asked in parallel.

The second reason for not compartmentalizing the subject and trying to see it as a whole is that the different levels of analysis, although distinct, are never totally separate. Many of the same concepts that have been used and tested at one level have been found to be useful at other levels. The behaviour of animals is, ultimately, due to the workings of nerves, muscles, hormones and sense organs. Equally, however well we understand the workings of those parts of the body at a physiological level, our understanding is incomplete unless we understand how they interact to produce complex behaviour. The most fruitful studies have been those in which the boundaries between the different levels have been breached and in which 'outside the skin' and 'inside the skin' have not been the demarcation lines of a particular investigation.

Some of the best examples of how our understanding has benefited from just such a multi-level approach have come from studies of behavioural development, where we can often begin to link up the building of a nervous system with the emergence of a behavioural repertoire. Thus although – quite arbitrarily – Tinbergen put development as his fourth question for ethologists, it is the one to which we choose to turn first.

SUMMARY

Niko Tinbergen's four questions about behaviour (function, evolution, causation and development) are as relevant today as when he first outlined the scope of the study of animal behaviour (ethology). By asking how a behaviour helps an animal to survive and reproduce, we study how natural selection has shaped its behaviour. By asking about evolution, we study how an animal's ancestors behaved in the past. By asking about causation, we study how the nerves, muscles, hormones and brains of animals are coordinated to produce the amazing complexity of animal behaviour. By asking about development, we study how genes, learning and other environmental factors shape the behaviour of growing animals.

Remarkably, we find that there are certain principles underlying behaviour that help us to understand the behaviour of whole animals as well as that of the fundamental building blocks of their nervous system, the neurons or nerve cells. These include inhibition, facilitation and feedback. Animal behaviour is diverse but our understanding benefits from a unified approach.

CHAPTER TWO

The development of behaviour

CONTENTS

Here we describe some of the principles of behavioural development, with different animals varying greatly in the extent of contributions from their genetics and their environment both physical and social.

- Young animals grow up
- Instinct and learning in their biological setting
- The characteristics of instinct and learning
- Genetics and behaviour
- Developmental changes to the nervous system
- Hormones and early development
- Early experience and the diversity of parental behaviour
- Play
- Imprinting
- Bird song development
- Conclusions

One of the most remarkable features of living organisms – plant or animal – is the way in which a single-celled zygote, the fertilized egg, is transformed by cell division, cell differentiation and cell movement into the adult form, far more complex and often millions of times larger. Development (often called embryology when it describes the progress from egg to a young but free-living stage) is a most challenging field of research and there has been some remarkable progress in recent years. New insights help us to understand how the sequences of development have been modified as a group of animals has evolved, for example as the vertebrates have evolved through fishes, amphibians and reptiles to birds or mammals. The initial stages must stay the same and modifications can be added only at the end. Thus all vertebrate embryos go through a stage which involves their developing gill clefts. Fish – of course – retain these and so do some amphibians but in reptiles, birds and mammals they close up as other structures pertaining to their future life on land are added on and replace them. Hence the old phrase, 'ontogeny recapitulates phylogeny' which retains its power if not taken too literally.

 ## Young animals grow up

The zygote contains all the information necessary to build a new organism, provided that it can develop in and interact with a suitable environment. When we study the development of behaviour, we must obviously concern ourselves with some aspects of embryology – for example, the way in which the basic framework of the nervous system is laid down – but we need to go far beyond this. A young animal's behaviour may continue to develop long after it is independent and it is perfectly reasonable to argue that in some animals behavioural development continues throughout life. Thus learning (which we shall concentrate on in Chapter 5) might well be regarded as a form of development and young animals sometimes learn a great deal as they develop. However, here we shall concentrate on other types of behavioural change which often take place both rapidly and dramatically during the early part of life.

At the outset, we must recognize that young animals are not simply part-formed creatures, inadequate stages on the path to adulthood: they have at all times to be fully functional animals capable of behaving effectively in their own world. During their early stages of growth some animals may be protected inside an eggshell or a uterus, or watched over by attentive parents, but others are free-living and have to look after themselves entirely. Young animals may emerge as miniature adults, gradually growing in size, but their behavioural responses must change as well to keep pace. Young cuttlefish (*Sepia*) begin and remain as carnivores, but at first they can kill only tiny crustacea which are ignored as prey when the cuttlefish have grown. They move on to larger and larger food and the behaviour patterns employed for detecting and catching prey have to change accordingly with growth towards the adult size. The behavioural and morphological changes can be even more dramatic, for some young animals lead a life totally different from that of adults. Tadpoles swim and breathe like fishes and are herbivores before metamorphosing into land-living carnivorous frogs or toads. Aquatic, filter-feeding rat-tailed maggots (*Eristalis tenax*) (so called because they breathe air through a long snorkel tube at their rear end) metamorphose into flower-feeding hoverflies (see Fig. 2.1). For these life histories, young and adult each require an almost totally separate repertoire of behaviour.

(a)

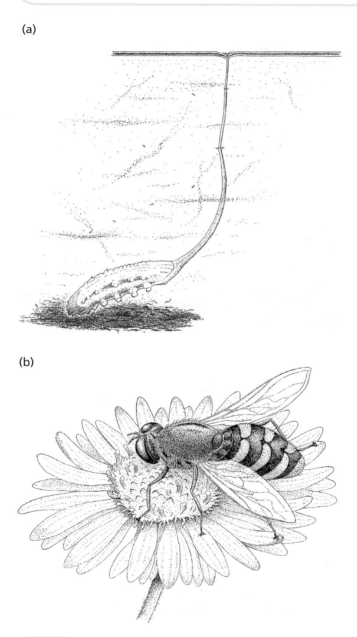

(b)

Figure 2.1 (a) A 'rat-tailed maggot', an inhabitant of stagnant water where it browses on detritus or the bottom mud, breathing by means of an elongated 'snorkel' linked to the insect's tracheal system. After a pupal stage it metamorphoses into (b) a flower-feeding hoverfly, whose environment and behaviour are totally different.

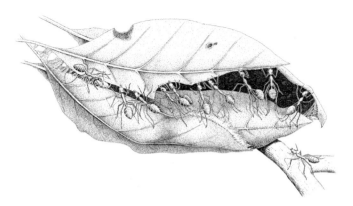

Figure 2.2 Collaboration between the generations! Weaver ants constructing their nest by bringing two leaves together and then binding the edges to join them firmly. Adult workers cling to the open edges with jaws and feet to pull them even closer together, whilst on the left another worker holds one of the colony's larvae and, by moving it from side to side like a living bobbin, induces it to spin silk which binds the leaf edges together.

Such changes mean that development often has to generate patterns which operate for only part of an animal's life and then disappear. Provine (1976) has described the particular synchronized movements by which cockroaches break out of their individual eggshells and then the protective case which packages a batch of eggs together. These movements involve a series of reversed waves of contraction along the body from tail to head and they are seen only on this one occasion. They appear, precisely timed, at the close of development in the egg stage and serve to launch the young cockroach nymph into its next stage of growth; this done, they are never elicited again. The feeding behaviour patterns of the aquatic larvae of hoverflies, just mentioned, must have a rather longer life than this, for they have to carry the larva through its whole growth period until it pupates, but then, like the cockroach hatching movements, they disappear. Animals with complete metamorphosis have, in effect, to be capable of organizing two different lives, carrying two sets of genes which organize two developmental programmes, interacting with two different environments. Just sometimes the two stages of life can combine. Figure 2.2 shows a nice example of 'cooperation' between young and adults within a colony of weaver ants (*Oecophylla* sp.). Using adult behaviour patterns, clinging on with jaws and clawed feet, worker ants bring together the two opposing edges of a leaf to form a sheltered nest space within it. They then seal the edges together by bringing up larvae and inducing them to spin silk across the gap – a pattern the larvae had to evolve to spin their cocoons prior to pupation. There are many other examples of the divergent lives of young and adult animals. Monaghan *et al.* (2008) provides an excellent review of these aspects of behavioural development.

We shall be describing in a later section (p. 57) some of the developmental processes in the nervous system which mediate the changeover from juvenile to adult behavioural systems. Here we may note that specialized infantile behaviour patterns do not always disappear but may return in a slightly different context. Baby meerkats (an African mongoose, *Suricata suricatta*) go limp and behave passively when an adult seizes them by the scruff of the neck. This reflex facilitates their being moved without injury. Adult female meerkats similarly relax when seized in a neck bite by the male during copulation. The neural basis for the reflex remains beyond infancy and we have other examples which reveal that even if they are not used again, some juvenile reflexes remain. As they hatch, domestic chicks break out of their eggshell using a series of unique movements – beating their beak against the shell and making strong thrusting movements of the legs which serve to rotate the body. In a series of elegant experiments, Bekoff and Kauer (1984) showed that these same movements will be performed again by young chickens several weeks after hatching if they are gently pushed back into a huge artificial eggshell and then placed in the posture of the chick at hatching. Probably one factor which contributes to the retention of the neural circuitry underlying this pattern is that elements of it are involved in adult locomotor behaviour. We shall later discuss other examples of this kind of 'neural re-cycling' from insects (p. 60).

The progression, sudden or gradual, which marks the transition from young to adult, is not always made along a single pathway. In this chapter we shall be considering a number of ways in which behavioural development can proceed and how far young animals can compensate for deviations from the normal path. However, few animals have one single end point, one fixed pattern of adult behaviour that all individuals show. Some animals become males, others females, and their behavioural repertoires may be very different. Female honeybees (*Apis mellifera*) develop into queens or workers, grasshoppers of certain species may stay solitary or become swarming locusts. In these cases, we know that there are clear-cut genetic or environmental triggers which send subsequent development along one of two or more alternative paths. The presence or absence of a Y chromosome in the zygote of mammals determines the sexual pathway, although as we shall see, subsequent 'sign posts' along the path are not genetic (p. 64). The length of time that royal jelly is provided to the growing larval honeybee determines whether it develops into a queen or a worker (p. 367). The desert locust (*Schistocerca gregaria*) exists in two forms, or 'phases' as they are termed; it is either a highly gregarious, black and orange locust or a solitary-living, pale green grasshopper. These two are so totally distinct in both behaviour and appearance that when first described they were put into different genera! The switch is brought about in part behaviourally, in part it can be through egg quality. Solitary nymphs change their colour and behaviour if they are crowded together, whilst the eggs of females which have been crowded, even if they started adult life solitarily, hatch into nymphs with gregarious tendencies (Roessingh *et al.* 1993; Islam *et al.* 1994; Fig. 2.3). This remarkable transformation is discussed further in the next chapter.

(a)

(b)

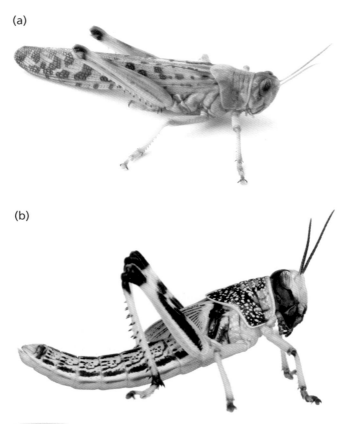

Figure 2.3 The solitary (above) and gregarious phases of the grasshopper locust (*Schistocerca gregaria*). The former is cryptically green to brown in colour with prominent eye stripes, and has five or six larval or 'nymphal' stages. The gregarious phase is yellow and black with less conspicuous eye stripes, and always completes its development in five stages. This is a fourth-stage nymph with incomplete wings. There are many behavioural differences between the two phases.

Not all triggers are so clear-cut. Males of some vertebrates fight to defend territories to which females are attracted and then mate with them. Other males may not fight and hold a territory of their own, but rather hang around submissively in another male's territory as so-called 'satellite' males. Such behaviour in the breeding season is found in animals as diverse as sticklebacks (Fig. 2.4), ruffs (*Philomachus pugnax*), gelada baboons (*Theropithecus gelada*) and white rhinoceroses (*Ceratotherium simum*). Although at first sight satellites may appear to be unsuccessful, close observations show that they certainly do sometimes manage to mate with females – often when a territory holder's attention is concentrated on fighting off other territorial males. Depending on the exact circumstances, becoming a satellite – at least for a time – may be an adaptive strategy. It is not always the case that there is just one fixed developmental pathway to adult life.

Figure 2.4 Two alternative mating strategies for male ten-spined sticklebacks (*Pungitius pungitius*). The black male in full nuptial colours has built the nest and led the female, swollen with eggs, towards it. He 'shows' her the entrance, inducing her to enter and spawn after which he will pass through the nest and release sperm to fertilize the eggs. But below is another male; cryptically coloured, he appears very similar to a female. He has moved stealthily around the territories of nest-building males and approaches with the female. He may slip into the nest after her and fertilize the eggs before the territory holder has a chance to do so. (Morris, 1958.) Morris recorded that this reproductive strategy, which he called 'pseudo-female', often resulted when competition between males for territorial space was intense. It was not always successful but probably offers more subordinate males a chance to breed and perhaps also a chance to inherit a territory if the dominant male is killed.

With both males and females, we now know of many cases where newly independent young birds or mammals have two alternative reproductive strategies to follow: they can disperse, find mates and reproduce themselves, or they can stay behind with their parents and help the latter to rear another set of offspring. We shall discuss this kind of choice in later chapters on evolution and on social behaviour.

Identifying the events, genetic or environmental, which bias such development one way or the other, is now an active area of research. In addition to genetic and physiological considerations, ecological factors often turn out to be crucial for such life-history 'decisions'. Nor will they necessarily be permanent decisions, for further change and development may be possible as ecological conditions allow, or as animals get older and individual experience enables them to adapt.

Instinct and learning in their biological setting

Viewing it overall, the role of development is to acquire a behavioural repertoire which fits an animal for success adapted to its mode of life. Whether we are looking at negative feedback processes in the muscular control of a limb movement or the nest-building patterns of a bird, whether we are observing young animals or adults, the beautiful match between form and function is striking. Of course animals do make mistakes and may appear clumsy at times, particularly when they are put into unnatural situations, but for the most part their behaviour is beautifully matched to their way of life. They respond appropriately to the features of their world and thereby feed themselves, find shelter, mate and produce offspring. How does behaviour acquire this near perfect match? How can it develop so effectively?

This question has fascinated people for centuries, because we have always been observers of animals. Of course, we have rarely been concerned directly with how their behaviour develops, but with their 'nature' as creatures which crucially share with us the 'spark of life'. However much we may exploit or ignore the interests of other species in favour of our own human wants, we cannot simply ignore them. This fact has led to extremely disparate attitudes. Sometimes animals have been regarded as gods. The Egyptians often portrayed their god of writing, Thoth, as a hamadryas baboon (*Papio hamadryas*) and kept a sacred bull, Apis. Conversely the Madagascan aye-aye (*Daubentonia madagascariensis*), a type of lemur, was persecuted until recently as the embodiment of a devil (Fig. 2.5). In Europe, cats and toads were often denounced as the 'familiars' of women accused of witchcraft. On the other hand, St Francis famously preached to animals as part of Creation and perhaps possessors of immortal souls. We can now dismiss superstitions, but how far animals are sentient beings like ourselves – if simpler – remains an important area of research which we shall discuss later. Certainly the majority of pet owners will – even if part in jest – endow their animal friends with some human qualities. At one extreme, we may end up with animals portrayed as if they share all our goals and feelings, not unlike Badger, Ratty, Mole and Toad in Kenneth Grahame's *The Wind in the Willows*. Such attitudes tend to stultify the objective study of animals for their own sake. We come closer to some semblance of biological reality with the highly influential views of the brilliant seventeenth century French philosopher, mathematician, and sincere Christian, René Descartes. He certainly recognized that animals had special characteristics of their own, but saw them as altogether below us in the scale of things, with human beings set apart from brute creation. Yet he had to recognize that the brutes do cope remarkably well with the problems of their

(a)

(b)

Figure 2.5 (a) The animal as god. The Egyptian god of writing, Thoth, who is also portrayed as an ibis or a human form with the head of an ibis, is here represented as a male hamadryas baboon with his cape of long fur compare Fig 7.22, p. 400. (b) The animal as devil. A nocturnal lemur, the aye-aye of Madagascar. If seen close to a village, it was killed and the village sometimes abandoned after such visitation by an evil spirit.

lives. How could they do this, lacking souls or the power of reasoning which enabled humans to survive and prosper? Descartes supposed that animals were able to respond adaptively because they operated using instincts endowed by the Creator, which automatically provided them with the correct response, requiring neither experience nor thought. We deliberately choose the term 'automatically' because Descartes did indeed view animals as little more than automata, blindly driven along pre-set but adaptive paths.

The concept of instinct or instinctive behaviour (the terms 'innate behaviour' or 'inborn behaviour' are effectively synonymous) is still a familiar one. It is part of common speech and may be used to account for human actions as well as non-human ones. Instinct is often described as patterns of inherited, pre-set behavioural responses which develop along with the developing nervous system and can evolve gradually over the generations, just like morphology, to match an animal's behaviour to its environment. It might be defined in a negative kind of way, as that behaviour which does not require learning or practice, but which appears appropriately the first time it is needed. This definition immediately suggests its converse and the other familiar way in which behaviour can become matched to circumstances. Animals may not have any pre-set responsiveness, but be able to modify their behaviour in the light of their individual experience. They can learn how to behave appropriately and perhaps practice or even copy from others to produce the best response.

Expressed in these terms, we appear to have come up with a clear dichotomy, rather as Descartes intended: **instinct** (nature), in which adaptation occurs over generations by selection, or **learning** (nurture), where the same end result – responding appropriately – is engendered afresh within the lifetime of each individual. But all animal behaviour moves along a developmental pathway and, as we shall see, studies of the way it does so do not support such a clear division and suggest that the categories themselves are inadequate. Nevertheless it is useful to survey briefly some of the extreme contrasts in behavioural development across the animal kingdom with its huge diversity of forms and lifestyles, for some patterns are discernible.

Pre-set behaviour which requires no learning or practice is obviously going to be advantageous for animals with short life spans and no parental care. Indeed it is completely essential for the many insects which lead almost totally solitary lives with no overlap between the generations and precious little social contact of any kind. A female digger wasp, *Philanthus*, hatches from its pupal case in an underground burrow long after its mother died the previous year. In flight she has one brief contact with a male to mate and then performs a series of elaborate patterns – details depending on the exact species – but often involving digging a burrow and forming a tunnel off which are excavated several 'brood chambers'. She then seeks out honeybees, captured on flowers or sometimes on the wing and paralyses them with her sting (Fig. 2.6). Usually two or three victims are placed in a chamber where the female wasp lays a single egg on the inert bodies. She moves on to repeat this sequence for each brood chamber, finally sealing up the whole burrow. Depending on circumstances, she may then go on to build a second and subsequent burrows before her brief life ends. Throughout, the wasp has to rely entirely on pre-set responses to the environment into which she emerges. There is no chance to learn from others or gradually shape up her behaviour by trial and error to end up with the best techniques. She has to get it right first time because for her – as indeed for almost all short-lived animals – there is a harsh rule of life, 'If at first you don't succeed – you don't succeed!'

But of course not all animal lives are so rigidly constrained. For the most extreme of contrasts we may compare the digger wasp's life history with that of African elephants (*Loxodonta africana*) (Moss and Colbeck, 1992). Their life span is comparable to our own and the key element of elephant society is the matriarchal group led by a mature female with her daughters and their offspring (Fig. 2.7). A baby is born into a group where

Figure 2.6 The digger wasp *Philanthus*, the so-called 'bee wolf'. The female is carrying a paralysed honeybee towards her burrow.

Figure 2.7 A 'family group' of African elephants. The old matriarch leads the group across dry grassland of the Amboseli reserve in southern Kenya. Behind her follow sisters and daughters with their offspring of all ages. Males will leave the family as they become sexually mature around 10–15 years of age, but they are unlikely to be able to mate with females until much later.

every individual knows all the others from long experience in their company. It is nourished and closely protected for several years by its mother and other relatives. Slowly it acquires the adult repertoire of feeding behaviour, learning how to select food and how the group migrates around its home range to match the seasonal changes of vegetation and water supply. Females do not become sexually mature until about 20 and puberty is even later for males, who leave the group when mature to lead more solitary lives. The behaviour of individuals and of groups varies considerably according to their history. For instance, Ian and Oria Douglas-Hamilton (1975) record how one group of elephants at Addo in South Africa are abnormally nocturnal in their habits and, whilst fearful, are also unusually aggressive towards humans. This behaviour can be traced back to an attempt to annihilate the Addo population by shooting, in 1919! None of the elephants alive at that time is still there, but their descendants have acquired and transmitted the behaviour which enabled a few to survive some 90 years ago.

The digger wasp which must rely on pre-set instinctive behaviour and the elephant which can learn at relative leisure and transmit information from one generation to the next represent two extremes on the behavioural scale. In fact, our descriptions over-emphasize the dichotomy, for the wasp can and must learn many things during its brief life – the exact locality of each of its nest burrows, for example, so that it can return to them after each of its hunting trips. The young elephant possesses some tendencies, such as those for feeding and reproduction, which are instinctive even though it may have to learn how to direct them. Certainly all animals whose complexity exceeds the annelid worm level show both types of behaviour and each has its own special advantages. One advantage of learning over instinct is its greater potential for changing behaviour to meet individual changing circumstances. Such a consideration is obviously more important to a long-lived animal than to an insect which lives only a few weeks. A further relevant factor may be body size, because highly developed learning ability requires a relatively large amount of brain tissue, insupportable in a very small animal. Usually body size and life span are positively correlated to some extent and most large animals live longer than small ones.

Apart from these physical constraints, it is clear that natural selection can operate to match different degrees of learning ability to an animal's life history. Two of the most advanced orders of the insects, the Hymenoptera (ants, bees and wasps) and the Diptera (two-winged flies), are comparable in size and life span. The Hymenoptera, in addition to a rich instinctive behaviour repertoire, show an extraordinary facility for learning, albeit of a specialized type, and this plays an important role in their lives. During her brief three weeks of foraging life, a worker honeybee will learn the precise location of her hive and the locations of the series of flower crops on which she feeds. She will move from one to another of these in the course of a day's foraging because she also learns accurately the time of day at which each is secreting the most nectar. Comparison with the Diptera yields a striking contrast. Certainly some flies can learn. Hoverflies learn something of the position of the flowers they visit, and houseflies tend to return to the same place to settle in a room. Using some ingenuity, it has proved possible to get the fruit fly (*Drosophila melanogaster*) to learn some simple discriminations. But for the most part the dipteran memory is brief and their learning powers very limited. Unlike the Hymenoptera, they do not reproduce in fixed nests to which they must return regularly. Their short lives are largely governed, with complete success, by inherited responses to stimuli which signal the presence of food, shelter and a mate.

The characteristics of instinct and learning

We mentioned earlier that instinct and learning are not much more than shorthand labels to be used at an early stage in studying what actually happens during the development of behaviour. Superficial considerations may suggest that whereas learning results in flexible patterns of response, instinctive behaviour is characterized by rigid, stereotyped responses and patterns of movement. Yet common observation will tell us that learnt patterns can be just as stereotyped as instinctive ones. In his delightful book, *King Solomon's Ring*, Lorenz (1952) describes how water shrews (*Neomys fodiens*) learn the geography of their environment in amazing detail. If at one point on a trail they have to jump over a small log, this movement is learnt with such fixity that they continue to make the jump in precisely the same fashion long after the obstacle has been removed. All digger wasps build their burrows in the same way and all kittiwake gulls (*Rissa tridactyla*) show the same 'choking' display (Fig. 2.8) on their cliff nest sites. Neither of these patterns requires learning for their development. But all chimpanzees (*Pan troglodytes*) of one community in Tanzania show one specialized mutual grooming pattern (Fig. 2.9) which is not seen in other communities and is certainly copied afresh by each generation of young animals. Rowley and Chapman (1986) describe how occasionally natural cross-fostering occurs between two very distinct Australian parrots, the galah (*Eolophus roseicapillus*) and Major Mitchell's cockatoo (*Cacatua leadbeteri*), when a female galah lays eggs in a cockatoo's nest hole (see Fig. 2.10). Galahs reared in such nests emerge strikingly abnormal in their behaviour. They retain some galah-like calls, but have also the cockatoo's calls; they have also acquired the cockatoo's manner of flight and choice of diet. So even such species-specific stereotyped behaviour patterns may develop through learning and practice in some cases.

Figure 2.8 Highly stereotyped behaviour – the 'choking' display of the kittiwake. The bill is wide open revealing the brilliant reddish-orange gape and tongue. As the body inclines forward the bird makes a rapid series of jerky up-and-down movements of the head and neck.

Figure 2.9 Chimpanzees of one Tanzanian community raise arms and clasp hands for mutual grooming of each other's underarms.

Such an example illustrates some of the problems that arise if one tries to force behaviour into categories, such as instinct or learning. There is a natural enough temptation to do this because instinctive behaviour is such a constant feature of an animal species: unless their environment is grossly changed, they all tend to develop a repertoire of the same patterns of behaviour and perform them in the same way. The cross-fostered galahs mentioned above must make us more cautious, but we know parrots to be long-lived and very intelligent animals. They probably are rather exceptional and we know many cases in which behaviour patterns typical of feeding, courtship and aggression do develop in the absence of learning.

Nevertheless, it has become clear that the dichotomy suggested by Descartes is not by itself going to be sufficient to account for the behaviour of animals. If the question is

Figure 2.10   A mixed brood of two species of Australian cockatoos about 45 days from hatching and ready to fly. Three Major Mitchells are on the left with their foster sibling, a galah, the latter's mother having laid an egg in the Major Mitchell's nest. All are marked with conspicuous wing tags which allowed them to be followed in the wild. The galah NO spent his first 2 years living with a flock of major Mitchells.

posed, 'where does this or that behaviour come from?' or, 'how can it be so well adapted to circumstances?' it has to be accepted that simple, either/or, nature/nurture answers are inadequate, however attractive they may seem. Our path to understanding is by an examination of how genetic and environmental influences interact to shape the behaviour of animals as they move from egg to adult.

One way of attempting this, which used to be popular amongst ethologists – see, for example, Lorenz (1966) – is the 'isolation' or 'Kaspar Hauser experiment'. Kaspar Hauser was a youth who appeared on the streets of Nuremberg, Germany, on 26 May 1828 with confused stories of his origin and upbringing. He was obviously mentally disturbed and displayed bizarre behaviour. He died in 1833 of wounds, almost certainly self-inflicted. Kaspar Hauser attracted the attention of various philosophers and occultists of the day and was sometimes regarded (for no fathomable reason) as an example of the totally untutored human being. He was claimed to be an isolate, revealing native human characteristics unchanged by experience. Consequently, his name is sometimes used as a convenient label for experiments in which animals are reared in isolation.

Such a procedure attempts to keep animals solitarily, out of all contact with others from as early an age as possible. When mature, their responses to a variety of stimuli are tested and compared with those of animals reared normally. As we have seen, the life histories of many insects are effectively natural isolation experiments. Rather few other animals have been tested under really rigorous conditions, but some fish and birds have been shown to perform various feeding, sexual and alarm patterns of behaviour quite normally following

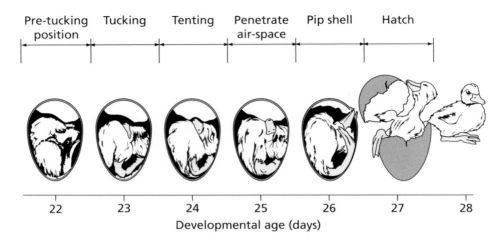

Figure 2.11 The final stages leading up to hatching in ducklings. They hatch after 27 days of incubation but for several days before this the bill first presses up against the air space in the egg (tenting) and then penetrates it. The duckling begins breathing air and vocalizing 2 or 3 days before hatching. It is these calls which affect the development of its subsequent responses to the mother's calls.

isolation. These are interesting data but, however easily isolation experiments enable us to eliminate the possibility that animals learn how to do something, they offer only a very restricted view of what happens during development. Isolation can tell us only what factors are *not* important for the development of behaviour; it tells us nothing of what is involved and some factors may not be at all obvious. We ought to begin our experiments by asking from what, exactly, are the young animals being isolated?

Gottlieb (1971, 1993) has carried out a series of investigations into the way in which young ducklings of various species come to respond correctly to the calls of their mother and how this choice is affected during their early life. One might regard the young bird, developing inside its eggshell, as another natural isolation experiment but this is too simple a conclusion. For example, young mallard (*Anas platyrhynchos*) hatched from an incubator respond more strongly to the call notes of mallard ducks than to other related calls and do so the very first time that they hear them, a preference we would call instinctive. But experiments reveal that one factor necessary for this instinct to develop is an embryo duckling's ability to hear the calls – quite unlike those of the mother – which it and other ducklings make while still inside the eggshell (see Fig. 2.11). This 'instinctive preference', once established in the newly hatched ducklings, remains stable in isolated birds, but if they are reared socially as would be normal in nature, then they become more responsive to other calls that they hear and may abandon their early choice. Bolhuis (1995) provides a number of other examples showing that various experiences, as apparently unrelated to vision as stroking, or being put into a running wheel – both in total darkness – can affect the visual preferences of young birds. It causes them, a few hours later, to express a preference for a realistic model mother bird (again we would call this instinctive) over a simple, artificial visual stimulus such as a wooden cube. Such experiments illustrate that instinctive responses can have a fairly complex developmental history. The details of the history may vary even between quite close relatives (something we shall also find for bird songs later in this chapter). Gottlieb (1983) found that young wood duck (*Aix sponsa*), which also hatch with a preference for the characteristic descending pitch calls of the mother, subsequently lose this preference unless they continue to hear the calls of other ducklings around them; their own calls are not sufficient.

Such examples force us to accept that behavioural development may involve processes and factors which are not easy to categorize, nor are they always easy to predict. It is useful to begin discussion of those processes by considering the information with which an animal's development begins.

Genetics and behaviour

Genes constitute one source of information which is present from the very outset of life. Certainly genetics immediately springs to mind when thinking about instinctive behaviour and it has been common enough to refer to such patterns being 'inherited', most especially when discussing their evolution. We have every reason to conclude that behaviour has often evolved alongside morphology under the influence of natural selection.

The concept of 'a gene' has been transformed by modern molecular genetics. Particular sequences of DNA determine which proteins can be synthesized within cells, but when, where and how much synthesis occurs depends on a whole range of other factors, both genetic and non-genetic. On its own, DNA can determine nothing. It plays its part in controlling how a cell differentiates and functions within the larger system of the animal itself. Noble (2008) provides an excellent commentary on how we should now reflect on 'genes and causation'. Such reflections are especially apt for those who study the genetics of behaviour. While there can be no doubt that genes are involved in the development of behaviour, it is never a straightforward matter to investigate how they act.

We have just been discussing some of the diverse factors affecting the way behaviour develops and genetic factors will be operating at all stages of such a complex system. At the outset of this chapter we referred to modern research into the way development itself evolves. This study – the evolution of development, now commonly referred to as 'evo-devo' – has yielded some astonishing insights about its genetic control. As the zygote begins dividing, the genes carried in the nucleus are distributed to all the daughter cells but they do not all become functional everywhere. As the embryo begins to form, some genes operate only in those cells which will make up the head, some only in other parts of the body. In insects, for example, specific genes organize those cell blocks which will make up the head, others the prothorax, others the mesothorax and so on into the future abdomen. These genes set up a kind of basic 'blueprint' of the body and then other genes, coming into operation later in development, are involved in the actual construction of limbs, wings, muscles, etc. in the appropriate field. The most remarkable discovery is that the genes which lay down the basic body structure are the same across a wide range of very different animal types. The genes for head and body segments of *Drosophila* are virtually the same as those for a mouse! The gene operating later in the future head of *Drosophila* which determines the formation of its eyes is the same gene as that operating in the head of a mouse to produce the same result. Of course, the eyes that develop are totally different in almost all respects, but the initial instruction, 'build an eye', is the same. The genetic basis of this instruction must have evolved very early in the evolution of complex animals, before the arthropods diverged from the chordates. The instruction has persisted, largely unchanged, for at least 600 million years to the beginning of the Cambrian geological period. It is perhaps the most compelling manifestation of the old rule, *'If it ain't broke, don't fix it!'* Natural selection has been operating to refine the numerous genes which influence the way eyes subsequently develop towards their diverse end points, but the instruction to set them going has continued to function and has been conserved. Carroll *et al.* (2005) provide a full account of the whole fascinating field of evo-devo research which now involves a very wide range of animals.

All this new work forces us to look afresh when we come to consider how genes may affect behavioural development. At first sight, behaviour genetics does seem to present some extra difficulties of its own because the 'distance' between the genes and the end product is so large and the nature of this end product is so diverse. Behaviour represents the functioning of the nervous system, none of it can be manifested until the nervous system is in place. Much of the behaviour we study is seen only when adult animals are out in the world and very far removed from their developmental beginnings (Rueppell *et al.*, 2004). However, we do now have some evidence that, as with the development of basic body structures, so with the development of some rather basic behavioural characteristics, the same or at least very similar genes operate across diverse animals.

For example, Sokolowski (2001) noted that *Drosophila* larvae as they fed in a culture medium seeded with yeast foraged in one of two distinct ways. Some stayed feeding in a small area immediately around themselves – 'sitters' – others foraged more widely, moving considerably longer distances – 'rovers'. This striking difference turned out to be largely due to different alleles of a single gene, named *for,* and the protein which it codes for has been identified, a protein kinase named PKG. It turns out that variations of the *for* gene are also found in nematodes and in honeybees and affect their exploratory behaviour in an analogous way. The tiny nematode worm, *Caenorhabditis*, feeds in a manner similar to *Drosophila* larvae and also has both 'rovers' and 'sitters'. Probably each method of foraging will be advantageous at different times and for different types of food. In the honeybee *for* becomes active at the time when the workers switch from being sedentary in the hive and leaving on excursions to forage. Here *for* becomes active at a particular stage in the life of all the workers. Like the gene which gives the instruction 'build an eye' the operation of the honeybee version of *for* also involves many other genes which control foraging.

Another remarkable example of a conserved 'instruction' for a behavioural trait is the gene labelled *FoxP2* which is known to be involved in pathways which affect the expression of human speech. Remarkably – and it can scarcely be a coincidence – the same gene operates in zebra finches, affecting those parts of the brain which are active when they are learning their songs (p. 94)! More examples of such genetic communality are given by Fitzpatrick *et al.* (2005).

We should continue to look for some kinds of 'behavioural framework' common to a whole range of animals and probably therefore very ancient. There are already some remarkable hints of the great conservatism of some basic genetic instructions. Tomer *et al.* (2010) describe similar suites of genes with similar patterns of action in the development of 'higher' brain structures from worms to insects to vertebrates! However, for our discussion here concerning the way behaviour develops within individuals, we shall also want to identify genetic factors operating on more detailed, specific behaviour patterns which probably evolved more recently in the history of a species. Morphological features such as coat colour, number of facets in a compound eye, length of tail, etc. are always there to be measured for any individual, but a threat display or an alarm call may not, indeed may never be, manifested during a lifetime. We can certainly identify genes which have effects on brain structure but inevitably they tend do so in a rather drastic manner which cannot tell us much about specific behaviour patterns.

The best examples come from animals such as *Drosophila* and mice which can easily be bred in the laboratory and whose genetics is thus familiar. *Bar* reduces the number of facets in the eye of *Drosophila* to a narrow strip; *no-bridge* causes a gap in the interconnections of the central complex, a major structure in *Drosophila*'s brain; *staggerer* mice have gross abnormalities to the Purkinje cells of their cerebellum, whilst those carrying the mutant *waltzer* have defective inner ear structure. In all these cases, and many others like them, we can observe changes to behaviour which correspond to the neural defects. Not surprisingly, they are rather gross changes for the mutations do act rather like hammer blows to a complex and sensitive piece of machinery. Nevertheless, we can gain useful information about how genes may operate to underpin the development of behaviour's neural basis. It may be obvious enough that *Bar* flies have defective visual responses and *reeler* mice – the name describes the behavioural effect – lack good balance, but in both cases we find that it is not just one target organ which is affected. *Bar* flies lacking eye facets also lack large areas of neural tissue in their optic lobes, the centres controlling visual information. The formation of eye facets triggers a range of developmental processes downstream, as it were, leading to normal development of the neurons which form the optic lobes. *Reeler* mice exemplify a different kind of developmental sequence. They seem to have the potential to form a normal cerebral cortex, but the cells that should migrate out to form connections fail to do so and remain piled up beneath the cortical layer (see Barinaga, 1996).

Common sense suggests that genes must be involved when we observe instinctive behaviour emerging fully fledged, as it were, at the first performance in every individual of a population. But, paradoxically, such constancy of development offers no rigorous way of investigating if and how genes are acting. This is because genetic analysis usually relies on the study of variation – setting up crosses between two different types and measuring how the differences are inherited. As we have seen, there are plenty of examples of behavioural differences between individuals being controlled by single genes. But behaviour patterns themselves will involve many genes controlling the neural mechanisms which in turn determine how the whole body operates to produce the behaviour. Our genetic analysis is usually confined to studying the development of individual differences and when there are variations in behaviour between individuals of the same species mostly it turns out to be quantitative. Animals commonly differ in how easy it is to elicit a behaviour pattern, or how frequently it is performed. Thus, given standard rearing and housing conditions, some strains of laboratory mice will instantly attack an intruder introduced into their cage and continue doing so, others are far less aggressive. Crossing the two strains gives mice with roughly intermediate levels of aggression which is pretty clear evidence that the behavioural differences are mediated by genes. However, the actual behaviour patterns the mice display in their fighting, sideways approach to the rival, arching the back, rattling the tail and so on, are always the same. So we can gain very little information on how genes operate in the development of the patterns themselves. Much of the conspicuous behaviour which makes up courtship and aggressive displays is very closely species-specific. We've just seen with the galahs reared with Major Mitchell's cockatoos that some such behaviour can be acquired as animals grow up, but it remains most likely that many of the subtle differences between species are due to inherited features of the nervous system – its structure and how it operates. The genetics of such complex features remains elusive for the most part.

We certainly have direct, if imprecise evidence of genes operating from some of those cases when it is possible to hybridize two distinct species. For instance, doves and pigeons develop their species-specific cooing song patterns in the absence of any experience, even when deafened so that they cannot hear themselves. Interspecies hybridization is frequently possible and, although the hybrids themselves are sterile, the males display and coo. When they do so, the patterns of their calling may be grossly distorted, full of cooing which is jumbled and fragmentary (Lade and Thorpe, 1964). The genes the hybrids inherit are mixed up and the songs are mixed up with them. The extent of the mix up probably depends on how closely related the parent species are. den Hartog *et al.* (2007) looked at the songs of hybrids between two doves which live close to each other in Africa and occasionally hybridize naturally. The hybrids' cooing is rather smoothly intermediate (see Fig. 2.12). It certainly demonstrates the genetic control of the behaviour but here the two sets of genes have been able to combine their action effectively, probably because they are very similar to begin with.

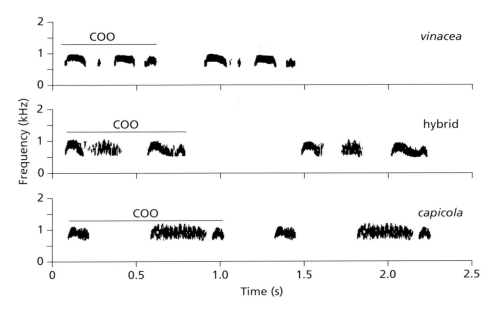

Figure 2.12 The cooing patterns of the two dove species, *Streptopelia vinacea* and *S. capicola* and that of the naturally occurring hybrid between them. These are sound spectrograms which show the sound frequency of the calls against time on the horizontal axis. The interpretation of spectrograms is discussed later in this chapter in relation to bird song development. The human ear interprets the cluster of three sounds which is repeated as a 'coo'. The clear point here is that the hybrid's 'coo', in the centre, is intermediate in form and timing between that of the two parent species.

We do have one good example where interspecific hybrids reveal a more subtle type of gene-based disturbance. Dilger (1962) worked with colourful small parrots of the genus *Agapornis*, of which there are a number of African and Madagascan species that readily breed in captivity. They all have a unique method for gathering nesting material, using their sharp bill to perforate large leaves parallel to an edge and then tearing off a strip of leaf, which they carry back to their nest hole as building material. At one point in this sequence of behaviour patterns there is a divergence between species. Fischer's lovebird (*A. personata*), carries leaf strips back to its nest singly, held in its bill; the peach-faced lovebird (*A. roseicollis*), interposes another stage. As each leaf strip is torn free, the bird elevates its rump feathers, turns round and tucks the strip amongst them, then lowering them so that the strip is held fast. The bird repeats this part of the sequence a number of times, finally flying back to the nest with several strips in place (Fig. 2.13).

Dilger hybridized these two species; as with the doves, the hybrids are sterile, but they perform all the usual nesting behaviour. In aviaries the parrots readily used newspaper instead of leaves as nesting material. The hybrids all showed normal piercing and strip-tearing behaviour, and we must assume that the genes received from each parent species were, at the very least, sufficiently compatible to allow normal development

Figure 2.13   Love birds (*Agapornis roseicollis*) tear strips of paper for nest-building. This species turns to tuck the strips into their rump feathers for carrying off to the nest site. Further explanation in text.

of the neural mechanisms behind this pattern. The sequence continued and it appears that peach-faced behavioural control now took over because a hybrid would always raise its rump feathers and, turning round, tuck in the strip. So far, so good, but at this point things went wrong. Instead of following through in the manner of its peach-faced parent, the hybrid did not open its bill at the end of the tucking behaviour. It hung onto the strip, more like the Fischer's parent pattern, and so, as its head returned to the front position, the strip was drawn back out of the feathers and the parrot was left standing with a strip still held firmly in its bill! It could then have reverted to Fischer's-style single strip carrying, but never did so. Instead it opened its bill, dropping the strip and then returned to the paper again and began another strip-tearing sequence, which ended in the same abortive fashion. Dilger found that it took months of repetition before some sequences began to be successful – always when the hybrids managed to inhibit dropping the strip at the end and flew back to the nest. However, they always went through the tucking pattern first. We know parrots are intelligent birds which can rapidly learn fairly complex tasks, but it was 2 years before they were successful in nearly all sequences and even then their heads would turn round slightly before they flew off with their strip – revealing a last persistent trace of the tendency to tuck.

One could scarcely ask for a more vivid demonstration of the force of genetic instructions operating on the development of particular neural mechanisms. The hybrids lack a very special part of the control chain and are thus persistently biased to perform particular behaviour patterns but fail to complete a crucial part of the sequence.

Even though the brains of advanced vertebrates are extremely complex and have huge numbers of neurons, the remarkable technical advances of modern molecular biology do make it possible to get some window into the way that genes are operating. It is now almost routine to isolate genes and generate multiple copies of them to discover what they produce in the cell. Not uncommonly, it is possible to reveal patterns of gene-controlled production in the cells of embryos and hence to deduce where and at what stage particular genes are acting. It might become possible to locate gene action within the developing nervous system, but the fine scale required is very demanding. However, given simpler material, it is sometimes possible to identify how genes operate on their neural substrate. Even if its development still eludes us, we may be able to understand how genes may regulate the way different behaviour patterns are brought into action at the appropriate time.

One step along the way comes from work on the big marine mollusc *Aplysia*, the sea hare. *Aplysia* is hermaphrodite and individuals usually mate when they encounter one another. Following mating, *Aplysia* switches its behaviour from feeding to egg-laying, which involves a series of specialized stereotyped patterns not seen at any other time. Using the techniques of molecular genetics, Scheller and Axel (1984) have been able to identify a gene and its product which, acting on specific nerve cells, causes them in turn to release 'egg-laying hormone'. The hormone targets several sites, neural and muscular, in *Aplysia*'s body and following the behavioural patterns involved in mating, activates particular neural circuitry which switches the behaviour towards egg laying. So a single gene here affects several aspects of behaviour but of course it can do so only by acting on a whole range of sense organs, nerves and muscle which have themselves already been affected by numerous other genes during development.

Until recently, it was at this point that analysis came to a halt but now we can sometimes proceed a few steps further as with some remarkable work on the genetic control of courtship behaviour in *Drosophila* (reviewed by Manoli *et al.*, 2006). The males of *D. melanogaster* perform a complex series of courtship patterns towards young females (which are illustrated in Fig. 2.14). Mutants are very well studied in *Drosophila* and some years back one labelled '*fruitless*' (*fru*) was identified which led to males whose courtship is incomplete. This is the recessive mutant form of a gene *Fru*, which turns out to be a complex gene some of whose protein products are absolutely crucial for courtship. Males who lack them show no courtship but their behaviour is otherwise apparently normal. The gene is expressed in about 20 clusters of neurons in different parts of the male nervous system – some of which had already been implicated in the performance of the behaviour. Of course, females will also carry the *Fru* gene, but it is not activated in their nerve cells. Using ingenious genetic techniques it has been possible to disinhibit, as it were, the expression of *Fru* in females and they then show some elements of male courtship! What is particularly significant is that this gene appears to operate very locally and in a very specific way to determine the functioning of the male behaviour. The activation of *Fru* can already be picked up in the developing neurons of the 3rd larval instar and in the pupa during its

(a)

(b)

(c)

Figure 2.14   The courtship of *Drosophila melanogaster*. (a) The male follows close behind the female and, extending one wing, vibrates it rapidly, about 30 cycles per second, creating sound pulses that are picked up by the female's antennae. Intermittently (b) the male will extend his proboscis to touch the female's genitals and attempt to copulate. If she accepts, the female parts her wings and enables the male to mount, (c). Copulation in this species lasts about 12 minutes.

metamorphosis into an adult fly. Thus *Fru* may well be involved in the laying down of the specific neural circuits for courtship. We must look forward to many other such examples before we can to begin to generalize, but there is the exciting possibility that the distinct and separate nature of some instinctive behaviour patterns may reflect a similarly dedicated genetic background. The complexities of gene expression in the developing nervous system which this would require are formidable but a wonderful field for research. Progress will most probably come from concentrating on those few species, e.g. *Drosophila,* zebra fish (*Danio rerio*) and mice, where detailed molecular genetics can be linked to interesting and complicated behaviour.

There are a number of textbooks dealing with the whole field of behaviour genetics, often concentrating heavily on evidence from humans because of its intrinsic and often controversial interest, e.g. Plomin *et al.* (2000), Plomin and Haworth (2009). Fuller and Thomson (1978) remains a valuable survey, especially for zoologists, because it covers more examples relevant to the evolution of behaviour. At the moment this is where a genetic approach can record more progress and we return to genetics in Chapter 6.

Development and changes to the nervous system

External events as well as genetic factors affect the way the nervous system develops. Gottlieb's work, described above, showed clearly that the auditory system of mallard ducklings requires experience – perhaps one might call it 'priming' – with certain types of sound if it is to develop normally and maintain its special responsiveness to the mother's calling. Specific external factors affect this behaviour through some effect on the developing nervous system. Obviously the development of behaviour must often run parallel to the complex process of building a nervous system. As with the study of gene mutations affecting the nervous system, sometimes we can deduce the changes underlying behavioural development only from observing changes in the physiological responses of neurons. Sometimes we can observe relevant structural changes to the nervous system directly.

To take a neurophysiological example first, it has been possible to study very closely the changes underlying the acquisition of visual responsiveness in some mammals. It turns out that the process shares some features analogous to the development of auditory responses in Gottlieb's ducklings. Thus the eyes of many mammals develop all their basic structure in total darkness, but do not develop their normal functioning unless 'primed' with certain types of visual experience. Blakemore and Cooper (1970) reared kittens in complete darkness except for brief periods in a very visually stimulating environment (Fig. 2.15a). From 2 weeks of age when their eyes opened they spent an average of 5 hours per day looking at a pattern of either horizontal or vertical stripes. When first brought out into the normal world at 5 months the kittens were, understandably enough, disorientated. However, they quite quickly managed to cope but their vision remained abnormal. This was revealed dramatically when a 'horizontal' and a 'vertical' kitten were put together. If the experimenters presented them with a rod held vertically and shaken the 'vertical' kitten moved forward to play with it, but the 'horizontal' kitten ignored it. When the rod was turned to the horizontal, the

(a)

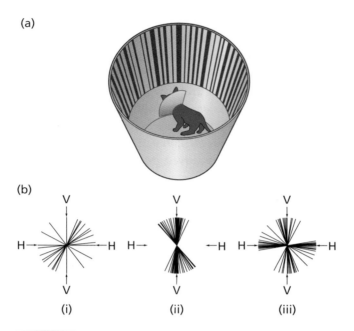

(b)

V · V · V

H → ← H   H → ← H   H → ← H

V · V · V

(i)   (ii)   (iii)

Figure 2.15 (a) Apparatus used by Blakemore and Cooper to restrict the visual environment of kittens. The kittens wore a black collar so that they couldn't see their bodies and stood on a glass plate inside a high cylinder whose walls carried vertical or horizontal stripes of varying thickness. (b) Diagrams – so-called polar graphs – of the preferred orientation of visual cortical neurons of kittens. Each line represents one neuron and they are arranged according to their preferred orientation around the marked vertical and horizontal axes. (i) Neurons from a kitten reared in total darkness; (ii) from a kitten after 33 h exposure to the vertical stripes in the apparatus illustrated in (a); (iii) from a kitten after 50 h exposure to both horizontal and vertical stripes but no other experience.

'horizontal' kitten immediately took interest and approached whilst the 'vertical' kitten no longer responded – it behaved as if the rod had suddenly disappeared.

Subsequently, Blakemore and Cooper used neurophysiological techniques to look at the functioning of neurons in the kitten's visual cortex. (This is the part of the brain which receives, in coded form, the visual information from the cat's retina.) Each of these neurons has a 'preferred orientation', which means that they fire when lines or edges at that orientation are presented to the eye. In a normal cat the orientations are distributed all around the clock, but vertically or horizontally reared kittens showed a very strong bias towards the corresponding orientation (Fig. 2.15b). Thus horizontal kittens had no units responding to vertical or to lines or edges within 20° either side of vertical, they literally cannot see a rod when it is held vertically.

Such experiments reveal clearly how the early environment, sometimes before hatching or birth, sometimes later, affects the physiological functioning of the nervous system and hence shapes the development of fully functioning behaviour. At the level of morphology too we can regularly observe that behavioural development takes place in parallel with normal growth processes both of the nervous system and the rest of the body. Thus the emergence of sexual behaviour in vertebrates depends on the growth of the gonads which begin to secrete hormones. Very often the nervous and muscular systems of young animals have to go through further growth or differentiation before they reach adulthood and there are attendant behavioural changes.

For example, young birds can often be seen making vigorous flapping movements with their wings while still in the nest and it is commonly supposed that they are practising flying. Similarly, human parents often support young babies on their legs and encourage them to 'practise' walking. In fact, there is no evidence that the early development of bird flight or human walking is affected in any way by such activities. Well over a century ago Spalding (1873) showed that young swallows (*Hirundo rustica*), reared in cages so small that they could not stretch their wings, flew just as well when released as normally reared birds. By the age of 18 months, when the majority of children are walking, their skills are very uniform whether they walked first at 9 months or at 15 months. In both cases it is the maturation of the central nervous system and its coordination with muscular development which count. Practice, of course, eventually adds all the finer points of skill – young fledglings are clumsy fliers – but the basic pattern emerges without it.

Because the developing embryos of amphibia and birds are relatively accessible, it has been possible to investigate more closely the growth of their nervous systems and behavioural changes. As development proceeds, the increasing complexity of an embryo's structure is paralleled by an increasing repertoire of behaviour, both spontaneous and in response to external stimuli. Oppenheim (1974) describes work on chicks showing how movements appear quite suddenly just at the stage when the requisite connections between growing nerve fibres are made. Sensory fibres grow in to the spinal cord from the muscle spindles and then can provide information on the state of contraction of a muscle (Chapter 1, p. 29). These fibres must form their central connections before that muscle can take part in coordinated movements with others.

The gradual nature of development in crickets enabled Bentley and Hoy (1970) to study one example of neural growth in great detail. Crickets go through a series of 9–11 larval or nymphal stages (instars) which become increasingly similar to the adult with each successive moult. The final moult involves the biggest changes, with the full development of wings and adult genitalia. Since they have no wings, cricket nymphs do not fly, but some elements of the flight pattern can be detected as early as the 7th instar. Nymphs at this stage will adopt the typical flying posture (see Fig. 2.16) when suspended in a wind tunnel. All the muscles necessary for flight are present from an early stage, albeit reduced in size. Bentley and Hoy inserted fine wire electrodes into some of these muscles of the thorax and recorded action potentials from them. Since the relationship between input from motor neurons and the output from the muscles is relatively simple in insects, the muscle activity gives a very accurate picture of the nerve activity responsible for it. Bentley and Hoy could detect signs of the rhythmic nerve

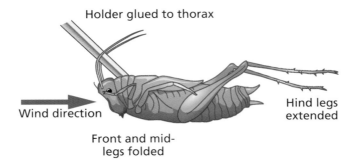

Figure 2.16 A 7th instar nymph of the field cricket (*Teleogryllus commodus*) adopts the flying position when suspended in a current of air, although it has no wings. The antennae point ahead, the fore- and mid-legs are drawn close to the body and the hind legs are held straight back parallel to the body. All the features are identical to the position of the adult in suspended flight.

impulses characteristic of flight from 7th instar nymphs, but it was incomplete and not sustained. It became more complete with each successive instar; new elements could be identified as becoming active, presumably as functioning synaptic contacts were established. Although timing varied between individuals, the units always developed in the same sequence until, by the last instar before the final moult, the entire pattern was complete. At this stage then the maturation of the nymph's nervous system is finished and it awaits only the development of wings. Substantially the same story is true for the cricket's song, another activity which requires the wings of an adult (they are rubbed together rapidly to produce sound) for proper expression. For both flight and singing Bentley and Hoy found that they could not elicit neural activity from the thoracic centres of nymphs unless they made lesions in the brain. This presumably removed inhibition, which normally prevents the nymphal muscles from being stimulated into useless activity until the rest of the body's development has caught up.

The crickets and their relatives show gradual development towards the adult form, which provides a gradually growing morphological framework in which to build up the neural basis of adult behaviour. Insects with complete metamorphosis have to reorganize their nervous systems quite radically during the pupal stage. In experiments similar to those on crickets, Kammer and Rheuben (1976) could detect patterns of activity in the developing muscles of the pupae of some large moths which indicated that the neural basis for both flight and warming up (shivering) movements had developed before half-way through the 21 days of the pupal stage. Elsner (1981) reviews these and other studies in more detail.

Studies on the changes to the nervous system of insects during larval growth and metamorphosis have revealed some remarkable parsimony in their development (Levine, 1986). Insects sometimes modify and reshape the same neurons for use in both larval and adult forms. Levine and Truman (1985) have shown that in the hawk

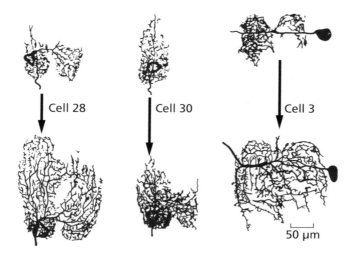

Figure 2.17   How three motor neurons of the moth *Manduca* are 'remodelled' during metamorphosis. Individual cells can be identified regularly in all moths and injected with a cobalt dye, which reveals the rich arborization of the dendrites. The top row shows the relatively simple dendritic pattern of each cell in the larva – the cell body and axon to the larval muscle are clear. After metamorphosis into the adult, each cell has a greatly enlarged dendritic field, which has grown into areas not penetrated in the larval state and the axons have, in fact, lost their target larval muscles and now innervate new muscles in the adult moth.

moth, *Manduca sexta*, all the motor neurons supplying the abdominal muscles are used again. Different as the two body forms are, the abdominal movements of the caterpillar and those of the adult moth may not be totally dissimilar. Some of the motor neurons can stay innervating the same muscles, but others lose their specifically caterpillar connections at pupation and grow new processes to innervate newly formed adult muscles (Fig. 2.17). Most larval sensory neurons supplying sensory hairs die away at pupation but one group come to innervate specialized trigger hairs on the abdomen of the pupa. They trigger a reflex, unique to the pupa, which causes it to flex suddenly in a movement which defends it against predators.

And so behaviour patterns mature, flourish and disappear, correlated with growth, modification and death of the underlying neurons. For instance, Truman *et al.* (1992) describe the way programmed cell death removes neurons which have become obsolete once larval life is over. It all looks very much like the running of a pre-set programme of development, but this is a description not an explanation and close examination of neuro-embryology can help us to understand what 'pre-set' implies. Such work certainly exemplifies how neurobiology is an aid to studying behavioural development itself with insects providing excellent material for studying the way in which a nervous system is built up. For example, in grasshoppers the axons of sensory neurons from the developing legs grow out from cells in the sensory hairs and make connections with the central nervous system. Bentley and Keshishian (1982) have shown that the first axons to grow this way – the pioneers as they call them – use special cells situated at intervals along the route as 'guide posts'. As the limb grows and lengthens, the guide posts are stretched apart and later axons do not use them but grow along the line already established by the pioneers. Thus, although the final result looks uniform – sensory axons running smoothly into the central nerve cord – the actual developmental histories of those neurons have not been uniform. There is also considerable flexibility in the way in which detailed synaptic connections are formed as sense organs develop and their sensory axons reach their central destinations. Which cells they connect with and how they do it depends upon what other cells are in the vicinity and what sort of inputs are arriving from the sense organs. If inputs are abnormal or the cells with which they would normally have synapsed are missing, neurons may be able to compensate, at least to some extent. There must be a genetically based programme for development but it is much more likely to lay down ground rules rather than a specific point-to-point blueprint. Murphey (1985, 1986), Stent (1980, 1981), Keshishian and Chiba (1993) and Katz and Shatz (1996) provide interesting discussions of such evidence which is certainly relevant to the development of behaviour itself and its genetic basis. We shall meet similar concepts of flexibility, predispositions and sensitive periods when we come to discuss the development of bird song (p. 94), where apparently similar behavioural end points may have been attained by markedly different paths.

Hormones and early development

All behavioural development will involve changes to the structure and functioning of the nervous system, sometimes quite clear-cut, as in the examples just given. More often we observe changes during development where a complex of factors is influencing the behaviour we are studying and the fascination rests in teasing this apart and trying to understand how the different influences are operating.

A good place to begin is with the development of sex, which certainly involves a whole range of interactions between structure and behaviour. As we mentioned earlier, sex represents a clear case of alternative pathways along one or other of which an animal's development, both morphological and behavioural, is switched by an early 'triggering event'. Because being male or female is such a pervasive feature for ourselves and the animals we are most familiar with, it is important to keep some broader biological perspective and remember that a good number of invertebrates are hermaphrodites and have only a single developmental pathway to becoming both male and female. Even more strange in some ways, a number of fish species retain the capacity to switch paths and behavioural factors are involved in this change (see Francis, 1992). For example, in the coral reef fish *Anthias* all individuals develop into females, but some transform later into males. They normally live in mixed groups and if a male dies, one

Figure 2.18 The coral reef fish *Anthias squamipinnis*; male in the foreground, female behind. Within 3–6 days of the removal of the male from a group, females begin to show colour and fin changes towards the male type.

of the females changes sex. This change is very rapid: it takes only a few days for the previously female fish to develop male features and begin making sperm in its gonads rather than eggs. But even earlier, within hours in some cases, its behaviour begins to change and it is treated like a male by other females in the group (Shapiro, 1979; Fig. 2.18). As if to demonstrate that all variations are possible in nature, the anemone fish (*Amphiprion*) – also a coral reef species with a remarkable symbiotic association with large sea anemones – does the opposite! It normally begins life as a male, mating with a larger, more dominant female. If she dies, or is removed, the male changes into a female and another, hitherto non-breeding male takes up the fertile male role (Godwin, 1994).

In most vertebrates the embryonic gonad can develop either way. It requires a switch of some kind to direct it towards becoming a testis or an ovary and the switch is usually permanent. We are most familiar with a genetic switch from the sex chromosomes, female mammals have paired X chromosomes, males an X and a Y. In birds and in snakes, among the reptiles, it is the male which has paired sex chromosomes, labelled W whilst the females are heterogametic, one W one Z. This situation in birds and snakes probably evolved later because in most other reptiles – and birds are quite recent reptilian descendants – the triggering event is environmental. Sex is determined by the temperature at which the eggs develop and there are some interesting variants here (Crews, 2003). In some groups high temperature during incubation leads to a preponderance of males, in others just the opposite. In the wild, female reptiles probably tend to lay eggs where shade or direct sunlight will vary across the clutch and so their progeny will be suitably varied!

The fish which can change sex demonstrate that they inherit the potential to perform both male and female behaviour patterns. The same must be true for mammals and birds because the two sexes share almost all their genes. The 'problem' for natural selection has been how to make them develop differently. In morphology we can see that sometimes it has not been worth differentiating between the sexes. Male mammals develop functionless mammary glands and nipples; some male marsupials develop the bones which support the female's pouch. For those aspects of morphology and behaviour where sexual differentiation is vital, the switch is a hormonal one. Once the gonadal sex – ovary or testis – has been determined, so is the pattern of hormone secretion. Short and Balaban (1994) review the complex story of sexual differentiation, which has many interesting variants across the mammalian groups. Here we must summarize the main events as they affect behaviour.

In mammals, the differentiation of the genitalia and of those neural mechanisms responsible for initiating sexual behaviour is not determined at fertilization, but much later in development. Both males and females inherit a brain that can mediate both masculine and feminine behaviour; which pattern becomes dominant depends on whether the brain receives a pulse of hormone. The embryonic or newborn testis, unlike the ovary, has a brief period of activity and its hormone is picked up by particular regions of the hypothalamus. This is a small region of the brain much involved in regulatory activities of all kinds, and which controls the functioning of the pituitary gland that lies attached to and immediately below it.

If the developing hypothalamus picks up hormone during a certain brief period, it is switched along masculine lines; if not, it becomes feminine. The structural effects of this early hormone pulse are quite obvious, particularly in the preoptic nucleus, a discrete cluster of neurons so-called because it is located just anterior to where the optic nerves pass into the brain. This nucleus is much larger in males than in females. If females are given tiny quantities of hormone at the right stage, their preoptic nucleus develops along male lines and their behaviour is masculinized. It is interesting to note that either the male or female sex hormones, testosterone and oestrogen, respectively, will serve to masculinize male or female embryos – they are both steroid hormones and closely related in chemical terms. Denying male embryos their pulse of hormone at the normal time or removing the testes before they secrete, or injecting a chemical which inhibits the action of testosterone, results in de-masculinization. We may note that this is a clear case of a critical period in development (p. 84) for hormone treatment earlier or later has no effect. The die is cast and, for example, a male de-masculinized by early castration will not have his masculinity restored by large doses of testosterone given later in life. When adult, masculinized females show changed sexual behaviour, some-times dramatically so. Thus in some rodents, such as rats and mice, masculinized females will mount and thrust on receptive females and eventually go on to perform all the behaviour patterns which are shown by males when they ejaculate; they also show masculine-type aggressive behaviour (Manning and McGill, 1974). In addition, such females are 'de-feminized' in that their vagina fails to open, they lack oestrus cycles and are consequently sterile. Female primates can readily be masculinized and, as with the rodents, it is not only specifically sexual behaviour which is altered. They are more aggressive and, as infants, show more of the 'rough-and-tumble' patterns of play which are characteristic of males. However, unlike rats and mice, such females show little sign of de-feminization for they remain fertile and can reproduce. Actually, it is rather difficult to generalize about the effects of early hormone treatment on behaviour, because the extent and nature of the critical period varies between species. In sheep, for example, it extends for some weeks during mid-gestation and there are different critical periods for masculinization of genital development and for sexual behaviour itself. By closely time-limiting a female foetus' exposure to hormones, it is possible to produce sheep whose apparent genital sex and sexual behaviour do not match up. In primates, it is also the middle period of gestation which is critical but, in rats and mice, it is very late in development and, in fact, can be extended until a day or two after birth.

Careful observations have shown that the masculinization of females is not simply an artificially induced condition confined to animal behaviour laboratories. Female mice produce 8–10 pups after a 3-week pregnancy in which the foetuses have been developing in a row along each horn of the uterus, rather like broad beans growing inside their pod. Each has its own placenta but it appears that some mingling of hormones must occur between adjacent foetuses. vom Saal and Bronson (1980) found that, at birth, male foetuses had three times the level of circulating testosterone in their blood as did females – this is presumably a result of the pulse of secretion by the testis which we know occurs around birth. Having delivered mouse pups at full term by Caesarian section, they noted the position each female had occupied in the uterus. They compared females which had developed between two male foetuses (2M females) with those who had had two female neighbours (0M females) and found a number of significant differences. Most interestingly, they found significantly more testosterone in the blood of 2M females than their 0M sisters and they could correlate this effect – presumably due to leakage between adjacent blood systems – with a number of morphological and behavioural differences which extended into adult life. 2M had slightly masculinized genitalia, had longer and less regular oestrus cycles, showed more aggression towards other females and in defence of their young, and were less attractive to males than 0M in a choice situation. This does not necessarily mean they are at a disadvantage. vom Saal and Bronson suggest that under crowded conditions, the increased aggression shown by 2M females may improve their survival and that of their young. It is probably a case of swings and roundabouts in fluctuating environmental conditions. Such naturally arising variation within a litter may in fact be advantageous for the parents. Crews and Groothuis (2009) discuss some of the possible behavioural consequences and natural selection will not necessarily act so as to insulate female foetuses better from their brothers' hormones.

And so it appears that some of the normal variation we expect to find between females may have its origin in the chances of foetal life. For centuries it has been known that if a domestic cow gives birth to twin calves of different sexes, the female calf very frequently has malformed gonads and uterus and hence is sterile. This so-called 'freemartin' is an androgenized female resulting from her sharing some blood circulation between her placenta and that of her brother in their mother's uterus. In human beings too there are clear effects when female embryos are exposed to male hormones during pregnancy. These are not detectable when dizygotic twins are of different sexes but happen when a single female foetus has a rare genetic abnormality which leads to a condition known as Congenital Adrenal Hyperplasia (CAH). The adrenal glands of the foetus develop abnormally and release a masculinizing hormone into the bloodstream. When CAH girls are born their genitalia are often somewhat masculinized but this can be corrected with surgery and they can grow up to be fully functioning females in every sense. However, it is fascinating that the effects of masculinization can be detected in their behaviour as children. Ehrhardt and Baker (1974) have shown that CAH girls are more 'tomboyish'. For example, they indulge in more rough play, prefer boys to girls as playmates and tend to reject dolls as playthings. Now we know for certain that such preferences can be greatly influenced by parents and the particular culture in which children grow up (see Money and Ehrhardt, 1973). However, it seems very probable that the CAH girls were indeed treated as girls by their parents and that the changes to their behaviour were spontaneous.

Ehrhardt and Baker are cautious in their interpretation of how early hormones may act. It might bias infants towards specifically male patterns of behaviour; but in humans these would always be interpreted by a developing child in ways which reflect the 'male role' as presented by its culture. Of course, it may not be a specific effect at all. Tiefer (1978) has suggested that early androgen may do no more than increase metabolic rate and hence energy levels. An especially active girl infant will be most likely to end up playing with other active infants and they will be boys; hence the other male-orientated patterns follow. We may note that although the behaviour of CAH boys appears qualitatively unchanged, they do have raised energy expenditure.

As mentioned above, it is the preoptic nucleus of the hypothalamus which is the target for the pulse of early hormone which organizes the brain along masculine lines and we tend to think of such developmental effects as part of the growth process which ends with the attainment of maturity. However, the preoptic region in adults is periodically reactivated during reproduction. Until recently, we would assume that the structure of the nervous system would be fixed by the time that maturity was reached and that sex hormones in adult life would only reactivate neurons already in place. This is because with a few familiar exceptions, such as the extensive powers of regeneration in urodeles – newts and salamanders – it has been generally accepted that the nervous system of vertebrates shows no further growth or cell division in adults. However, new evidence has come from an unlikely source – the neural control of bird song – which suggests otherwise. Most of the work has been done with canaries (*Serinus canaria*), a species in which males acquire new song patterns in each new breeding season. Certain brain nuclei, which we can identify as those involved in the control of singing, are much larger in males than in females. Nottebohm (1989, 2005) and his group discovered that these nuclei wax and wane each year, growing in the spring when the

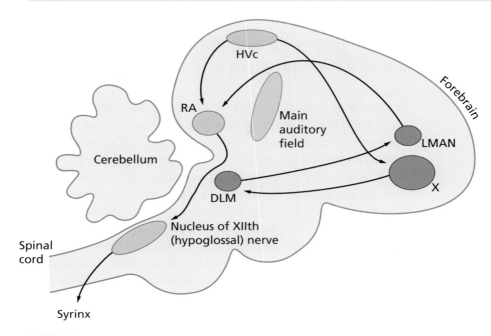

Figure 2.19 A simplified diagram of the main neural centres concerned with the control of song in a bird such as the canary. The brain is depicted in longitudinal section, anterior to the right. The dark-shaded nuclei are those, referred to in the text, controlling the descending motor pathway for song production. The unshaded nuclei, especially those of the forebrain, are known to be involved with song learning. Main nerve tracts linking the nuclei are indicated by arrows. The neural pathways between the 'learning areas' and the 'production areas' are quite complex and their operation is not fully understood. Hence one of the learning areas is, for the present, simply called 'X'. The 'production areas' are far larger in males than in females and it is especially in HVc where neural cells are replaced on a cyclical basis in canaries.

males are acquiring new songs only to regress at the end of the breeding season (Fig. 2.19). Their painstaking work has proved that neurons die as the singing centre regresses and that the next spring new neurons develop from neuroblast cells, just as they do when a young bird's nervous system is developing in the first place. At first, the world of neurobiology required some convincing that new cells were developing in the adult brain, but in the end the experimental results were clear. This work with canaries illustrates the value of a comparative approach which focuses on the diverse specializations of brain and behaviour which can be found if one ventures outside the conventional set of laboratory animals. In fact, we are now obtaining increasing evidence that, in mammals, neural changes, like behavioural changes, do not cease abruptly when animals become adult (Gould *et al.*, 1999; Johnson *et al.*, 2005). It is becoming more difficult to mark any clear division between the early behavioural development of young animals and changes which occur as they move into maturity and go through their lives.

Early experience and the diversity of parental behaviour

In a review of work on behavioural development and life history strategies, Monaghan *et al.* (2008) begin, 'The environment is not merely "permissive" of development, but to some extent also guides, or even induces it.' We are familiar with the idea that animals are particularly sensitive to events early in life, many of which may have lasting effects. However, if we look across the animal kingdom, it is obvious that the extent of such influences varies enormously. As outlined earlier, many short-lived invertebrates have a life history which runs through a pre-set programme, well adapted to the normal environment, but very limited in the extent to which it can adapt to changed circumstances. The moth *Manduca*, described earlier, has effectively two pre-set programmes: the first equips it for life as a caterpillar, this then shuts down and in the pupa a second programme develops for its life as a moth. *Manduca*'s environment must be supportive, but it does little to shape the insect's behaviour at either stage of life. Certainly there are no influences from its parents for many, although not all, insects develop as orphans! All their parents can do for them is lay eggs in a suitable place and containing enough yolk for growth.

With longer-lived animals there will usually be overlapping generations, more time and often more sensitivity to the kind of influences Monaghan is referring to. Direct parental influences are obviously going to be a key factor but before turning to this we should consider some striking examples of sensitivity to external events in the early life of mammals and birds, where most of the studies have been centred.

Gottlieb (1997) reviews his work on the early environment of birds which, as we discussed above, although apparently isolated are affected by experiencing sounds heard while still in the egg. Their responsiveness may have surprising results for a clutch of eggs in the nest can already act to some extent as a group. Vince (1969) showed that embryo quails respond to the clicking sounds made by the other members of the clutch. Remarkably, this communication between embryos serves to synchronize their hatching. Slow developers are accelerated by hearing the calls characteristic of advanced embryos and, to a lesser extent, advanced embryos are slowed down by the calls from less advanced neighbours. In nature this synchronization may be important, because it will reduce the dangerous period when there are both chicks and unhatched eggs in the nest before the mother quail can lead the whole brood away.

Unsurprisingly, young mammals prove to be very sensitive to circumstances soon after birth. Numerous studies with laboratory rodents have explored the behavioural sequels of a variety of treatments. For example, mild electric shock, cooling, or even the simple act of lifting babies from the nest and then replacing them can affect their subsequent behaviour as adults. Rats which have been handled during infancy mature more quickly – their eyes open sooner and they are heavier at a given age than unhandled controls. In addition, they appear less 'emotional' in frightening situations. By this we mean, for example, that when placed in a large, brightly lit arena they move around and explore more than unhandled rats, which spend more time crouching and cling to the edges of the arena if they do move. Such effects and the way in which early experience produces them are not fully understood, nor is it a very easy subject to follow because, as with many complex questions, different studies do not always yield consistent results.

Much of this kind of work forces us to recognize the wisdom of the ethological approach to behaviour – be mindful of the natural environment in which the species evolved. Almost certainly the only reason we can detect any effect at all from such apparently trivial manipulations of early life is that we are measuring it against the background of the laboratory-reared animal, which is deprived of so much stimulation. Rather than emphasizing how early handling 'adds' something we should accept that laboratory rearing takes something away. In the wild, it is common for a mother rodent to abandon her nest and shift the litter to a new site, sometimes more than once. Several mothers may share a nest at times and the environment around the nest, when the young begin to emerge at about 12–15 days of age, is far more complex than a laboratory cage. A number of studies have found that rats and mice which grow up in 'enriched', i.e. more natural, environments show improved learning in a variety of situations. For example, Joseph (1979) reared rats under two conditions, one group were kept in standard laboratory cages, the other group were given a variety of 'toys' – cups, ladders, corks, bottles, etc. – and these were changed regularly. Also their cages were moved around within the animal-house room every other day, whereas the control group's cages remained in one position as is normal practice. The enriched rats proved to be markedly superior in maze learning and they also explored new environments more actively. There are a number of such studies and we know something of the neural changes which underlie such effects on behaviour. It has been possible to show that the brains of enriched animals are slightly heavier than those of unstimulated controls, they have a richer pattern of interconnections and they contain higher concentrations of key enzymes concerned with neural functioning (Rosenzweig, 1984).

But, of course, with these rodents, as for all mammals, a key factor in their development is the behaviour of their parents, especially the mother. In a few mammals, a father may also be involved in protection and bringing food, but gestation and lactation can involve only the mother who is bound to be the dominant. In other vertebrates, other life histories have different outcomes. Where fertilization is external as in most fish, it is often the male who cares for the eggs and the developing young, such as with sticklebacks and seahorses. Parental care is not common in reptiles, typically young emerging from their eggshells are on their own, but crocodiles, for example, show quite elaborate maternal care both before and after hatching (Fig. 2.20). Birds almost always lay their eggs in some kind of constructed nest and one or sometimes both sexes must attend their incubation. The majority of the familiar perching

(a)

(b)

Figure 2.20   Crocodilians are unusual among reptiles for providing considerable maternal protection to their young. (a) A tiny young crocodile takes shelter in its mother's mouth but (b) even quite large young may travel on their mother's head as in these alligators.

birds – the passerines – require insects to raise their young and usually both parents are needed to feed the developing brood. They act as a monogamous pair although, as we shall discuss in Chapter 6, broods very often have more than one father! Passerine young hatch as helpless nestlings which require warmth and all food to be brought to them in the nest until they have fledged. The young of game birds or ducks are much more independent at hatching. Their mother, sometimes their father too, can lead them to feeding areas and show them food, but they feed themselves from the outset.

We described above how the calling of young game birds whilst still in the egg helps to synchronize the hatching of a whole brood. Here, we might say, the developing young are acting to improve their own survival. But in some birds we now know that the mother can affect the development of individuals from the outset. Modern techniques have made it possible to detect minute levels of hormones in blood, urine, saliva or, as in this case, the yolk of eggs. Females sometimes vary the amount of maternal androgens deposited in the yolk of their eggs, doing so in a systematic way across the successive eggs of a clutch, usually laid one per day. Even if incubation is delayed until the last egg of the clutch is laid, the first eggs nearly always get a head start in development. They are first to hatch and begin feeding, thus the last hatched may find themselves in competition with larger siblings. Canary mothers 'compensate' for this disadvantage by depositing more androgen into later eggs. The effect of this tiny hormonal boost is not to influence later sexual behaviour, as in the mammals we have just been describing, but it does raise the overall vigour of such nestlings and, in particular, increases their rate of food begging just after hatching. The parents respond to this and the late hatchers are helped to catch up. We can understand the advantage to parents under normal conditions with good food supplies – the number of successfully fledged young will increase – but interestingly not all parent birds can accept this. Like some owls, cattle egrets (*Bubulcus ibis*) regularly lay more eggs than they can normally rear and they begin incubating early so hatching asynchrony is marked. Commonly they lay three eggs and only two young are reared. Unless food is unusually abundant the last hatched nestling is killed by the others. Female cattle egrets do the exact opposite of canaries. There is more androgen in the first laid eggs so the differences in strength and success are emphasized. Schwabl *et al.* (1997) suggest that it benefits the parents to have competition in the nest rapidly resolved whilst still allowing a third nestling to be raised if conditions are very good. These are examples where the mother selectively androgenizes particular eggs – variation within a clutch – but studies on black-headed gulls show there are also variations between clutches of eggs within a colony. Pairs which have larger territories have overall more testosterone in all their eggs, although they still favour later eggs within their clutch, like canaries. Perhaps this is a measure of extra vigour in the females on such territories and it may possibly give their offspring some advantage (Groothuis and Schwabl, 2002; Crews and Groothuis, 2009).

We might regard these as examples of parents taking direct action to optimize their success in reproduction, but, of course, all these varied styles of early life among the vertebrates demand appropriate behavioural reactions from the developing young. They cannot just be passive recipients of parental attention. Always they must respond effectively in their own best interests, interacting with their parents, so as to emerge as fully functioning adults. In Chapter 5 we discuss how the interests of parents and offspring are by no means identical – conflict may be rife! However, at this point we can concentrate on the more positive aspects of parental behaviour.

A great deal of the experimental work has involved mammals, where maternal behaviour is so dominant. Much of the so-called 'early experience' or 'enrichment' interventions described above breaks into the normal intense and almost continuous communication system that operates between a mother rat or mouse and her growing litter. The pups provide a variety of stimuli and, in particular, have a range of calls, the most important of which are ultrasonic and far above the limits of our hearing. Disturbing the pups in any way – as when they are handled – leads to an increased rate of calling and the mother responds (see Cohen-Salmon *et al.*, 1985, for an introduction to such work). A litter of pups may get scattered and if, for example, some towards the outside of the nest get chilled, their increasing outburst of calling will accelerate the mother's retrieval of them back to the nest cup. As the pups get older their normal rate of calling decreases, they move around voluntarily and the responses of the mother are less intense. The behaviour of both mother and developing young is reasonably balanced so as to promote successful weaning. Elwood (1983) and Fleming and Blass (1994) have reviewed a range of work on parent–offspring interactions in rodents.

Maternal influences do not just begin at birth. In mammals, the embryo is more insulated from the external world but, of course, more directly dependent on its mother's physiological state. If female mammals are stressed during pregnancy this may have significant effects on the development and behaviour of their offspring. They may become more fearful in their approach to novelty and their sexual development may be affected (see Sachser and Kaiser, 1996, for an example from rodents and Clarke *et al.*, 1994, for one from primates). There is now much interest in exploring how far in responding to her current environment a pregnant mammal may 'equip' her developing offspring to be ready to cope with the world outside. A female who is restricted in food supply may have smaller young who may be better able to cope with a poor diet. Conversely, abundant food may result in large young not only given a head start, as it were, but well adapted to deal with a rich diet. All this will increase fitness in each case but problems will arise if the environment changes during pregnancy and, accordingly, the young are not well matched. Bateson *et al.* (2004) discuss a number of examples, many of them from humans, where we have an abundance of evidence, although its significance and interpretation are still uncertain.

Neel (1999) has suggested that our long human history as hunter-gatherers has programmed us to cope with food shortage interspersed with periods of abundant food – a huge animal carcass, for example – upon which you must gorge in order to get through the next lean period! Now at a time when nutritious food is easily available (for fortunate people!) there is too much gorging. There is no doubt that the developed world is currently experiencing what amounts to an epidemic of obesity, quite often associated with the development of diabetes. In its simplest form, the hypothesis that we have inherited eating habits from our evolutionary past has not been supported (see Speakman, 2008). For a start, it underestimates the diet of hunter-gatherer societies, which is often very good. However, it is notable how attractive abnormally high sugary and fatty foods are, not just to humans but other species which, in nature, will rarely come across them. Hill sheep, able to subsist on poor vegetation, are powerfully attracted to sugar when, as at human picnic sites, it suddenly becomes available! Presumably, selection will have favoured sensitivity to the richest patches in an otherwise low-grade nutritional environment.

(b)

(a)

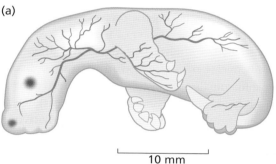

10 mm

Figure 2.21 The maternal style of kangaroos – a marsupial. (a) A newly born young. It is minute in relation to the size of its mother and is effectively still an embryo. The clawed forelimbs help it to haul itself up through its mother's fur into her pouch. The prominent blood vessels run just beneath a moist skin and at this stage the young kangaroo may well breathe as much through its skin as through its embryonic lungs. (b) Many months later the young (the 'joey'), although now sub-adult, still retreats into its mother's pouch when alarmed. One of the long hind limbs sticks out awkwardly over its shoulder.

Although natural selection may have determined how the physiology of gestation affects young mammals, we must be struck by how varied are parental influences and parental contact in different species once they are born and separate from their mother's body. The details of the developmental pathways towards behavioural independence must be equally diverse. For example, marsupial infants are born after a very brief gestation period and they are effectively no more than externalized foetuses at first, fused to the mother's nipples. Their sensory capacities are so restricted at this stage that mother–infant interactions can only be through the milk supply, but even when they become more independent they are constantly with the mother, carried on her body and often in a pouch (Tyndale-Biscoe, 1973; Fig. 2.21a, b). This represents a far higher degree of contact than the part-time attention received by baby rats and mice. At the other end of the scale are rabbits (*Oryctolagus*) where Zarrow *et al.* (1965) showed that the mother visits her litter only once per day and then for only a few minutes. Milk is pumped into the babies, they are briefly groomed and then the nest is covered and left; the mother herself has another nest chamber elsewhere. Presumably this behaviour helps to minimize the chances that predators will find the nest by picking up the mother's scent. A similar explanation is suggested for the yet more extreme maternal behaviour of tree shrews (*Tupaidae*), a rather distinct group of small insectivore-like creatures probably with some distant linkage to the primates. Martin (1968) (see also D'Souza and Martin, 1974) observed pairs breeding in captivity and found that they build two separate nests, in one of which the adults sleep while the female produces her litter, usually of two, in the other. The young are visited briefly on every *second* day; they suckle rapidly and become distended with milk. The mother then leaves them with the very minimum of grooming and typically does not return for *48 hours*. Martin discovered that the young are capable of grooming and licking themselves from the day of birth. They stay alone in the nest until they emerge at 33 days of age.

For a further contrast, we may compare the tree shrews, whose maternal attention, although brief and intermittent, extends over 4 or 5 weeks, with the extraordinary specializations of some seals. These large mammals come ashore only to breed and for some solitary Arctic species their time on land, or more often on floating pack ice, is very hazardous. Ice floes are always liable to break up and move off in sea currents and even if they remain stable, mother and pup are threatened by foraging polar bears. Perry and Stenson (1992) and Riedman (1990) describe how hooded seals (*Cystophora cristata*) have adapted to this situation. The female hauls out onto an ice floe and almost immediately gives birth to a single pup, which is born with a thick layer of blubber beneath the skin. She remains alongside it continuously with frequent and intense bouts of suckling (Fig. 2.22) and the pup grows at a phenomenal rate: the seal's milk is some 60% fat. The mother then abruptly abandons the pup to return to the sea and mate; this whole process from birth to desertion all takes place within *four days*!

This is a very remarkable type of development for a mammal and it forces us to recognize that natural selection can produce a range of specializations in the pattern of development which are matched to other aspects of a species' life history. The influence of the mother on the earliest stages of behavioural development must be very different in the marsupial, rat, rabbit and tree shrew. Even close relatives can vary enormously in their stage of development and independence at birth: we can contrast the helpless blind naked young of rats and rabbits with the active young, fully furred and with eyes open, born to guinea pigs and hares. The young of most ungulates struggle to their feet and move off with their mother within minutes of birth – in most cases it would be fatal to stay put. As we have seen, some seals are suckled for a few days only but Californian sea lions stay close to their slow-growing offspring for 6–9 months. There are no predators menacing them on the rocky islets where they breed and mothers can go off to feed between bouts of suckling.

Figure 2.22 A hooded seal mother with her pup resting on Arctic pack ice where she gave birth. The pup is born with a layer of blubber and, after 4 days of intense lactation, the mother will leave it to fend for itself.

Not surprisingly, it is with long-lived, highly social animals that parental behaviour is most richly developed. Our closest relatives, the primates, have some of the longest and closest associations between parents and offspring, lasting for 6 or 7 years and beyond in the great apes, as in ourselves. The relationship between a primate mother and her offspring is absolutely crucial for almost every aspect of behavioural development. Separation, particularly in the early months, can have very severe effects on the infants. In the experiments of Harlow and his group (Harlow and Harlow, 1965) young rhesus monkeys (*Macaca mulatta*) were isolated at birth and reared artificially. Although they grew well enough, they were behaviourally crippled and, when subsequently put with other monkeys, showed almost none of the normal social responses. In particular, they showed totally inadequate sexual and parental behaviour. Male isolates did not respond normally to receptive females; if female isolates became pregnant and gave birth, they totally ignored their offspring.

This is another illustration of the subtlety of behavioural development. If we watch a young monkey with its mother, it is not easy to identify in its behaviour those processes which are contributing crucially to the infant's social development, yet they must be there. The work of Hinde's group, well described in his book (1974), followed in great detail the normal development of rhesus monkeys reared by their mothers living in small groups. They could trace the gradual growth of independence due in part to the infant, in part to the mother. At first the infant is scarcely ever out of contact. The mother rarely allows her infant to move beyond arm's reach and even when it leaves its mother to explore it returns frequently, using her as a secure base from which to investigate new objects and new situations. Gradually, in parallel with the infant's increasing independence, the mother becomes less solicitous and even begins to reject some of the infant's approaches (Fig. 2.23).

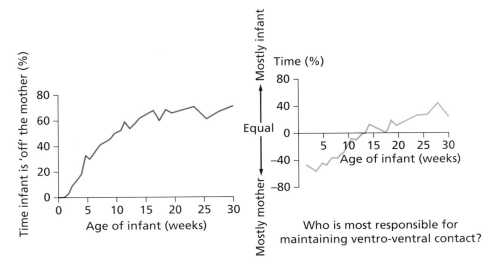

Figure 2.23 Two measures Hinde and his group used to define the manner in which the mother–infant relationship in rhesus monkeys changes as the latter grows up. (a) The increasing proportion of time an infant spends off the mother. This means not in contact with her body, although they are often very close; (b) is a more complex measure. Ventro-ventral contact (V-VC) means the infant is clinging beneath its mother and often clasped by her. Here the observers record for each bout which of the two is most responsible for initiating and maintaining this close contact. V-VC diminishes with age but clearly more and more often it is the infant who seeks to cling in this way when contact with the mother is made – often when the infant seeks reassurance after exploring away from her. Further details are given in Hinde (1974).

Knowing the normal course of development, Hinde and his co-workers then went on to study how it was disturbed by forced separation of mother and infant. It is now universally acknowledged that the deliberate isolation of baby monkeys from their mother is unacceptable. We can gain a better understanding of the normal processes of development by arranging levels of maternal deprivation far less drastic than those imposed by Harlow. In a typical set of observations, a baby rhesus and its mother, living in a group, were watched regularly for some weeks. Then, when it was 6 months old – well able to feed itself – its mother was removed from the group for a few days. Far from being isolated, the baby was usually 'adopted' by other females of the group and given a great deal of attention – 'aunting' as it is called. Nevertheless, its behaviour showed a marked change, distress calls increased, it moved around less and spent long periods in a very characteristic hunched posture. When the mother was returned, there was usually an instant reunion and the infant spent much more time clinging to her than just before separation. The pattern of their relationship was different from that normal for a 6–month infant and it took several weeks to recover.

Several fascinating conclusions, with obvious human parallels, have emerged from systematic studies of this type. For instance, the infants worst affected by the separation were those whose relationship with their mother before separation appeared to be less good. From our point of view, in this discussion of early experience and development, we may note how profoundly the normal smooth adjustment of mother–infant relations was affected by even a brief interruption. The effects on the young monkey were both general and persistent, for Hinde could detect even years later that monkeys that had been separated were more fearful in strange environments.

Just as within other mammal groups, so observations on primates in the wild have revealed striking differences between species in what we would call child-rearing practice. Thus, mother langur monkeys are quite permissive in that they allow other females to take their babies from time to time. Rhesus monkeys are far more restrictive, scarcely allowing their babies to leave their side for several weeks and controlling outside contact for much longer. However, a close relative of the rhesus, the Barbary macaque (*Macaca sylvanus*) of North Africa, behaves quite differently. Mothers allow other individuals, males as well as females, to hold their babies even within a few days of birth. Deag and Crook (1971) described how young babies can be seen moving about in the group not obviously under the care of any particular adult. It is a common observation for all primates that newborns and young generally are objects of intense interest to other members of a group, particularly young females. 'Aunting' behaviour has frequently been reported from primates and is also an important part of parental behaviour within the matriarchal groups of African elephants. They too have a long infancy and a wonderfully rich social environment (Moss and Colbeck, 1992). Lee (1987) has shown that young elephants do best if they grow up in a group where they have not just their mother's attention, but that of other females and juveniles as well.

Even within a single species mothering styles can vary substantially, depending on circumstances. Altmann's (1980) long-term study of baboons describes the behaviour of young adults, and suggests how differences in the social, exploratory and feeding behaviour of young animals relate to the type of mothering they have received. Are there then 'good' and 'bad' mothers and ought not natural selection to eliminate the latter? Altmann's conclusion is that in a varying and uncertain environment, restrictive or permissive mothers may sometimes be 'good', sometimes not. Perhaps natural variations in the pathways of behavioural development are not, over a lifetime of reproduction, disadvantageous but quite the opposite, a point we have already made (p. 65) when considering variations in the development of females within a mouse litter.

As hinted above, we need to recognize that the interests of parents and young do not always coincide. Trivers (1974), in an influential review, referred to 'parent–offspring conflict'. The latter always wish to get the maximum help from their parents before they are forced to face the world alone. This will match the interests of parents of those species which reproduce only once – many insects, for example. The more they invest in their young the more descendants they will leave. But many birds and mammals, for example, live on to reproduce more than once in their lives. They have to balance their investment in the current set of young with how such investment will leave them placed to invest in the next clutch or litter. They may well need to husband their resources and, as we have seen, some of their behaviour may be directed towards accelerating the independence of their offspring and even to culling their numbers if times are hard. We return to this issue in Chapter 6, for it has important implications for the evolution of behaviour.

Play

For anyone who has enjoyed watching kittens or puppies grow up, the most conspicuous aspect of their behaviour is play. They spend long periods playing with each other and with objects but, although adults may continue to play intermittently, it does tend to fade out with age. It is commonly suggested that play has a role in the development of adult behaviour. Indeed, there is an enormous literature on the importance of play in infancy and childhood for the normal development of human beings (e.g. Bruner *et al.*, 1976). It is certainly a compelling aspect of the life of young animals where it does occur but we must recognize that, looking across the animal kingdom, it is actually very rare. Fagen's book, *Animal Play Behaviour* (1981) still offers the best comparative survey (see Fig. 2.24). We have good, undeniable records only from mammals and a few birds. Some observers suggest that some play-like behaviour occurs in reptiles and fish – see for example accounts and illustrations in Burghardt (2005) – but not everyone will be convinced. When it comes to invertebrates, play becomes almost unimaginable and it is valuable to consider why we feel this. Partly it is time available, so many invertebrates live very short, rigidly programmed lives, but probably also because implicitly we think of play as requiring some level of intelligence, flexibility and social interaction.

This may be so, but we must also think in objectively functional terms too. If we observe that animals spend significant periods of time engaged in a certain activity, we would, as a first assumption, conclude that it is important for survival. Play is a very energy-consuming activity in ourselves, but there is some debate as to how far this is true of other species. Estimates range from between 5 and 20% of the 'surplus' energy available, i.e. that not required for metabolism and for growth, is expended in play. Even at the lower figure, this is not a trivial amount and play clearly does impose a cost. The cost is not just in energy; young desert bighorn sheep (*Ovis canadensis*) cavorting in play through their harsh environment are often pierced by cactus thorns; young gelada baboons (*Theropithecus gelada*) playing on the steep cliffs of Ethiopian gorges have been seen to fall 5 or 10 m and limp painfully away! Another significant cost arises because vigilance is bound to be reduced when young animals are playing and this will make it potentially dangerous. For example, Harcourt (1991) finds that young fur seals (*Arctocephalus australis*) spend about 6% of their time in play swimming with others. They are preyed upon by sea lions and over 86% of such casualties take place when they are playing in this way. Caro (1995) noted that their play certainly made cheetah cubs more conspicuous to him as a human observer and presumably also to lions and spotted hyaenas which are known to kill them.

It is among the ungulates, carnivores, whales, elephants and primates that play becomes very conspicuous. (We may note that these groups are certainly those with relatively large brains and they often live socially.) Each group has its own typical play repertoire; ungulates butt each other, chases and mock fights are common among primates, whilst carnivores show elaborate stalking, leaping and all aspects of prey-catching behaviour. The behaviour is often repeated again and again and has an exaggerated form, often switching from one kind of behaviour to another. Play is certainly easier to recognize than to describe just because it is so variable in form and may change from second to second. Adults may join in and, in some animals, there are special postures which indicate that an individual wants to engage in play (see p. 149). All this may give the impression that play is clearly distinguishable from 'real' behaviour, but this is not always true. It can be quite difficult to identify play with certainty, particularly as animals get older, and we have all observed that animals themselves may have the same problem, for mock fights may change rapidly into real fighting as one partner really gets hurt.

When discussing parental care we had to accept that, although there could be little doubt that the interchange between mother and offspring contributed to the healthy development of the latter's behaviour, it was difficult to identify exactly how it operated. The same problem is met even more forcibly with play. As we watch mock fights among kittens or the amazing high-speed chases and wrestling bouts of young monkeys, it is difficult not to believe that such behaviour plays a part in the development of skills which are going to become vital for survival in adult life. Apart from the obvious physical skills, numerous other functions have been suggested for play – gaining knowledge about potential prey species, gaining knowledge of the social group and one's position within it, exploration of the environment and so on. Spinka *et al.* (2001) believe that the very chaotic nature of so much play may be its most important feature. It prepares animals for the unexpected. Suddenly you are flung onto your back, or slip from a branch or detect an opportunity to switch from the defensive to the offensive. All this, they suggest, builds up flexibility of responsiveness in every sense and hence will improve survival. There is some support for this idea for play in young mammals takes place at a time when crucial maturation events are occurring in the brain and muscles which lead to the differentiation of synapses in the cerebellum and the control of complex movements (Byers and Walker, 1995). Of course, we cannot be certain of the significance of this association in timing, but it could be that play provides some kind of behavioural template around which such neural and muscular development can take place.

However reasonable suggestions for a function of play may seem, it is certainly not easy to show convincingly the effects of play itself on subsequent adult behaviour. Although we may not understand how parental care operates, we can at least

Figure 2.24 Animals at play! Some examples to illustrate its delicious variety. Animals often play alone with objects (g) but social play between young animals is particularly common, (c), (d) and (f) although some adults also play (h). Mothers, especially carnivores, frequently play with their offspring (a). Play is not totally confined to mammals, (i) and (j) are playing ravens. The lower of the two birds in (i) is upside down, preparing to grapple with the other. (j) is a drawing from observations made by Heinrich and Smolker (1998).

(f)

(g)

(h)

(i)

(j)

Figure 2.24 (*cont.*) The bird repeatedly slid on its back down a snow slope. It had to go head first because of the lie of its feathers. Reaching the bottom it flew back to the top to slide down again! (b) shows two species which normally have no social contact racing round a field together. Play can cross species boundaries. Finally (e) illustrates one of the subtleties of play. The large dog weighs twice as much as the smaller. It could easily pull the latter along or take the rope, but it 'self-handicaps' itself in order to keep the game going. Further discussion of all these issues in the text.

demonstrate the effects of its deprivation, as in some of the work just described. However, it is very difficult to deprive a young animal of the opportunity to play without also depriving it of other things, such as social contact, which we know have drastic results anyway. Sometimes field observations over several years may offer some useful comparisons. One such natural experiment comes from Lee's (1983) work with vervet monkeys (*Cercopithecus aethiops*) in East Africa. Her observations extended over several seasons and included periods of severe drought. She noted that play which was such a conspicuous feature of the infants and juveniles during normal seasons, and which occupied a great deal of their time, virtually disappeared during drought. All animals, including the young, managed to survive only by constant searching for food. When Lee compared young adults which had grown up during the dry period – and hence had been deprived of normal play opportunities – with the rest, she could detect no differences in their behaviour. In Chapter 7 we describe some of the complex behaviour of African meerkats, a strongly social species of mongoose. Young meerkats play a great deal and in a very through study Sharpe (2005a, b, c) was able to quantify the variation between individuals, who else in the groups they played with and how successful they were in play fights. She went on to measure several features of domin-ance and cooperation between adults but was unable to find any correlation with their play behaviour when young. We shall probably look in vain for any general conclusion and, for example, with a very different animal, North American brown bears (*Ursus arctos*), Fagen and Fagen (2004) used a different measure of play's outcome. They found better survival in brown bear cubs which had played more in their first season and they were convinced that this could not be explained by other aspects of fitness, food, maternal care and the like. Again, Chaloupkova *et al.* (2007) and Dudink *et al.* (2006) have shown convincingly that good opportunities for social play in litters of young pigs (*Sus scrofa*) reduce signs of stress when they are removed from their mother to be weaned and reduce aggression when they are feeding as a group. Indeed, the occurrence of play in domestic animals is often taken as a sign of good welfare – that they are content.

Research on play is unlikely to yield clear and easy answers. However domestic animals such as pigs, cats and dogs do offer some opportunities for a more experimental approach. The very conspicuous play behaviour of kittens makes them good subjects and, additionally, they develop rapidly and their period of most intense play is over by about 3 months. Bateson and his collaborators devised a simple type of enclosure which enabled them to observe the behaviour of a mother cat with her litter and to quantify developments in play as the kittens grew up. They could distinguish two fairly distinct phases of play, the first from about 4 to 7 weeks was largely 'cat-orientated' play with each other and with their mother. From 8 weeks onwards, more attention was directed towards objects in their environment, playing with ping-pong balls, a small stuffed dog, etc. It was noticeable that in single-sex litters male kittens showed more object play than females, but this difference tended to disappear when brothers and sisters grew up together.

Mothers often take part in play sessions and their influence can be detected indirectly as well. If early weaning is forced upon a mother by administering a drug which suppresses milk production or simply by keeping her on short rations, the kittens respond by eating more solid food. This is an obvious result, but the signalling of early weaning by reduced milk supply also leads to the kittens accelerating the onset and increasing the amount of their play (Martin and Bateson, 1985a, b; Bateson *et al.*, 1990). This certainly suggests that play is not just a casual activity for the kittens, but something they need to do in order to function well later as independent animals.

Perhaps the kittens' object play is related to special feline skills. With some members of the cat family there is one very conspicuous type of play which does seem particularly directed towards the development of prey-catching. We have all noticed that domestic cats bring home live prey which they proceed to 'play' with, letting them loose only to pounce on them again, often repeatedly and to our distaste. Cheetah mothers may bring back a crippled gazelle fawn. In the natural context it makes grim functional sense for mother cats to loose live prey close to their litter. It will familiarize them with suitable prey types and also gives them repeated opportunities to practise capturing relatively crippled prey. Hunting is a difficult skill and we know that there is high mortality amongst young carnivores after weaning and cheetahs are a good example. Caro (1995) has made extensive observations of the play of young cheetahs, which occupied about 3–4% of their time, but again it was not possible to identify any unequivocal results as they grew up. However, he noted that sometimes the most important predatory skill a mother can impart to her offspring is restraint. The play of young cheetahs can render useless their mother's careful preparatory stalking.

Dogs have had a very different evolutionary history. Many of the domestic breeds have been selected to keep the strong social organization of the ancestral wolf and retain social play into adult life. Ward *et al.* (2008) describe the way elements of play change as puppies mature. As they become adult, their play patterns change. Dominance relationships develop but the top dogs increasingly 'self-handicap' themselves so that a chase or a play fight can keep going (Fig. 2.24e).

This is one feature among several which suggest that play does extend beyond immediate concerns of survival. For instance, we find numerous examples of different species playing together which otherwise never interact socially. We are familiar with this among domestic animals, puppies and kittens will play together, dogs always play with their human owners, but more remarkably we see it in the wild. Chimpanzee and baboon groups both forage on the shores of Lake Tanganikya in Tanzania. Their young mingle and play together even though adult chimps will sometimes hunt and eat baboons. Fig. 2.24b shows play leading to another striking relaxation of the normal 'rules'. Then again the self-handicapping referred to above, which is a regular feature in animal play, might be regarded as a sophisticated type of behaviour. Some would consider that it suggests a recognition, at least to some degree, of the world of another individual outside oneself – a concept we shall return to in Chapter 5 when discussing the possibility of animal consciousness.

More detailed comparative evidence will help us to obtain a reasonable assessment of the role of play in behavioural development. With any animal it is never likely to have a single function, and the balance of different functions is bound to vary between species. Later, in Chapter 5, we discuss how far animals share our intellectual and emotional lives. We must accept that most people watching dogs playing together or ravens (*Corvus corax*) sliding on their backs down a snow bank (Heinrich and Smolker, 1998; Fig. 2.24j) will conclude that they do so *because they enjoy it*, just as we would! Even if we could ever prove this still would not answer the biological puzzle. The key question of ultimate function remains. The short-term reinforcement for play may involve some kind of internal 'reward' (as it does for sex or feeding) but play certainly has costs, as we have seen. Natural selection would not have enabled it to continue unless its benefits outweighed them. Apart from Fagen (1981), Martin and Caro (1985), Bekoff and Byers (1998) and Burghardt (2005) also provide extensive reviews of this perplexing subject whose investigation, whilst perpetually fascinating, remains so difficult.

Imprinting

The effects of early experience can be fairly general in their nature, affecting a wide range of behaviour. Certainly, some of the examples we have been discussing come into that category. However, there are circumstances under which particular early experiences have rather sharply defined results. Of course, whether we call the results of an experience 'general' or 'specific' depends to a great extent on how closely we look for effects. It is very easy to label an effect as specific if only a few aspects of behaviour are studied. Nevertheless, some examples are quite striking. Playing a tape recording of song to a young bird may affect the song it sings when it becomes sexually mature months later (p. 98), but as far as we know nothing else in its behaviour is changed by this experience. Some of the results of 'imprinting' appear to be equally specific.

'Imprinting' is a term in common parlance now and it has taken on a meaning of finality, of something stamped in – as by a printing press – which suggests permanence. In animal behaviour it has less precise implications but nevertheless describes some striking phenomena. Here we use the term to refer to various behavioural changes whereby a young animal becomes attached to a 'mother figure' and/or a future mating partner. It was Konrad Lorenz who introduced the topic to most behaviour workers. He had a colony of tame geese and watched the goslings as they hatched. There is usually a brief period when they huddle together – in nature, of course, kept warm under the mother bird – but then they become very obviously attracted to conspicuous objects around them, especially if they move. Again, naturally this will be the mother who will lead the brood away from the nest site. They approach and keep close to her, but Lorenz showed clearly that in the mother's

(a)

(b)

Figure 2.25 Konrad Lorenz, one of the founding fathers of modern ethology, followed by a brood of imprinted goslings. In (b) the mature Lorenz is attended by mature geese which were also imprinted as they hatched.

absence almost any conspicuous object would be approached and followed assiduously. He arranged that it should be himself and such broods followed him closely and, as far as possible, continued to treat him as their mother figure (Fig. 2.25). A good deal of the early work on imprinting used birds such as geese, ducks and game birds which have precocial young (i.e. those that have good vision, can walk and feed themselves from hatching and do not stay in the nest like passerine birds). This type of response to a mother figure is usually called 'filial imprinting' to contrast it with 'sexual imprinting', not measurable at the time but only later in life when we observe how its early experience affects an animal's choice of sexual partners at maturity.

Imprinting is distinctive because it occurs so soon after hatching or birth and commonly results in an attachment which is difficult to change. This justifies its inclusion in a chapter on behavioural development, although clearly it does involve learning which we shall be considering in more detail in Chapter 5. Lorenz originally described imprinting as irreversible and being possible only during a brief 'critical period' just after hatching. Because it takes place before anything else has been acquired, it does form a very clear and identifiable event which has made it useful for studies of the neural basis of learning and memory. Few now would wish to consider the actual process of imprinting as different from other types of associative learning, but here it is more useful to concentrate on the role which it plays in development. Bateson (1990) and Horn (1998) review these issues further. Clearly, its function is to ensure that the vulnerable young brood get the maximum protection by staying close to their mother, but what they inherit is a tendency to approach and follow, not any specific template of a mother figure.

Certainly, a very broad range of objects can elicit approach and attachment. Birds have been imprinted upon large canvas 'hides' inside which a person can move, down through cardboard cubes and toy balloons to matchboxes. Colour and shape seem to be equally immaterial, see Fig. 2.26 for example. Auditory stimuli are also attractive to many young birds. For example, wood-ducks (*Aix sponsa*) nest in holes in trees and the mother calls to her newly hatched brood from the water below the nest hole. They are induced to approach before they have ever seen her properly.

For all the wide range of potentially attractive stimuli, we know that some young animals do not come to the imprinting situation with no biases at all. However, as we saw with the development of sound preferences in young mallard ducks (p. 48) it appears that sometimes imprinting bias emerges only following some earlier rather non-specific experience. Hampton *et al.* (1995) have found that domestic chicks reared in the dark will rapidly develop a preference for a realistic mother hen over other models but only if they are handled or exposed to the calls of a mother hen; non-exposed chicks do not develop it. In nature, one supposes, some of the relevant experience will almost always be there to bias the chicks towards approaching and attaching to their mother, rather than other conspicuous objects around them. Similar biases, however acquired, may be involved in sexual imprinting situations, as we shall discuss later. This is another example of how genetic predispositions are expressed only following distinct environmental inputs, a complex and intriguing developmental pathway discussed further by Bolhuis (1991) and Hogan and Bolhuis (2009).

Imprinting-like phenomena are also involved in the social development of mammals. It is common knowledge that the younger we take and rear a litter of wild mammals,

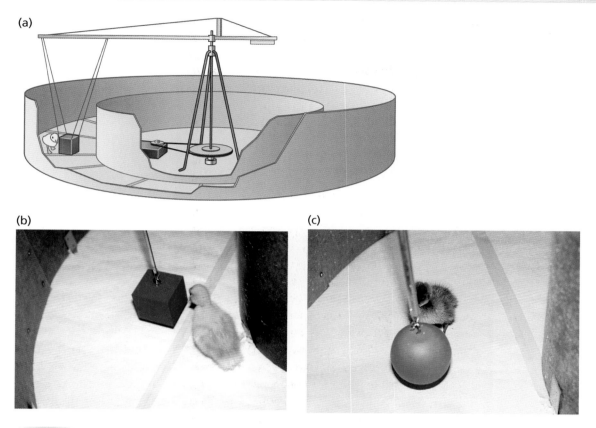

Figure 2.26 (a) Apparatus used to study the following responses of young birds. Different models can be attached to the long arm, which rotates slowly, moving them around the circular runway. (b) and (c) The end point of such a procedure: young ducklings firmly imprinted on wooden models. They stand close to them at all times; if the models are moved the ducklings follow.

the easier it is to tame them. Orphan lambs reared by humans follow them about and often show little attraction towards other sheep ('Mary had a little lamb ... !'). Whilst visual and auditory stimuli are dominant in birds, mammals are very much dominated by their sense of smell and it is not surprising to find that their early olfactory experience is crucial. Many baby mammals are confined to a nest or den, blind and helpless for some days, hence there is no following response to measure, but they are certainly acquiring knowledge of their mother's scent and that of the nest itself. The mother also 'imprints' on the scent of her offspring and must do this rapidly if she is to become attached to them and allow suckling. The time around birth requires a crucial switch in the responses of the mother, who must accept her newborn and start lactation. We know with sheep and goats that the mother must be able to smell and lick her newborn

Figure 2.27 Twin orphan lambs of the Soay breed fostered on to a Blackface ewe by tying half of her dead lamb's skin over each of the orphans.

within an hour or two of birth if she is to accept them. Beyond this period, all young are rejected. Farmers have long understood that if they are quick, an orphan lamb can be cross-fostered to a ewe whose lamb has died. They present the orphan covered by the skin of the dead lamb, which the ewe had already 'accepted' during her sensitive period (Vince, 1993; Fig. 2.27). For the young this may not be just a filial attachment; a form of sexual imprinting (see below) sometimes occurs and zoo staff know to their cost that hand-reared animals are often useless for breeding when they are mature. Armed with better ethological knowledge, steps are now taken to reduce the problem, see Fig. 2.28.

Both of the criteria which Lorenz claimed made imprinting a unique form of learning – irreversibility and restriction to a brief critical period – have had to be modified in the light of later research (Bateson, 1990). It is possible to get some reversal of preferences in an experimental situation but, of course, in nature such instances will be rare and fidelity to the parent figure is the rule – a brood of ducklings stays with its one mother. Regarding critical, or perhaps more accurately, sensitive periods, most would agree that imprinting manifests itself only in young animals, but

(a)

(b)

Figure 2.28 Young whooping cranes hatched from an incubator need to be kept from seeing humans or they will imprint on them and subsequently refuse to mate with other cranes. The chick is fed with a glove looking like the head of an adult crane so that it does not see a human.

the reasons for this are not entirely simple. Once a young animal has learnt the characteristics of its mother object it begins to discriminate against novel ones and in this way the imprinting process itself could bring about closure with the appearance of the ending of a sensitive period. Alternatively, a sensitive period may arise because imprinting may resemble certain types of discontinuous embryonic development, where we know that certain events must take place within a critical period if they are to take place at all. Once the critical period is past, embryonic cells may no longer be competent to respond. (This is the rationale behind the huge attention now paid to stem cell research, where we may be able to harness the persistent equipotentiality of these cells and carry it over for repairs in adult life.) The development of behaviour probably exhibits both continuous and discontinuous qualities and much will depend on a particular animal's life history. With short-lived animals, which must grow rapidly and are sexually mature at a few months of age, behavioural development is highly compressed and events like imprinting come to take on an all-or-nothing, irreversible appearance. (When we come to discuss the development of bird song on p. 94 we shall certainly find evidence of sensitive periods beyond which experience has no effect.) But the infancy of some mammals is very prolonged and development is far more a continuous process – we have only to think of the work with rhesus monkeys which was discussed earlier on p. 75. The behaviour of the young monkey gradually develops over months or even years. Experience has effects at all stages, some transient, some permanent, but rarely appearing as irreversible since there is usually time for later experiences to change the direction of development.

We can certainly find instances of sensitive periods of behavioural development in young mammals. Scott and Fuller (1965) describe some dramatic examples from their extensive work on dogs. They found that there is a period from about 3 to 10 weeks of age during which a puppy normally forms social contacts. If isolated beyond the age of 14 weeks they no longer respond and their behaviour is very abnormal. A very short contact with human beings at the height of this sensitive period is sufficient for them to form a normal relationship with us. Dogs, like some sexually imprinted birds, seem perfectly capable of accepting both humans and their own species as social partners. In such cases, we may try to distinguish between a change in responsiveness that occurs as a result of previous experience and a change which occurs anyway by some form of growth or maturation process, but this may well be a false dichotomy. We know that behavioural experiences during early life can affect the structure and functioning of the brain. There is now a great deal of research on this with rodents and primates, particularly humans (see Johnson *et al.*, 2005 for a varied set of collected readings). Effects on the brain will, of course, affect in turn responses to subsequent events. Now, following the pioneering work of Nottebohm with canaries mentioned above, it is well recognized that the central structural effects of experience may not cease, even in adult life.

Imprinting has always attracted the attention of psychiatrists because there is no doubt that human infants are extremely sensitive to early experience of many kinds. In particular, Bowlby (1969, 1973) developed a theory of the attachment of a baby to its parent which drew extensively from ethological work. He suggested that the period from 18 months to 3 years was especially important and that separation from, or lack of an adequate parent figure at this time led to a greatly increased risk of psychological disturbance in adolescence and later life. However, the idea of a restricted sensitive period in human development, with its implicit suggestion of irreversibility, has been widely criticized and replaced by a much more dynamic view of developmental processes (see Rutter, 1991, 2002). Nobody would deny that maternal separation in childhood may have long-term effects on human children, but inevitably the situation is not as clear-cut as with simpler animals. Separating a human baby from its mother will involve a whole series of changes to its life, almost always deleterious, and it is difficult to link up cause and effect.

Sexual imprinting

Lorenz found that the early experience of his young geese and ducks affected their choice of sexual partner when they were mature and subsequent work with a wide variety of birds and mammals has confirmed this. As with filial imprinting, there seems to be virtually no limit to the range of objects which can provoke attachment. Male and female turkeys have been sexually imprinted on humans, cockerels will imprint on plain cardboard boxes which they court and attempt to mount. We have already referred to problems which arise when mammals born in zoos are reared by human foster-parents.

The early observations of such imprinting, however striking, were not well controlled. The first extensive experimental evidence which demonstrated the full extent of the phenomenon came from the work of Immelmann (1972) with small Estrildine finches, in particular the zebra finch (*Taeniopygia guttata*) and the Bengalese finch (*Lonchura striata*). They are particularly useful because they breed readily in small cages and have a very rapid life cycle. They fledge and become independent at about 5 weeks from hatching and will breed themselves soon after. Immelmann carried out a range of investigations on the effects of cross-fostering between these two species. For example, in one set of experiments he placed a single zebra finch egg in a clutch of the Bengalese finch and allowed the Bengalese parents to rear the whole brood. Subsequently, cross-fostered zebra finch males were isolated until they were sexually mature and then their sexual preference was tested. A cage was divided into three parts by two transparent partitions with a continuous perch running through them. Each male was placed in the central part and presented with a choice; there was a female zebra finch on one side and a Bengalese female on the other. The results were unequivocal: unlike zebra males reared normally, the cross-fostered males directed all their courtship towards the Bengalese female. This preference was all the more striking because when a zebra finch male was put in with the two females, the zebra female usually responded at once with all the usual conspecific greeting calls and perched as close to the male as the partition allowed. The Bengalese finch female was, at best, neutral and usually showed avoidance when approached (see Fig. 2.29). Further, such males retained their preference for Bengalese females even if they were subsequently confined with and eventually paired up with zebra females.

Such results are very dramatic and they parallel some of Lorenz's original observations where, for example, hapless cockerels attempt in vain to attract female mallards which they avidly court from the bank whilst the ducks are swimming some way off shore! From a developmental point of view, one wants to ask questions similar to those raised for filial imprinting. Is sexual imprinting as irreversible as it may appear from the above observations? Does it too have a sensitive period and, if so, is this coincident with that for filial responses and brought about in a similar way? Closer observation reveals

Figure 2.29 A male zebra finch (on the left), reared by Bengalese finch foster-parents, shows full courtship song and posture, with raised feathers, to a Bengalese finch female.

more about the developmental processes involved. Certainly, a young bird's experience during rearing by foster parents and its experience when sexually mature but before the choice test between own species and foster species may both have effects which reduce the strength of preference for the foster species which we outlined above (Immelmann *et al.*, 1991; Kruijt and Meeuwissen, 1991.) The cross-fostered males in Immelmann's first experiments had had the experience of taking part in courtship with foster-parent females. It appears that this serves somehow to 'stamp in' or consolidate the effects of the sexual imprinting, which is initiated during the period of dependence as a nestling. If they had no such early choice but were given only zebra females then their subsequent preference for Bengalese was much reduced. We must recognize, then, that adult birds are still susceptible to experience which affects their sexual preferences and there is not the rapid closure of a sensitive period.

Few other types of bird or mammal have been tested properly, but sexual imprinting on parents certainly occurs in pigeons, geese and ducks (as Lorenz showed), in gulls and probably in parrots. Note that it cannot occur in cuckoos and cowbirds, whose young are always reared by foster-species. Such nest parasites must develop their sexual preferences in other ways, presumably by some inherent recognition of 'own species' characteristics, which may well include vocalizations as well as appearance. Overall it appears that experience affecting choice of sexual partners usually develops gradually over time and, here as in so many places, we can see interactions between inherited tendencies and their environment as animals are growing up.

The term 'imprinting' rather suggests that, however arrived at, there is a clear model type an animal becomes attached to or chooses to mate with. With sexual imprinting, for example, one may recall the old Music Hall song, 'I want a girl, just like the girl that

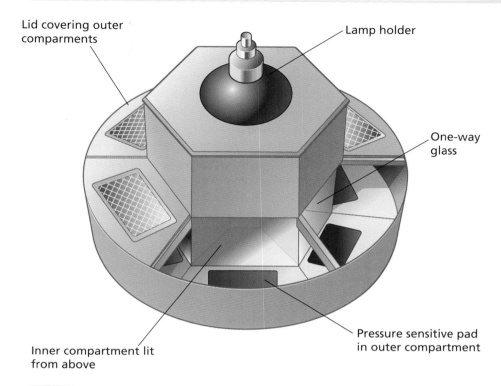

Lid covering outer comparments

Lamp holder

One-way glass

Inner compartment lit from above

Pressure sensitive pad in outer compartment

Figure 2.30   Apparatus used for testing courtship preferences of adult Japanese quail. Stimulus birds of various types are put singly into the inner compartments, which are brightly lit from above. Their outward-facing windows are fitted with one-way screens, so that although the stimulus birds cannot see into the unlit outer part of the apparatus, the bird making its choice can see through into each compartment. A sensitive platform in front of each window records automatically when a bird is standing on it. Those familiar with that city will understand why Bateson called this 'The Amsterdam Apparatus'.

married dear old Dad!' – clearly such a detailed identification will go too far. Animals will need to ensure that they mate within their own species, but not too closely. It would seem likely that the ideal mate will be similar to your parents and siblings but not identical with them and in one ingenious experiment Bateson (1983) demonstrated that was so in a small game bird – the Japanese quail (*Coturnix japonica*). He reared the young birds in a family as normal and when they were mature offered males a choice of three types of female to approach and court. These females were either from the male's family, or a variety with different plumage or thirdly with more normal plumage but from a different family. Bateson found that the preferred choice was for the latter type of female, in other words, the males chose mates which were neither too similar nor too dissimilar from themselves. The experiment is illustrated in Fig. 2.30 and Bateson referred to sexual imprinting here leading to the choice of 'optimal discrepancy' from oneself.

Given the extreme dominance of their sense of smell, sexual imprinting in mammals is bound to offer some contrasts to that of birds. However, in a remarkable way, these findings in quail match very well to evidence about optimal mate choice in mammals. This field has burgeoned over the past decades because of the importance of a set of some 80 genes, quite closely linked, which make up the so-called 'major histocompatability complex' (MHC). These code for proteins which, attached to the surface of cells enable animals to distinguish between 'self' and 'non-self' when they come into contact with substances circulating in the body. This is the very heart of the vertebrate immune system and the major weapon against bacteria and other parasites. Detection of non-self brings into action specific killer cells which are mobilized to engulf and neutralize the foreign proteins. The behavioural interest relates to the extreme variability which is found in the MHC with all the genes having many alleles. In many ways a diverse MHC gene set is a good thing, coding for a multiplicity of proteins, but there will be a limit – in other words an optimal discrepancy. Thus it is fascinating to discover that mate choice in rodents seems to operate to achieve just this and that it is based on smell. The MHC gene set results in a characteristic smell detectable in urine or on the body of the carrier and mating partners respond to this. It is often found that female rodents choose males which smell different from themselves in preference to closer relatives. They must become familiarized with the smell of their mother and siblings when in the nest and are able to compare this with strangers they encounter later. Sometimes males have been found to be less discriminating than females but not all results are consistent. We might well expect on purely behavioural grounds that mammals would discriminate for or against potential mates using smell and certainly work on the MHC and mate choice is confirming this. Brown and Eklund (1994), Penn (2002) and Milinski (2006) review this whole active field.

 ## Bird song development

Imprinting has just provided us with an example of how an inborn tendency to approach conspicuous objects is linked with learning of their characteristics. We have also mentioned previously how a motor pattern, such as bird flight, may mature to a basic level in the absence of practice, but that all the finer skills of flying are added later with practice.

Some of the most beautiful examples of such interweaving of inherited and learnt components during development come from studies of bird song. Song can be regarded just like any other behaviour pattern – it is a controlled sequence of muscular activity, which in this case we perceive as sound. Detailed analysis of song is, in fact, easier than for most motor patterns, because it can be recorded and then played into a sound spectrograph to yield a chart on which marks of different densities show how much energy was emitted at the various sound frequencies over time. We have already shown

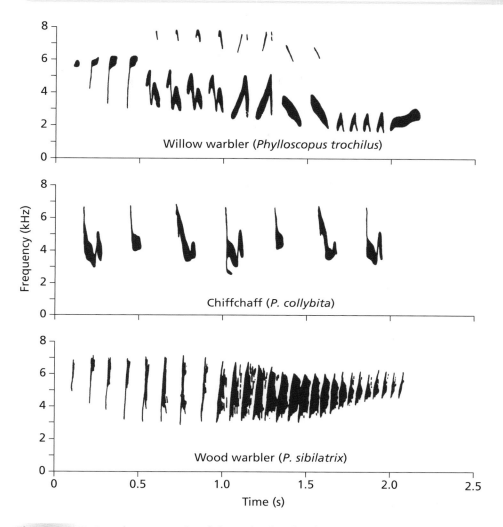

Figure 2.31 Sound spectographs of three closely related species of European warbler. the three are very similar in plumage and share the same general habitat, hence the songs probably serve, in part, as a mechanism of species isolation (see p. 345). These warblers were first clearly distinguished as separate species by the eighteenth century English naturalist Gilbert White, who published *The Natural History of Selborne* in 1789. He separated them by their songs.

some examples of dove cooing patterns in Fig. 2.12. Some more typical examples of sound spectrographs of bird song are shown in Fig. 2.31. With practice they can be interpreted rather like a musical score and they show up variations in sound pattern which are hard to detect by ear alone, and impossible to describe in words.

Bird vocalizations exhibit an astonishing variety, from the briefest of call notes through to elaborate songs lasting several minutes. Song, as we popularly understand it, is characteristic of one of the most abundant and advanced groups of birds, the Oscine families of the Order Passeriformes or perching birds. This order includes some of our most familiar birds, the finches (Fringillidae), warblers (both Old World – Sylviidae and New World – Parulidae) and thrushes (Turdidae). The ways in which they acquire their songs have proved to be a rich source of material for investigation. In their diversity they provide us with opportunities to study some important functional and evolutionary aspects of behavioural development. Catchpole and Slater's (2008) excellent book provides the best introduction to what they refer to in their subtitle as 'biological themes and variations'. We can touch on only a few of these themes, but studies of bird song have enriched our knowledge on all of Tinbergen's four questions, not just that of development.

A good example to illustrate some of the basic features of song development comes from the pioneering work of Thorpe (1961) and his collaborators with the European chaffinch (*Fringilla coelebs*), begun in the 1950s when the first sound spectrographs became available for scientific work. Chaffinch males produce a 2–2.5 s burst of sound consisting of several sequences of repeated notes of different types. As we listen to them singing we notice immediately that individuals sing several variants of the song – all clearly recognizable as being 'chaffinch' but differing in detail. Neighbouring males often have several songs in common but each bird will usually have its own individual mix of song types. Figure 2.32 illustrates some of these variants from a study by Slater and Ince (1979).

Such variability at once raises a developmental question; what is its origin? It has been known for some time that chickens and pigeons produce all their characteristic calls when isolated and even if deafened. Could the song of chaffinches be similarly resistant to environmental influences; could the variants even be genetically based?

Thorpe set out to find an answer. Young birds were taken from the nest immediately after hatching and reared alone in sound-proof chambers. The next spring when the males began singing – still in isolation – their song was recognizably chaffinch but was much simpler in form and showed little variation either within or between individuals (Fig. 2.32, bottom). Following our discussion of the validity of isolation experiments on p. 47, we may note that here such an experiment gives a very clear answer to a first stage question. The different song variants are not genetic but due to some influence from outside. It was an obvious guess that this external source was the song of adult birds which the young bird heard both in the nest and early in its first spring, and this proved to be the case.

One very striking result, which allows us to go into more detail, came from an early study by Marler and Tamura (1964) on an American finch, the white-crowned sparrow (*Zonotrichia leucophrys*). This bird has a wide distribution on the Pacific coast and, like the chaffinch, its songs vary. But they vary in a more systematic way, in that birds from a given geographical area tend to sing similar song variants so that there are clear local 'dialects' (Fig. 2.33). Quite a number of birds have geographical dialects of this sort.

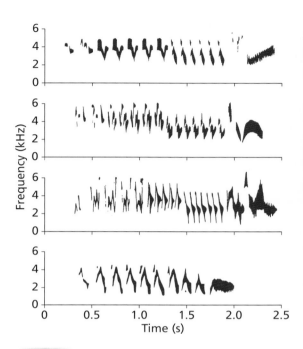

Figure 2.32   At the top, three examples of chaffinch song. They share the same basic pattern but are very different in detail although they are all unmistakably chaffinch to the human – and certainly to the chaffinch – ear. One male will commonly have several such variants in his repertoire and share some of them with birds in neighbouring territories. The lowermost song is that sung by a chaffinch brought up from the chick stage in complete auditory isolation from others. It has chaffinch timbre and rhythm but is obviously far simpler in form and, in particular, lacks the 'terminal flourish' so typical of the normal songs.

How sharp the boundaries are and the degree to which they persist from year to year depends on the life histories of the species concerned. Some are very sedentary – the western white-crowned sparrows are – others tend to disperse more and their song types do not vary in the same way, they are not so related to a particular area (Catchpole and Slater, 2008). Such natural differences allow us to analyse the development of song more closely. They suggest that young males may be picking up songs from what they hear around them and Marler and Tamura (1964) showed this was certainly true of their sparrows. If young males were taken immediately after hatching and isolated, as were the chaffinches, then no matter which region they came from, they all eventually sang very similar and simplified versions of the normal song. Obviously then, in a manner analogous to how humans acquire accents, they must pick up the local dialect by listening to adult birds and modifying their own simple song pattern accordingly.

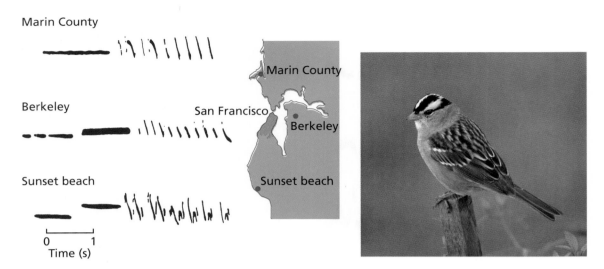

Figure 2.33 The song 'dialects' of the white-crowned sparrow around San Francisco Bay, USA. The examples given are typical of their area and, unlike the chaffinches of Fig. 2.32, there is much less song variation within geographical areas than between them. The use of the term 'dialect', as in human geographically based dialects, seems justified.

Marler and Tamura found that this learning process usually takes place during the first 3 months of life and thus before the bird has ever sung itself. Males captured in their first autumn and reared alone begin to sing for the first time the following spring, and produce a recognizable version of the local dialect. There are clear indications of a sensitive period for learning dialect characteristics. Up to 3 months of age, isolated males can be 'trained' to sing their own or other dialects by playing tape-recorded songs to them, though the results of such experience do not show until the birds begin to sing themselves some months later. Beyond 4 months of age the birds are less receptive to further training by tapes and their songs, when they begin singing, are less affected. Here, then, we appear to have a simple, inherited song pattern which is sensitive to modification by learning but only during early life. The young birds 'carry' the memory of the songs they hear and reproduce them when they first sing.

Experiments by Konishi (1965) took the analysis still further and have enabled us to qualify the conclusion that simple song is inherited. Remembering our earlier reflections on what isolation experiments can tell us about development, we must note that there remains one song which the isolated sparrow can still hear – its own. If a young fledgling is deafened just as it leaves the nest it will subsequently sing, but it produces only a series of disconnected notes. These are quite unlike the song of isolated birds which, though simple in form, would still be recognizable as 'white-crowned sparrow' to an ornithologist. The bird has to be able to hear itself in order to produce this simple song pattern. Consequently, it is more accurate to say that it is not the capacity to produce the simple song which is inherited, but rather some kind of neural 'template' representing this song, against which the bird matches the notes which it produces and adjusts them to fit. It requires auditory feedback if it is to realize this inherited potential.

In many songbirds, the way they first begin to sing in spring perhaps reveals this feedback control in actual operation. For example, chaffinches begin the breeding season singing a 'subsong', a soft rambling pattern of notes varying in pitch and length (Fig. 2.34). As this is repeated, the notes become louder and less variable, always approaching closer to a final pattern which presumably corresponds to one held in the template. Such a song is often said to be 'crystallized', a term which indicates that it does not change further.

With the white-crowned sparrow, Konishi found that if he deafened young birds after they had been 'trained' during their sensitive period with normal song but before they had themselves sung, their subsequent song resembled that of birds deafened as fledglings. They need to hear themselves in order to match up the song they produce with that which they have stored in their memories. Presumably the songs they have heard as fledglings modify their inherited template so that now it conforms to the more complex characteristics of normal adult song. Once the birds have matched their own output with this and sung the adult song, they can go on singing normally even when deafened. At this stage song development comes to an end in the white-crowned sparrow; after its first spring the bird is no longer susceptible to further experience and keeps much the same song pattern for the rest of its life. However Catchpole and Slater's sub-title, 'biological themes and variations' is well borne out here, for other birds can acquire new song variants into their second or subsequent years. Some, like Nottebohm's canaries, mentioned earlier in this chapter, effectively develop a new song set each year, involving new neurons in the process!

The results of these experiments are summarized in Fig. 2.35. Another experiment illustrated there must now be mentioned. Playing tape recordings of *other* bird species to the young white-crowns during their sensitive period has no effect and the birds' subsequent songs sound like those of isolated males. It appears that young birds can be highly selective in what they will learn. Marler and Peters (1977) have analyzed in detail how in two other closely related American finches, the swamp sparrow (*Melospiza georgiana*) and the song sparrow (*M. melodia*), young birds hearing tapes for the first time will unerringly pick out the their own species' song to copy.

Certainly, as Kroodsma (1982) points out, such selectivity must be important in the song development of many species. A young male will have to concentrate on the song of its own species, often selecting it out from a variety of alternatives which can be heard in the nesting territory during its sensitive period. We might take this to be another example of a predisposition to recognize and learn particular things, in this case the characteristics of one's own species' song. However we should not think solely in terms of inherited responsiveness, we should also examine the nature of the learning situation.

Tape recordings offer only a sound stimulus and, as we have just described, they are remarkably effective within certain limits. But provide a young bird with a live

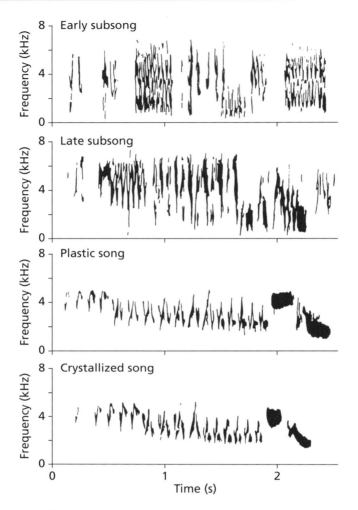

Figure 2.34   The development of song within a young male chaffinch in its first breeding season. Subsong is quiet and rambling, often quite lengthy. As the bird continues, some signs of phrasing are heard and louder and more structured songs begin – so-called plastic song. It justifies this name because it precedes a final version, which is completely structured with mixed length and pattern. After this stage, no further change takes place and the song is said to be 'crystallized'. Each male usually develops two or three different crystallized songs (Fig. 2.32) and, after its first season, does not change this repertoire, although early in spring it will run rapidly through these preliminary stages before crystallizing again into the familiar forms. Note how subsong is completely different from the song of isolated birds (Fig. 2.32, bottom). This latter is preceded by subsong during development and is, itself, a crystallized song.

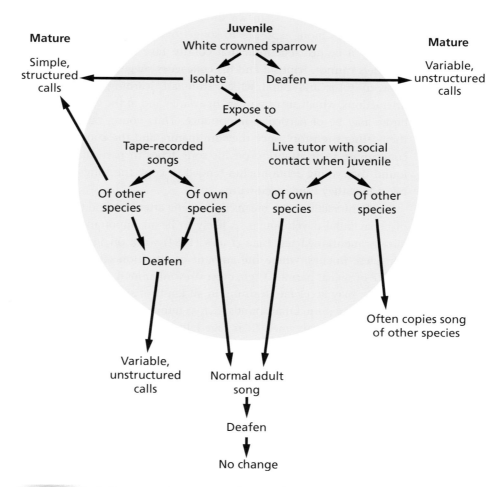

Figure 2.35 A diagrammatic summary of some of the results obtained by Marler, Tamura, Konishi, Baptista, Petrinovitch and others working on song development in white-crowned sparrows. Events inside the circle represent the juvenile phase before the bird has begun to sing itself. The results when it is mature and is singing are shown outside the circle.

song-tutor in the same or an adjacent cage and these limits are easily extended. For instance, isolated white-crowned sparrows have been shown to be capable of learning from live tutors well beyond the 50–55-day sensitive period for tape-recorded tutoring and, more strikingly, they will acquire the songs of other species of caged birds close to them and in sight of them, e.g. Lincoln's sparrow (*Melospiza lincolnii*) and the strawberry finch (*Amandava amandava*). Nor are such results confined to aviaries, for wild white-crowns have been recorded singing Lincoln's sparrow song in areas in California where the two species co-exist (Baptista and Morton, 1988). They do not learn these songs from tapes but again we should consider the learning situation. Perhaps the pupil receives more exposure to song from a live tutor than from a tape recorder, nor may it be just repetition that counts. Live tutors can interact with pupils, tape recorders cannot. Baptista and his colleagues suggest that it is the social interactions between tutor and pupil which stimulate learning and that aggressive, stressful interactions, which are often seen in aviaries and in the wild between adults and young birds, may be of particular importance. Thus young males are often driven out of their father's territory once they are mature and the consequent stress may serve to concentrate their attention on his song types in particular. Jones and Slater (1996) found that young zebra finches tended to copy the songs of the more aggressive of two adults they were caged with.

Social interaction, as a means of focusing attention and thereby affecting the course of behavioural development, is likely to be an important general concept. We have already mentioned ten Cate *et al.*'s (1984) work on imprinting between zebra and Bengalese finches, where the amount of attention given to nestlings affected their choice of sexual partner. High, even stressful arousal, directing the attention of young animals, may accelerate learning in all kinds of social situations. Parrot owners have commonly reported that sounds such as human cries or speech which the bird hears at stressful times, e.g. when their cage falls to the floor, may rapidly be imitated. (This notion is used to chilling effect by Osbert Sitwell in his macabre short story, *The Greeting*.)

One cannot but be amazed by the astonishing variety of bird song across species and the heterogeneity of its development and function. Many birds have songs which are far more elaborate and less proscribed than those of the finches we have been discussing, either in content or in time for learning and modification. In particular, the thrushes often have songs of great beauty to us, which consist of a large number of elements, strung together in variable order to make up a song bout which may last for several minutes. These elements are renewed with each successive breeding season, some being retained, others jettisoned, whilst new elements are added. Many elements are copied from other species and even from non-avian sounds in the birds' environment. It is well known that some birds are superb mimics – starlings, mocking-birds, mynahs and lyre birds are famous for incorporating other birds' calls into their own songs. The developmental processes involved here must be significantly different from those we have been describing above, although the canary – a finch – also develops fresh song elements each year and the remarkable neural changes underlying this process are probably shared by the thrushes and other elaborate songsters. We know that one variant of song development offers some parallels with the acquisition of human speech and which involves rather the opposite of learning new elements and incorporating them into new songs. Human infants go through an early 'babbling' phase before they can speak and analysis reveals that all the sounds involved in the diversity of human languages can be heard there. As speech development proceeds infants begin to eliminate sounds from this total repertoire and incorporate only those sounds from the language they hear spoken around them. They lose the ability to pronounce foreign sounds easily and, in fact, may no longer be able to distinguish them. Some birds, for instance American swamp sparrows (Marler and Pickert, 1984) and sedge warblers (*Acrocephalus schoenobaenus* – see below) begin singing in a similar way, using a very wide range of elements which they produce without any kind of prior experience, but gradually narrowing down their repertoire over their first season, probably in response to those songs they hear around.

Presumably, for all the variations on the theme of development, the process of matching your own song to those of adults in the surrounding area is in some way reinforcing or rewarding to a young bird. In one unusual example, we have some direct evidence of the nature of such a reward, and it is the young male's potential mate who provides it. Cowbirds (*Molothrus ater*) are brood parasites and therefore young males do not hear the songs of adults of their species during early life. West and King (1988) studied the origins of two of the geographical dialects in the United States, one eastern, the other southern. They discovered that, counter to expectations, the songs of males reared in isolation are more complex than those of normally reared birds. Beginning with elements of both dialects, males crystallize down not to the song pattern of other males that they are exposed to, but to match the dialect of the area of origin of the females with which they are housed! The explanation for this remarkable result comes from close observation of pairs when the young male begins singing to a female. When the male sings particular elements from his repertoire, she responds with a brief 'wing-stroking' display – a rapid shuffling of the wings, similar to the copulation invitation display of a number of passerine birds. The male responds in turn by increasing the frequency with which he produces these rewarded elements and eliminating those others with no effect on the female. In such a fashion, the female's preference for her local dialect comes to be matched by the males of the locality. Song development in cowbirds is thus a natural form of the 'operant conditioning' much studied by experimental psychologists and which we shall discuss further in Chapter 5.

We have left rather to one side the question of why so many passerine birds develop their songs through this often elaborate process of learning. Some birds have completely unlearnt songs – pigeons and doves, for instance – why not all? There are a number of possibilities, the balance of which will vary with each species' life history. One is that through such learning, with its inevitable imperfections, a group of adjacent territory holders will come to acquire a range of song variants, some of which will overlap, others being personal and distinctive. This gives the opportunity to distinguish individuals by their song repertoire. We can observe that the songs of neighbours with whom boundary disputes have been resolved tend to be ignored ('It's only old so-and-so'), but a new song variant attracts a rapid response ('I must repel this interloper'). Catchpole and Slater (2008) discuss the evidence for this, and other possible functions.

Again, developmental variability probably reflects a varying balance of the functions which song serves in different species; carving out a territory, repelling rivals, attracting a mate, and so on.

For example, it is noticeable that in some species male birds cease singing as soon as they have attracted a mate and formed a stable pair, others continue singing more or less uninterrupted throughout their breeding season. These latter may require song to hold their territory intact, but we know that some male birds will pair with more than one female, so mate attraction may continue as a function of their song. (We discuss these mating strategies further in Chapter 6.).

Old World warblers of the genus *Acrocephalus* are particularly interesting in this regard for close relatives differ in their pattern and timing of singing. The sedge warbler (*A. schoenobaenus*) and the marsh warbler (*A. palustris*) cease singing altogether once they have paired with a female, the reed warbler (*A. scirpaceus*) continues singing once paired though at a reduced level with peaks at dawn and dusk. Even more striking are the developmental contrasts found in this genus. Both the sedge warbler and the marsh warbler have amazingly complex songs – perhaps the most complex of any bird – which last for 20–30 s. The sedge warbler's song is made up of a selection from anything up to 75 different elements, which it employs in a fairly structured way, but varying between successive songs (see Fig. 2.36). All of this astonishing variety is built up from an inborn repertoire, none of the elements are copied. The marsh warbler's song, equally complex, may not contain a single inborn element! About one half consists of an amazing mixture of sounds copied from a variety of other species living around it during the breeding season in northern Europe. The other half of the elements were, at first, assumed to be 'native' marsh warbler, but later studies by Dowsett-Lemaire (1979) found that no, they were copies of the sounds of its African neighbours heard in the bird's winter quarters! Catchpole and Slater (2008) suggest that since males of both these species cease singing completely once paired, young male fledglings never hear their parental song before migrating. Both have to equip themselves with an attractive song as soon as they arrive back on the breeding grounds for their first spring, but they do so in totally different ways. We can only speculate about the origins of these two contrasted developmental pathways.

One factor which we do know has been involved in the evolution of these complexities is that the song of *Acrocephalus* warblers is really an auditory equivalent of a male peacock's tail, shaped by sexual selection (Catchpole, 1980). In Chapter 6, p. 342, we describe how the size of a male's song repertoire relates to his success in obtaining a mate. Probably the most extreme example is provided by the great reed warbler (*A. arundinaceus*). As is usual, males arrive back from their winter quarters before females and immediately begin carving out territories in the reed beds that are their breeding habitat. Not only is their song complex in form, with many different elements, they sing extremely loudly (close by, it is almost too loud for comfort to the human ear!) and they sing almost incessantly for several days – certainly 23 hours out of the 24 – and continue to sing whilst actively hunting and feeding! Such behaviour must certainly represent a very large expenditure of energy. Is this part of the male's strategy for repelling other males and carving out a good territory, or is it to attract the maximum

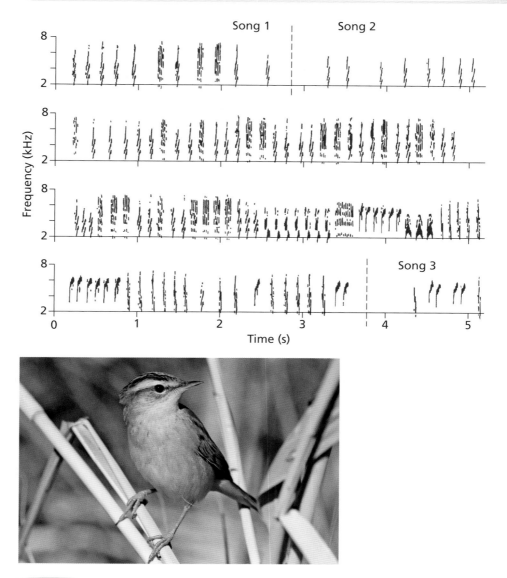

Figure 2.36   The astonishing complexities of sedge warbler song. One whole song lasting about 18 s is shown here (Song 2), together with the end of the song which preceded it and the start of that which followed. All songs consist of a selection of 'syllables' which are usually repeated several times before the bird switches. Each bird may have up to 75 elements in its repertoire and no two songs are exactly alike. The result is a song of dazzling complexity whose effect is amplified sometimes when a male delivers it in a looping display flight.

attention of the females when they arrive? Could both functions be interconnected? We shall return to discuss the topic of sexual selection and the evolution of displays in Chapter 6.

Conclusions

So, at the conclusion of this chapter on behavioural development, we link ideas on development with those on evolution and reasonably enough as we reflect on the question we began with: how does behaviour become so well adapted to requirements? Throughout we have had to emphasize the interactions between information provided in the genes, an inbuilt programme which becomes modified according to circumstances by information acquired in diverse ways from the environment, both physical and social. Bateson and Gluckman (2011) provide an excellent survey of the intricacies of such developmental pathways. Sometimes learning is involved and we must recognize that development may persist throughout life. Animals never come to novel situations with a blank mind. The title of a review by Gould and Marler (1987) makes this point very neatly; they call it *'Learning by Instinct'*. In many ways, it would be logical to move on next to consider learning in its own right – it is, after all, a form of behavioural development. However, it is impossible to discuss it without some detailed reference to the factors that determine how animals learn to modify their responses to new circumstances. Accordingly, we shall return to the topic of learning after chapters on stimuli and motivation, the external and internal components of Tinbergen's causation question.

SUMMARY

Behaviour, like any other feature of an adult animal, must develop from the interaction of genetically determined predispositions with the environment in which the young animal grows up. With behaviour there is often a temptation to label its origins as being either inborn – inherited – or acquired as the result of experience. Certainly there is a great diversity of life histories across the animal kingdom in which we can find extreme examples which might seem to support such a division. However, closer examination reveals a whole range of factors, some of them unsuspected, which contribute to behaviour's development.

Here we give a short survey of the diverse pathways along which animals grow up and then move on to examine how they are mapped out. We begin with genetic factors, discussing what we mean by using terms like 'genetic control' and describe some of the new discoveries of how genes affect the building of a nervous system and thus may have specific as well as non-specific effects on behaviour. Gene expression in nervous system development and in behaviour is modified by the young animal's environment and factors both internal – such as hormones – and external may have dramatic effects. Sex determination and all the behavioural differences it involves being a conspicuous example.

Most invertebrates develop their behaviour largely isolated from any social contact because overlapping of generations is rare, but for many vertebrates parental attention is absolutely crucial although, again, there is wide variation in its nature and extent. We discuss a range of examples of how parents and offspring interact, noting that, while much parental behaviour is supportive, offspring may try to get more from parents than the latter are prepared to give. We examine two types of behaviour which are particularly associated with young animals – play and imprinting. Imprinting may be regarded as a form of early learning which has major effects on how young birds and mammals attach to a parent figure offering protection, and rather later in development may affect the young animal's choice of a sexual partner. Play, which is such a conspicuous feature of young mammals in particular, nevertheless presents ethologists with big problems concerning its function and how it relates to the behaviour of adults.

We end the chapter with one particular example of behavioural development. How do birds acquire their song? This familiar behaviour, perhaps surprisingly, turns out to offer remarkably detailed information on how genetic and environmental factors interact. Also, because there is such diversity between bird species in the way their song develops, we can gain rich insights into how natural selection can operate to change developmental pathways.

Stimuli and communication

Behaviour can be so well adapted to an animal's way of life that the animal seems to 'know' what to respond to, what to do and when to do it. Thus partridges and small rodents 'know' that the most effective defence against a hawk flying overhead is to flatten themselves on the ground and remain motionless since the hawk, which is very sensitive to movement, is least likely to see them. Most of us have experience of wasps that know about sweet drinks or mosquitoes that know where to find exposed human flesh.

Using the word 'know' in this context does not necessarily imply that animals are consciously thinking about what they are doing (although they may be, as we discuss in Chapter 5). Animals may 'know' things about their environments in the same way that a heat-seeking missile knows how to find its target or a computer knows how many mistakes you have made in the course of a game. In other words, quite simple unconscious mechanisms are capable of giving rise to what we might refer to as knowledge. When we ask causal questions about behaviour, we are asking what this knowledge is and how the animal uses it. For example, if we see a small bird or mammal taking evasive action in the presence of a hawk, we could ask, 'How does the animal discriminate between a dangerous hawk and a harmless gull flying overhead?' or 'When it has detected a hawk, how does the animal decide whether to race for cover or to remain motionless where it is?'

Such questions about immediate causation are obviously closely linked to the ones about developmental history we discussed in the last chapter, but they are also usefully separated from them because they require rather different kinds of information to answer them. For example, two different species may both recognize a hawk by its distinctive shape, but one may have an innate (unlearned) ability to recognize it, while the other may have learnt by watching the responses of its parents to different birds flying overhead. The causal mechanisms might be similar but the developmental routes quite different. Conversely, two species may each have an innate capacity for recognizing hawks but use quite different cues to do so.

In this chapter, we begin our study of how behaviour is caused and controlled by looking at the way animals initially detect what is going on in their environments. Do they see, hear, smell as we do? Can they touch and taste? Do they have sixth or even seventh senses that we know nothing about?

But we also have to go further and ask how animals then make sense of all the information they are receiving. Is the patch of red light that has just been detected a piece of food, a mate, a predator, or just the setting sun? Such objects may have some similarities, but the appropriate behavioural response for an animal to make will be quite different depending on which one it is, so the animal must be able to distinguish between them. We might think that the best way to do this would be to have more and better information and we do indeed find that animal sense organs are often superbly good at picking up fine details in the world. But this in turn leads to yet another problem. Paradoxically, sense organs are often too good in the sense that they provide so much information that the brain may not be able to process it all and an animal may have to ignore or throw away some of the sensory information provided by its sense organs. For example, sense organs can potentially provide information about many objects or events in the environment, such as grass waving in the wind. If these carry no significance for survival, the animal will take no notice at all while continuing to respond strongly to other types of event, such as an approaching predator.

The choice of what to respond to is not, however, always as clear-cut as this. The animal may be receiving many different sorts of information at the same time, most of which may have some significance for its survival and reproduction. It may be receiving information about food and water as well as about its predator, the fly that has landed on its back, and so on. It then has to choose between the different sorts of information its sense organs are providing it with and selectively ignore the rest, sometimes changing which information it makes use of at any one time. When there are no predators around, information about food may be responded to very strongly, but once the predator appears, sensory information about food now has to be ignored and the animal must start responding instead to the more immediate threat to its life. Decision-making of this sort is an inevitable outcome of receiving a large amount of sensory input, so that this present Chapter 3 on the sensory worlds of animals will lead naturally into Chapter 4 on motivation and decision-making. These two chapters together show how animals perceive, evaluate and then respond to their environments – the key parts of what constitutes an explanation of the causation of behaviour. We should not underestimate the complexity of the mechanisms involved. Animal bodies are far more complex than any man-made machine yet made and it is worth noting that scientists who design robots are increasingly turning to animal behaviour for ideas about how to make their machines perform tasks such as finding their way around their environment, tasks that animals successfully perform all the time.

We thus begin our study of how animal behaviour 'works' by looking at the way in which animal sense organs detect the host of smells, buzzes, flashes and other changes taking place in the animals' physical environment. Then we will look at how their sensory nervous systems and brains process this unpromising raw material into objects or events that the animals can recognize and respond to. In doing this, we shall see that many of the most important objects in an animal's world often turn out to be other animals and many of the most significant stimuli to which they respond are in fact signals from other organisms. Understanding how animals detect and respond to stimuli will therefore, by the end of this chapter, have taken us into one of the most fascinating of all areas of animal behaviour: the study of animal communication.

What stimuli are and how they act

The term 'stimulus' means literally 'little goad' from the Latin verb *stimulare* – to goad, incite or arouse – and this describes very well what the term currently means when applied to examples of animal behaviour. An alarm call by one member of a flock of birds incites all the rest to take off. In this case, the effect is immediate, the goad of the alarm call prodding the other animals into action. But at other times the action can be delayed and the effects of a stimulus can seem to accumulate gradually, changing the responsiveness of the animal over a period of time, so that the stimulus does not so much 'prod' as 'arouse'. The courtship of male doves leading to induction of hormone secretion in the female, which we discuss in more detail in Chapter 4, is a good example of this. The sight of a male dove courting leads to hormonal changes in the female which make her more ready to take part in nest-building. The stimulus of the male's courtship is only effective, however, if it is repeated many times. One bout of courtship has little effect. It requires several hours of courtship behaviour by the male, usually spread over a number of days, to bring the female into reproductive condition.

An even more dramatic example is the way in which desert locusts change from being solitary and inactive to being highly sociable and active as described in Chapter 2 (Fig. 2.3). This dramatic change is brought about by the stimulus of being touched on the hind legs (Simpson *et al.*, 1999). One touch does not change the behaviour but repeated touching does. As locust numbers are beginning to build up they inevitably begin to jostle each other. These contacts act cumulatively and begin to alter their behaviour so that within 4 hours a locust that previously avoided others now actively seeks them out and by touching them becomes itself a stimulus for making even more locusts gregarious.

Yet another effect that stimuli have on animals – which again differs from a straightforward prod – is to orientate their behaviour in particular directions. Animals respond almost all the time to the basic physical qualities of their environment – light, gravity, air, water currents, and so on – and position themselves to be in correct relationship to them. Blowfly maggots crawl directly away from light when they move out of their food source to pupate. Fish rest with their heads facing into a water current. From an early age, young rats, even in total darkness, show a 'righting response' which keeps them upright with respect to gravity. In each of these cases, the stimuli act continuously over a long period of time to orientate the response in a particular direction.

Orientation can also refer to the way in which stimuli control responses to more transitory features of the environment. Male ducks direct their courtship displays according to the position of the females on the water. Some displays are given when lateral to the female, others when the male is directly in front of her. In Chapter 2, p. 42, we described the nesting routine of the female beewolf, *Philanthus*, a solitary wasp. The experiment shown in Fig. 3.1 shows how *Philanthus* uses landmarks such as pine cones to orientate and locate her burrow which she provisions with honeybee victims as food for her larvae. She makes several trips to do this and as she leaves the nest on a foraging trip, she makes a brief circling flight to orientate herself and clearly learns landmarks surrounding the entrance. If these are moved while she is away, she still uses them to orientate, or rather mis-orientate, her response on the return trip, since she will go to where the cones have been moved rather than to the real tunnel entrance.

Stimuli can even come from inside an animal's body. The external stimulus for the release of milk by the female rat is her young sucking at her nipples. This mechanical stimulation by the young induces the hypothalamus of the mother's brain to release

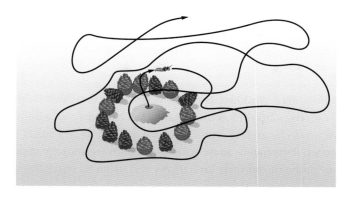

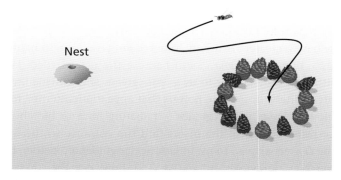

Figure 3.1 The female digger wasp builds a nest burrow in sand. While a wasp was in her burrow, Tinbergen (1951) placed a circle of pine cones around the entrance. When she emerged, the wasp reacted to the new situation by an orientation flight (a) before flying off. When she returned with prey (b) she looked for the nest in the centre of the circle of pine cones, which Tinbergen had moved while she was away.

the hormone oxytocin from the posterior pituitary gland. This is carried in the blood and acts as an internal stimulus to the mammary glands to let down milk. There are many other internal stimuli that are crucial to the efficient working of the body. There are receptors for detecting changes in blood pressure, the tension in muscles, the amount of salt in the blood, and so on. As we saw in Chapter 1, effective movement itself depends on information from within the body about the relative position of different parts of the body. There are receptors in joints, muscles and tendons, and, in fact, coordinated behaviour would be impossible if animals could not detect the relative position of their limbs or what was going on in different parts of their bodies. So the term 'stimulus', far from being just a simple 'trigger' that always leads to the same simple response, covers a wide range of different ways in which factors both inside and outside the animal's body are detected and evaluated over many different time scales.

Diverse sensory capacities

Vision

Every animal inhabits a world of its own whose character is inevitably shaped by the information it receives from its sense organs. As observers and interpreters of animal behaviour, we would be greatly handicapped if we had to rely solely on the evidence of our own senses. Fortunately, as we will see, modern technology makes it possible for us to overcome some of our own limitations and begin to enter the sensory world of other species.

For example, many animals are able to see ultraviolet light, i.e. light of much shorter wavelengths (below 400 nm) than our eyes are sensitive to. This means that bees foraging on flowers respond to colours and patterns that are invisible to us (and vice versa since we have a greater sensitivity to red or long wavelength light). Bees fly to the flowers of white bryony (*Bryonia dioica*) because the petals reflect large amounts of ultraviolet, although they appear to us as pale green, and provide little contrast with the leaves. Interestingly, not all the flowers that appear unpatterned to us do so to the bees. Many flowers have 'nectar guides' – radiating lines or blotches of contrasting colour on the petals that help the insect find the entrance to the flower where the nectar is. Some flowers appear, to us, to lack nectar guides altogether, but photograph-ing them through an ultraviolet system (Fig. 3.2) shows that they do have patterns, but adapted to the eyes of bees, not humans.

Fish and birds, too, are also sensitive to ultraviolet light (Stevens and Cuthill, 2007). The plumage of birds such as gulls and snowy owls, which we think of as white or dull grey, reflects ultraviolet light, possibly making them very conspicuous to other birds (Burkhardt, 1989). In Chapter 7, we see how the ultra-violet feathers on the heads of great tits function as 'status badges' and indicate the birds' dominance status. The mouths of nestling birds such as blackbirds (*Turdus merula*) and blue tits (*Parus caeruleus*) are made even more conspicuous to their parents than they are to us because they are rich in ultra-violet colours (Fig. 3.3).

At the other end of the visual spectrum, some snakes can 'see' infrared radiation, that is, the heat given off by their warm-blooded prey (Newman and Hartline, 1982; Bakken and Krochmal, 2007). Pit vipers (Crotalinae) and the distantly related pythons (Boidae) have pit organs or infrared 'eyes', small cavities on their heads (Fig. 3.4), which contain thousands of heat-sensitive nerve endings. They can detect warmth from the body of a mouse and the snake is alerted to the presence of potential prey. Rattlesnakes are so good at detection with their infrared eyes that they can accurately strike and kill prey even in complete darkness.

Finally on visual capacities, some species go beyond us by being sensitive to the plane of polarization of light. Light reaching us from areas of blue sky is vibrating

Figure 3.2 Flowers seen in two different lights. Photograph (a) shows what the human eye sees; the flower appears to have no obvious nectar guides; (b) is photographed through an ultraviolet system to show approximately what a bee sees; a striking pattern is now seen.

Figure 3.3 In addition to the yellow/red colours that we can see, the mouths of these nestling blackbirds (*Turdus merula*) show striking ultraviolet colours that we cannot see but the parent bird can. The UV colours make the gapes contrast maximally with the dark background of the nest (Hunt *et al.*, 2003).

Figure 3.4   The head of a python showing the heat-sensitive pits on the side of the snout. A further series of pits are found along the lower lip, below and behind the eye. Each pit has a slightly different field of 'view' and its lining is highly sensitive to heat radiation.

predominantly in one plane, known as the e-vector. This vector changes in a regular fashion with respect to the apparent position of the sun as it moves across the sky. Thus, animals that use the sun for navigation can potentially know where the sun is even if all they can see is a patch of blue sky (Brines and Gould, 1982). von Frisch (1967) showed that bees can indeed use the pattern of polarization of light to locate the sun's position even when it is obscured by clouds or a screen. Bees have a small dorsal region of the retina which has a row of analyzers, each of which is maximally sensitive to a different e-vector direction. When a bee looks at a patch of polarized skylight, the perceived direction of the e-vector is determined by which analysers in the array produce the largest output (Rossel and Wehner, 1986). Mantis shrimps are also sensitive to the plane of polarization of light (Land, 2008), as are squid, which appear to detect fish by responding to the pattern of polarized light that is reflected off their shiny bodies (Land, 1981).

(a)

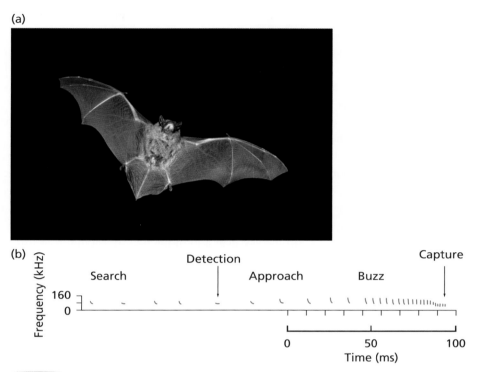

Figure 3.5 Echolocation sequence of a *Pipistrelle pipistrellus* as it captures an insect. It begins with a *search* phase, with relatively long time intervals between pulses. *Detection* marks the point at which the bat detects the insect. In the *approach* phase, the intervals between pulses rapidly shorten. Then during the *buzz* phase, the number of pulses per second rises dramatically just before the insect is captured.

Hearing

We are familiar with various sound-emitting devices which are sold to summon pet dogs or repel rodent pests. These produce high frequency sounds, at above 20 kHz and thus well outside the sensitivity of human ears, frequencies that we label 'ultrasonic'. Many animals hear well above this and in bats, this capacity is specialized to a remarkable extent. About 70% of bat species, such as the pipistrelle bats (Fig. 3.5), feed on insects and most of them use a form of sonar that enables them to detect flying insects even in complete darkness. Dolphins and porpoises, too, use sonar to track fish in dark or murky waters, although they also pick up the noises made by the fish themselves (Gannon *et al.*, 2005). Both bats and dolphins use their sonar to build 'sound pictures' of the environment around them. Just as with human sonar, they emit pulses of sound and then detect the returning echoes bounced off objects in the environment. Unlike the relatively slow sonar pulses of a submarine, however, bats can emit up to 100 pulses a second and then, remarkably, listen to the echoes coming back between each pulse. Different objects distort the sound in different ways so that by comparing the original sound it made with the distorted echoes that come back to it, the bat gains information about the size, shape, speed of movement and surface properties of what is around it (Simmons and Stein, 1980; Jones and Holdereid, 2006).

However, because most of the insects hunted by bats are small, they have to use very high frequency (i.e. short wavelength) sound because echoes only come back from objects when the wavelength of the sound is approximately equal to the object's diameter. (We expect to hear an echo if we shout at a cliff face, for example, but not if we shout at a needle.) To obtain a reasonable echo from an object the size of a moth, bats have to use sound of very short wavelength, which means frequencies of 50–80 kHz. Because of the higher speed of sound in water, dolphins have to use even higher frequencies of sound – up to 300 kHz.

Young people can usually hear a high-pitched squeaking from hunting bats. This is the lowest harmonic of their sonar calls and represents only a tiny fraction of the sound energy they put into the calls. These calls are extremely loud but, of course, we cannot hear them at all. However, some of the bats' prey can hear them and, in fact, some moths and bats engage in a life and death battle on a nightly basis. Owlet moths (Noctuidae), for example, have ears on their thorax, consisting of tiny thin tympanic membranes connected to sensory cells. These simple ears can pick up the echolocating calls of hunting bats and trigger evasive action. If the bat sounds are very faint, indicating that the bat is still some distance away, the moth simply flies off in the opposite direction. But at close range, the moth goes into a series of erratic manoeuvres – flying in spirals or dropping suddenly to the ground – which makes it more difficult for the bat to catch it (Fig. 3.6; Roeder, 1967; ter Hofstede *et al.*, 2008).

Figure 3.6   Evasive action taken by a moth that has detected the echolocating call of an approaching bat. The moth soars upwards (broken line) just as a bat (the long white streak) swoops by on a capture path.

'Ultra-sounds' are too high in frequency for us to perceive. At the opposite end of the sound spectrum are 'infra-sounds' – sounds that are too low in frequency for us to hear. Whales and elephants both use infra-sound to communicate over long distances. Elephants use low frequency 'rumbles' (10–35 Hz) to keep in touch with other members of their group (McComb *et al.*, 2003). These are very powerful and clearly effective over distances of 1–2 km. Although we cannot hear the lowest of these sounds, they can be 'felt' as vibrations in the chest by a person standing close to elephants. There is a lot of energy in the air movements produced by such huge animals. In the oceans, with even larger body sizes and less attenuation of sound energy in the dense medium of water, whales can communicate over hundreds of kilometres using low frequency sounds (Payne and McVay, 1971).

Some bats and owls also hunt in complete darkness not by emitting sounds of their own but by being extremely sensitive to sound made by their prey. The barn owl's hearing is so good, for example, that it can locate a mouse in complete darkness simply by homing in on the sound of the mouse rustling through leaves or even just chewing (Payne, 1971). On hearing the sound, the owl turns its dish-like face (Fig. 3.7) towards the source, until it is pointing directly towards it, and then launches itself towards the

Figure 3.7 The dish-like face of the barn owl (*Tyto alba*) is edged with short stiff feathers that help to deflect sound into the ear openings (not visible). The ears are asymmetrically placed at the rim of the mask, the right opening slightly below the level of the eyes, the left slightly above.

prey. The owl uses the intensity differences between its two ears to home in on the sound, since when the two ears are being equally stimulated, the sound must be coming from a point directly between them. This gives it information about where the mouse is in a horizontal (left–right) plane. In addition, barn owls have asymmetrical positioning of the ear openings, with that of the left ear being above the mid-point of the eye and that of the right one below it. This gives them the ability to pin-point the mouse's position in the vertical plane as well (Knudsen and Konishi, 1979).

Some animals can also pick up patterns of disturbance caused by objects moving near them without possessing what we would conventionally call ears. Fish possess a 'lateral line' organ – effectively a row of little pressure sensors called neuromasts down each side of the body – that allows them to pick up mechanical vibrations in the water (Fig. 3.8). Lateral line organs are most developed in fish that live in dark caves or the deep ocean where there is no light (Dijkgraaf, 1962), but even fish with normal eyesight can, in

(a)

(b)

Figure 3.8   (a) An Atlantic cod (*Gadus morhua*), showing the lateral line running along the length of its body. (b) Close-up of lateral line.

complete darkness, collect detailed information about the presence of other fish in the school, predators, and so on. Blind fish can keep up with the rest of the school but fish without a functioning lateral line system tend to crash into each other (Partridge and Pitcher, 1980). Mexican blind cave fish, which live in dark caves and have, over evolutionary time, lost their eyes altogether, can still find their way around and detect landmarks (Sutherland et al., 2009). If a novel object appears in their environment, they accelerate so that they can better detect the pressure waves in the water caused by their own movements pushing water against the object (Burt de Perera, 2004). Pressure changes are also important in air, too. Some caterpillars have hairs that can detect a flying wasp as much as half a metre away (Markl, 1977) and, as we saw in Chapter 1, cockroaches use the tiny air movements made by the movement of a toad's tongue as it prepares to strike to enable them to escape.

It has been known for a long time that spiders will attack a tuning fork touching their web. This is because the vibrations of the tuning fork resemble the vibrations produced by prey caught in the web and struggling to get free. Spiders' webs are very good transmitters of information about both the nature and the location of a mechanical disturbance and a spider can use the vibrations both to communicate with its mate (Maklakov et al., 2003) and to home in on its prey. Movements of as little as 1 nm (10^{-9} M) may be picked up by the spider (Masters and Markl, 1981; Masters and Moffat, 1986). Some mud-feeding shorebirds can detect their prey by 'touch' but without actually touching them. Red knot (Calidris canutus) have touch-sensitive corpuscles in the tips of their beaks that are so sensitive that all the bird has to do is dip its bill 0.5 cm into the mud and it can detect mollusc prey several centimetres away (Piersma et al., 1998). The bird probes the mud up to 10 times a second and can detect the increase in pressure that builds up each time the bill penetrates the mud in the direction of a solid object.

Perhaps the most extraordinary tactile organ of all is that of the star-nosed mole (Condylura cristata), which has 22 mechanosensory appendages surrounding its nostrils (Fig. 3.9). One pair (the 11th, in the middle, are particularly sensitive. When some of the other appendages have detected something, the mole turns the 11th pair onto the object of interest to investigate it further, almost like turning the eye to focus on the sensitive fovea. The speed of the mole's search of the environment is astonishing. It touches a different area 13 times a second (Catania and Remple, 2004).

Chemical senses

We are generally aware that the chemical senses – taste and smell – are often far more highly developed in other animals than in ourselves. The insects are famous in this regard. On emerging from their cocoons, female silk moths (Bombyx) emit a scent that is extremely attractive to any male within a wide radius. The males locate the females entirely by their odour and are able to pick up the scent of a female when it is diluted to only 100 molecules of sex odour substance per cubic centimetre of air. The antennae of the male (Fig. 3.10) resemble minute test-tube cleaning brushes. Each branch is divided

Figure 3.9 The bizarre face of the star-nosed mole, with its set of tentacles surrounding the nose.

Figure 3.10 Head of a male silk moth showing the feathery antennae that are finely tuned to pick up the scent of the female.

into sub-branches whose finest divisions are all covered with scent receptors. This gives the antennae exquisite sensitivity and they are 'tuned in' only to the scent of females and insensitive to all other chemicals. Similarly, as we discuss further in Chapter 7, worker honeybees have an extreme sensitivity to the 'queen substance' secreted from surface glands of a healthy queen.

Chemical senses are also important in almost all vertebrates. For instance, male garter snakes can follow chemical trails left by females and can even determine the direction in which the female was going. As the female moves, she brushes against objects on her trail, thus leaving more scent on one side than the other. The male tests each side of the objects with his forked tongue (in reptiles and amphibia there is little distinction between the senses of smell and taste) and thereby recognizes which side the female has pushed against and can follow her (Ford and Low, 1984). Males can also determine whether a female is ready to mate from her scent (O'Donnell *et al.*, 2004).

The scent detection of dogs is legendary and it is nice that, for once, controlled scientific investigations have amply borne out popular belief. Dogs have a huge area of olfactory epithelium on the inside of the nasal cavity where densely packed sense cells pick up scent molecules and transmit nerve impulses to the olfactory parts of the brain. A large dog has over 15 times the area of epithelium of a human and 100 times the density of sense cells per unit area (Pihlström *et al.*, 2005). This is a huge investment in the sense of smell and we know how far this sense dominates the life of dogs, for example running down prey whose flight path is detected by smell, the identification of companions and discrimination of strangers. Pigs and bears have made an even greater investment and use their sense of small to detect roots and bulbs or burrowing rodents well below ground (Marshall and Moulton, 1981; Wyatt, 2003).

Objective tests have shown that bloodhounds are indeed the most sensitive of all dog breeds and can accurately follow trails made by a garment that, having been worn by a person, is then dragged across a field. Even a day or two later such trails can be picked up if the intervening weather has not been too wild. Bloodhounds will discriminate easily between the scent trails of different people with one exception – identical twins not only look alike, they also smell alike! A dog can still discriminate between such twins if the scents of both of them are given together and it can compare simultaneously, but if only one twin's scent is presented first, then later the smell of the other, the smell of the other twin will be accepted and followed (Kalmus, 1955; Hepper, 1987).

It is difficult for us to judge how such an amazing sense of smell must operate. Certainly it means being able to detect, discriminate between and remember smells probably several million times more accurately than us. Yet dogs are not overwhelmed by the world of smells any more than we, dominated by our sense of sight, are overwhelmed by visual things. As we discuss below, animals have to filter out and ignore a high proportion of the stimuli which are impinging on them from instant to instant.

House mice develop complex communication networks based entirely on smell (Hurst, 1989). Without even encountering each other directly, mice know the sex and also the social status of others just from the scent left behind in their urine. Even more remarkably, mice have recently been shown to use smell to distinguish mice that are genetically similar or genetically different from themselves in one particular genetic region. The major histocompatibility complex (MHC) is a group of genes that were originally thought to be entirely concerned with recognition of 'foreign' bodies or tissues. It has now been established that odour cues associated with differences in the MHC affect not just cellular

Figure 3.11   A duck-billed platypus (*Ornithorhynchus anatinus*) hunts for prey in the bed of a stream, which it disturbs with its 'bill'. Around the edge of the bill are electroreceptors capable of detecting the minute electrical activity which results from the contracting muscles of the prey – worms aand crustaceans – as they try to escape.

recognition but behaviour as well. We have already described some of these effects in Chapter 2. Which companions a mouse nests with and which mate it chooses are both affected by its genetic similarity to these other mice at this particular genetic region (Lenington, 1994). By smelling out mates that differ genetically from themselves, mice have a built-in way of avoiding inbreeding, a subject we return to in Chapter 6.

It has recently been discovered that similar effects may operate in humans, suggesting that our sense of smell may be more sensitive than we realize (Thornhill *et al.*, 2003). Women asked to rate the smell of T-shirts that had been worn by six different men reported that the smells were pleasanter from men who proved to be most different from them in the MHC region and least pleasant from men who were similar (Wedekind *et al.*, 1995). What is more, the smell of the MHC-dissimilar men reminded the women most of their own actual or former mates, which suggests that they may have chosen such partners accordingly in the first place.

Other senses

Completely outside our own experience, sixth and even seventh senses are a reality for some animals. Duck-billed platypus (*Ornithorhynchus anatinus*; Fig. 3.11) are able to detect the electric fields generated by muscle activity in animals such as the crayfish and shrimps on which they feed (Scheich *et al.*, 1986). Even though playtpus seem to shut off the 'conventional' senses and close their ears, eyes and nostrils while they are underwater, they have another way of detecting prey. Around the edge of their 'bills', they have electroreceptors that enable them to detect their prey whose muscle activity generates minute electrical fields. Platypus will even approach an electrode placed in the water if it shows electrical pulses like an 'artificial shrimp' (Manger and Pettigrew, 1995). Sharks and rays are even more sensitive to electrical fields generated by muscle movement and, almost unbelievably, can detect fields as small as $0.005\,\mu V/cm$. Electric fish go one stage further and create their own electrical environment. Using

Figure 3.12 The electric fish *Gymnarchus niloticus* creates an electric field around itself and then detects objects by the distortions they make to this field.

specially modified muscle tissue, they set up an electric field around themselves (Fig. 3.12). Objects in the field that are either good conductors or poor conductors of electricity distort the electrical field in characteristic ways, and the fish use the distortions to identify the objects and locate them. Electric fish tend to live in very turbid waters where vision is not much use and they use their electric sense to detect obstacles and to find prey. They also carry on a great deal of their social life by communicating through electrical signals (Hopkins and Bass, 1981).

The Earth's magnetic field can be detected by certain bacteria which contain pieces of magnetite (Blakemore, 1975), by some fish, by amphibians such as the red-spotted newt (Phillips *et al.*, 2002) and even by migrating birds (Akessen, 1994). The exact mechanism by which birds detect the Earth's magnetic field and how they might use it is, however, still a matter of active debate. A major problem is that the Earth's field is extremely weak and so it is difficult to see how it could be detected at all. One clue is that magnetic orientation appears to be possible under short wavelength (blue or green) light but not under red light (Wiltschko and Wiltschko, 2001). This has given rise to the idea that magnetoreception might be dependent on molecules called cryptochromes that are blue-light photoreceptors found in plants, insects and birds, and have a response to light that is affected by applied magnetic fields (Rodgers and Hore, 2009). Despite 40 years of research, magnetic detection in birds still remains a considerable mystery.

In concluding this brief survey of animal sensory systems, we can begin to appreciate how much the behaviour of different animals may be affected by the kinds of sensory information they use. For some species, the ones that use vision and hearing, their sensory worlds may seem relatively easy for us to understand, because we too use our eyes and ears to learn about our world. Even here, however, some animals may take these 'ordinary' senses into realms it may be difficult for us to comprehend. We can hear sound but can't use it to hunt with the precision of bats or owls. And with animals that use electrical signals or magnetism, we encounter truly alien worlds that can only begin to enter by using a machine to do what they have been doing for millions of years. But understanding what senses an animal uses is only the start of understanding its behaviour.

The problem of pattern recognition

Information about what stimuli animals are capable of detecting is clearly an essential preliminary to understanding their behaviour. But it is only the first stage, and it would be easy to make the mistake of thinking that all we need to know about how animals respond to stimuli is to understand how their sense organs transform energy from the environment (light, sound, magnetism, etc.) into nerve impulses, which can then be interpreted or recognized by the brain. However, after the sense organs have responded, the brain still has to interpret the stream of incoming sensory messages and this process of interpretation is extremely complex and far from fully understood.

To understand why the interpretation of information supplied by the sense organs poses such difficulties and in turn to appreciate some of the 'solutions' that different animals have adopted, we can use the very simple example illustrated in Fig. 3.13. A common object like a cup is easily recognizable when seen from different angles and distances even though its apparent shape varies quite dramatically. Thus, even though our retinas are stimulated in different ways, we have no difficulty in saying that we are looking at the same object. However, we can also discriminate the cup from a saucer made from the same material even when, just going by outline, the cup seen from above looks more like the saucer seen from above than it does the cup seen from the side.

The 'problem' of pattern recognition can thus be broken down into two sub-problems: that of recognizing the similarities between patterns of stimulation that are very different but, in fact, come from the same object seen from different viewpoints or under different conditions; and that of discriminating between patterns that are very similar but, in fact, come from different objects. Exactly the same problem arises in the sound domain as with vision. If someone with a high-pitched voice and someone with a bass voice both say the same word, 'tell', the actual sound pattern reaching our ears will be very different and yet we can recognize them as the same word spoken in two voices. Yet if two words 'tell' and 'dell' are said by the same person, the sound patterns are much more similar but we recognize them as distinct words.

The difficulty is that the two sub-problems of pattern recognition – detecting similarities, on the one hand, and discriminating differences, on the other – tend to have mutually incompatible solutions. Thus, one way of recognizing the similarity between the cup seen from above and the cup seen from the side would be to give the label 'cup' to anything of roughly the right size that had at least one curved edge. This would successfully classify the various views of the cup together but it would also erroneously include the saucer as a cup. In other words, successful detection of similarities between the different views of the cup would lead to complete failure of discrimination between cup and saucer. Pattern recognition involves a variety of different solutions that animals have adopted to circumvent this dilemma. All involve compromises and are prone to some inevitable errors that reveal a great deal about what the worlds of animals are like.

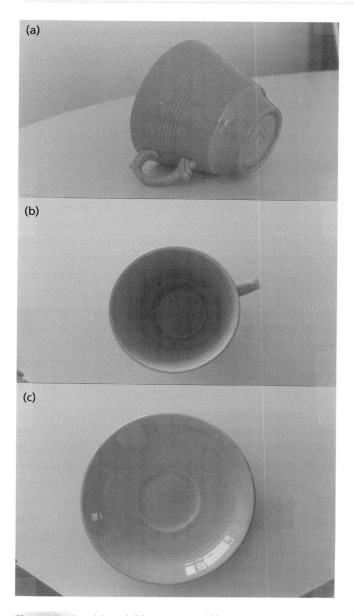

Figure 3.13 (a) and (b) Two very different views of the same object – a cup. We have no difficulty in recognizing them as the same object seen from different viewpoints even though the outline of a cup seen from above is very different from that seen from the side and much more similar to that of a saucer (c) than it is to (a).

Sign stimuli (key features)

One common solution we find is that animals respond to only a limited part of the array of stimuli presented to them. Their pattern recognition is thus accomplished by picking out one or a few key features that all instances of whatever they are trying to recognize have in common. Dragonflies attempting to lay eggs on the shiny metal of a car (picking out 'shiny surface' as the key feature of the water they normally lay their eggs in), or the male robin (*Erithacus rubecula*) described by Lack (1943), which flew down and attacked a bunch of red feathers on the lawn (robins have red breast feathers), are two examples of animals responding to 'sign stimuli'. The animals are responding to restricted features of their environment, apparently ignoring others. By using a small number of key features, they can recognize as similar patterns that might be much more difficult to recognize as the same if larger numbers of features were being used. Male robins present very different stimulus patterns (at different distances, view-points, etc.) if a complex analysis were done, but what they have in common is that have red breasts. Since isolated bunches of red feathers are a rare occurrence, using this one key feature makes recognition easy, simple and view-invariant; a few false alarms are a small price to pay for rapid identification. Even humans use sign stimuli in some situations. People were found to be three times as likely to put money into an 'honesty box' to pay for their drinks if there was a picture of an isolated pair of eyes on the wall than with a control stimulus (Bateson *et al.*, 2006).

There are some dramatic examples of the advantages and disadvantages of relying on sign stimuli. Turkey hens (*Meleagris gallopavo*) which are breeding for the first time will accept as chicks any object which makes the typical cheeping call. On the other hand, they ignore visual stimuli in this situation and deaf turkey hens kill most of their chicks because they never receive the auditory sign-stimulus for parental behaviour (Schleidt *et al.*, 1960). Roth (1948) describes how male mosquitoes respond with high selectivity to the sound of their females' wings which beat at a characteristic frequency, different from their own. The males can, in fact, be attracted to a tuning fork vibrating at the correct frequency. African elephants will actually run away from a playback of the sound of disturbed wild African bees (King *et al.*, 2007). Sign stimuli thus provide a rough and ready way of recognizing patterns in the environment that nicely illustrate the problems we discussed in the last section. They give easy recognition despite variation in exactly how a stimulus is seen or heard. But they can lead to a failure of discrimination if unusual circumstances (such as the presence of a human with a tuning fork) appear.

One of the most famous examples of animals responding to a sign stimulus is Tinbergen's (1951) description of male three-spined sticklebacks (*Gasterosteus aculeatus*) displaying aggressively to a passing red mail van they caught sight of through the windows of their aquarium. Normally, males direct their aggression to other male

(a) (b) (c) (d) (e)

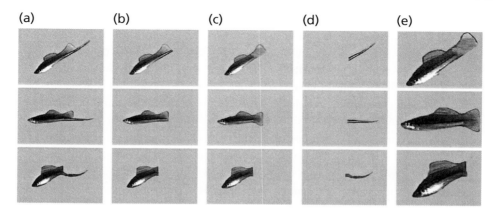

Figure 3.14 Response of female swordtails to video images of the courtship display of a male. The same video sequence was manipulated so that it showed either an intact male with or without a sword or just the sword. The female spent most time with the intact male, but that a sword on its own was almost as effective.

sticklebacks, which are also coloured red in the breeding season. These fish have good eyesight and can certainly discriminate between a fish and a mail van but in this case they reacted to the sign stimulus of red and ignored other characteristics. Fortunately, we don't have to rely on such accidental discoveries to tell us what animals are responding to.

For visual stimuli, we can use models of various sorts, including video and computer-generated images. A quite simple dummy female stickleback moved around on a stick in front of a male will elicit sexual behaviour from the male, but much more effectively if it has a 'head-up' posture than if it has a more level posture (Rowland, 2000).

With video and computer-altered images, we potentially have even more control over what we present to animals. Green swordtails (*Xiphophorus helleri*) are small, colourful South American fish, the males of which have a long 'sword' on their tail fin (Fig. 3.14a) and females have a sexual preference for males with the longest swords (Trainor and Basolo, 2000). Not only will females respond to video images of males, they will also respond to moving images of disembodied male tails (Fig. 3.14b). Using amount of time spent near the image as a measure of attraction, females found a moving tail almost as attractive as an intact male making the same movements (Rosenthal and Evans, 1998). The same technique of altering a video image can be used to discover even finer details of what the female is responding to. Altering the extent and position of the black stripes that occur on the swords with Adobe Photo-shop shows that the amount of black on the sword is important to a female and that she also responds more strongly to tails with black on the end than to the same amount of black nearer the body (Trainor and Basolo, 2006).

We have to be somewhat cautious with relying to heavily on video images, however, because animals may not respond to them in exactly the way they respond to the real world, any more than we do. In real life, female zebra finches show more sexual behaviour to unfamiliar males than they do to familiar males, and although they do respond with sexual behaviour to video images of males, they do not show the same preference for unfamiliar males seen with real birds (Swaddle *et al.*, 2006).

In fact, birds may be particularly difficult to completely 'fool' with video because their eyes have a higher flicker fusion frequency than ours and they may see flickering lights not moving images with ordinary (cathode ray tube) video as we do (D'Eath, 1998). They respond better to TFT (thin film transistor) images (Galoch and Bischof, 2006). For this reason, making alterations to live animals may sometimes be the best way to replicate animals in their natural environment. In a quite remarkable field experiment, Andersson (1982) glued extra long tails onto male widow birds (*Euplectes progne*) and then released them back into the wild. The normal male has extraordinarily lengthened tail feathers, probably larger in proportion to his body size than in any other bird (Fig. 3.15). Andersson cut the tail feather off males and glued even longer feathers onto some of them. As control, he glued shorter tails onto other males and the same length tails onto yet a third group. In this way, he could separate the effects of catching the birds and cutting their tails from the effect of changing the tail length. Males with artificially elongated tails attracted more females to their territories (Fig. 3.15).

This is a particularly dramatic example of what have come to be known as 'supernormal stimuli'.

'Supernormal' stimuli

A very curious phenomenon that has emerged from studies on sign stimuli, such as the swordtails and the widowbirds, is that it is often possible to produce a model which produces a greater response from an animal than does the natural object. Examples of such 'supernormal' stimuli are found in all sorts of animals, but some of the oddest have come from the incubation behaviour of birds. Some of the early observations made by ethologists in the field showed that with the herring gull, the greylag goose and the oystercatcher, the larger an egg is, within broad limits, the more it stimulates incubation. Figure 3.16 shows an oystercatcher struggling to incubate a giant egg in preference to its own.

Male sticklebacks preferentially court females which are large and have distended abdomens indicating that they have a lot of eggs. Rowland (1989) showed that when presented with two dummy females, a male would first direct his courtship to the largest and fattest one, even when the 'fat' female had an abdomen distended far beyond the normal range for female sticklebacks.

Herring gull (*Larus argentatus*) chicks respond to very crude models of their parents (Tinbergen and Perdeck, 1950). The adult herring gull has a red beak with a yellow spot on the lower mandible at which the chick pecks. This causes the parent

(a)

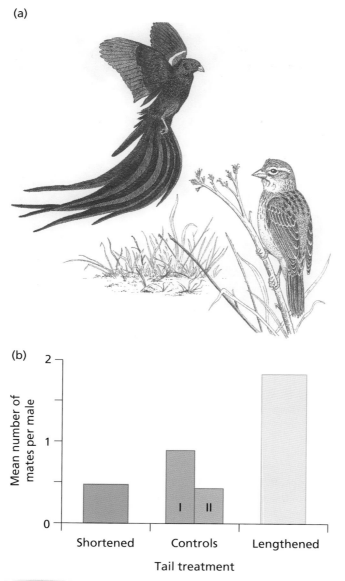

(b)

Mean number of mates per male

2

1

0

I   II

Shortened   Controls   Lengthened

Tail treatment

Figure 3.15 *Above*: A male long-tailed widow-bird (*Euplectes progne*) displaying to a female. *Below*: Results of an experiment in which the tails of males were artificially lengthened or shortened (with controls having artificial tails of the same length). Males with elongated tails attracted more females to nest in their territories.

to regurgitate food. However, the chick will not only peck at a headless beak: it will peck even more vigorously at a long thin red knitting needle, particularly one with yellow bands on it. Nor is it only our animal relations that respond in such ways. It is obvious that lipstick makes human lips into a super-normal stimulus

Figure 3.16 An oystercatcher attempting to brood a giant egg in preference to its own egg (foreground) or a herring gull's egg (left foreground). The bird's original nest was equidistant between the three 'test' eggs.

Figure 3.17 Human lips are made strikingly supernormal by the addition of lipstick.

(Fig. 3.17) and much of the advertising industry is based on persuading us to respond to exaggerated versions of everyday situations.

At times, the response of animals to supernormal stimuli can confuse both animals and their human observers. Burley *et al.* (1982) put tiny coloured bands on to the legs of male zebra finches solely, so they thought, to be able identify different males. To their astonishment, the males that happened to have been given red leg bands turned out to be more attractive to females than males given green, blue or other colour leg bands and they also had more surviving offspring. Female zebra finches are normally attracted to males with bright red beaks and respond to the sign stimulus of 'red'. Males with more red (even though it is in the wrong place) are 'supernormal'. Once mated to a supernormal male, a female works harder than usual at rearing young by making more food-gathering trips and so her well-fed young survive better than average. Without understanding how the females were responding to the identification bands, quite the wrong conclusions could have been drawn about which males were genetically the best fathers.

At this point, we can ask why – in an adaptive sense – animals respond to supernormal stimuli at all. It would appear that by responding to exaggerated versions of natural stimuli, animals are behaving 'stupidly', and against their own interests. Why has natural selection not acted against animals that behave in such ways? It is here that we see clearly the importance of asking different sorts of question about animal behaviour and of understanding its evolution as well as its causation. We have just asked why, in a causal sense, animals respond to sign stimuli and supernormal stimuli and have seen that such responses arise from the need to solve the difficult problems of pattern recognition. We have seen that sign stimuli help solve some of these problems, but they are certainly not perfect solutions. But why, in an evolutionary sense, have the animals not done better?

Many of the examples we have discussed of animals responding to sign stimuli and supernormal stimuli have been of animals put into very unnatural situations, i.e. ones that would rarely, if ever, occur in nature. For example, ostrich eggs are not a feature of the natural environment of oystercatchers, nor are knitting needles generally found in herring gull nests. Equally, red mail vans are not a major natural hazard for male sticklebacks and red leg bands are not worn by wild male zebra finches. Consequently, simple 'rules of thumb' such as responding to the biggest or brightest or longest object around will usually serve the animal perfectly adequately and may be the easiest to evolve given the structure of its sense organs. False positives will be rare or non-existent and probably confined to interference from ethologists.

Using a simple rule of thumb, such as courting the largest fattest female around for a male stickleback, will normally result in a male having the greatest number of eggs laid in his nest, since he will be directing his courtship towards the most fecund females. However, the fact that the fattest females may sometimes not be full of eggs but full of parasitic worms shows that there are dangers in solving a pattern recognition system by relying solely on simple sign stimuli. Where mistakes are relatively rare, these may persist ('the rare enemy' effect as Dawkins, 1982, describes it). But where mistakes or false positives are common and/or dangerous, we find that natural selection has favoured animals that do not make them.

The habit of brood parasites such as cuckoos of laying their eggs in the nests of other birds provides powerful selection pressure on the potential foster-parents *not* to rely on sign stimuli and to discriminate between their own eggs and those of other species. The brood parasite benefits if it can 'deceive' a host into caring for its young, but the cost to the host can be very high. A young cuckoo (*Cuculus canorus*), for example, kills the

Figure 3.18 Having pushed this reed warbler's own eggs out of the nest, this cuckoo chick has the undivided attention from its foster parent, which continues to feed it even when it becomes bigger than the parent itself.

Figure 3.19 Different races of cuckoo specialize in parasitizing different host species. Each race has evolved eggs that closely mimic the eggs of their own host.

host's own young by pushing all existing eggs out of the nest. So the host – such as a reed warbler – not only loses the eggs she has already laid, but loses again through working hard to feed the rapidly growing cuckoo (Fig. 3.18) rather than laying a replacement clutch itself. The host appears to respond to the 'sign stimulus' of an open gape, but more careful analysis shows that there is an evolutionary 'arms race' between the cuckoo chick to be accepted and the host to recognize it. The selection has been particularly severe at the egg-recognition stage.

Each individual female cuckoo specializes in just one species of host and lays eggs that match, often closely, the eggs of her host (Fig. 3.19). Females from different strains of cuckoos (known as gentes) somehow choose to lay in the nests of species whose eggs resemble their own. Wagtail-cuckoos produce white spotted eggs and lay them in wagtails' nests and pipit-cuckoos produce brown spotted eggs that they lay in pipits' nests. Females deposit a single egg in the nest of their chosen host and take one egg away. The cuckoo egg develops quickly and usually hatches before any of the host's clutch, which gives the young cuckoo the chance to destroy all the other eggs.

Even though cuckoos breed throughout Europe and northern Asia, they parasitize only about 10 host species. It is the difference between the small number of species that are parasitized and the much larger number that are not which provides us with the evidence that egg recognition has evolved under the evolutionary pressure from brood parasites.

Davies and Brooke (1989a, b) investigated the responses of 24 species of songbirds to eggs of the European cuckoo by placing model cuckoo eggs in their nests and recording whether they were accepted or thrown out. They found that most of the species that are currently parasitized by cuckoos, such as reed warblers, meadow pipits (*Anthus pratensis*) and pied wagtails (*Motacilla alba*), showed some degree of discrimination in that they would accept model eggs if they were similar to their own but not if they mimicked those of other species. However, the other species that are not currently parasitized by cuckoos, such as spotted flycatchers (*Muscicapa striata*) and reed buntings (*Emberiza schoeniclus*), rejected all model eggs, irrespective of type. These species would seem from their diet and nest locations to be highly suitable as cuckoo hosts, which raises the intriguing possibility that they were once parasitized by cuckoos until their egg discrimination became so good that cuckoo eggs could no longer survive in their nests.

By contrast, other species, such as greenfinches (*Carduelis chloris*), linnets (*C. cannabina*), great tits (*Parus major*) and swallows (*Hirundo rustica*), accepted all the model eggs, whatever they looked like, but these species, because of either diet or nest site, have probably never been suitable hosts for cuckoos and so have never been selected for sophisticated egg discrimination. Meadow pipits in Iceland, where the cuckoo does not occur, were much less discriminating about model eggs in their nest than the same species in Britain, where the cuckoo is a threat. This is a clear illustration that the ability to discriminate is itself under selection pressure (Langmore *et al.*, 2005).

 ## Neuroethological basis of sign stimuli

Sometimes we find that the sense organs themselves are responsible for the selective response to a particular sign stimulus through being relatively insensitive to anything else. One dramatic example we have already encountered is that of the male silkmoth (*Bombyx mori*), whose highly sensitive antennae pick up the scent of the female in very low concentration, but are unable to smell almost any other substance. Another example are the calls of male green tree frogs (*Hyla cinerea*) that have two peaks of sound energy – a low one at 900 Hz and a high one at about 3000 Hz (Gerhardt, 1974). The ears of the females are tuned to pick up these two frequencies in particular: in the female's auditory system, the so-called 'amphibian papilla' is most sensitive to frequencies between 200 and 1200 Hz while the 'basillar papilla' responds best to sounds of about 3000 Hz (Capranica and Moffat, 1975). Narins and Capranica (1976) showed a remarkable sex difference in the hearing of a related species of tree frog (*Eleutherodactylus coqui*), which accounts for the fact that males and females respond to different sign stimuli in the call of the male – a double 'co-qui'. By playing back tape recordings of the males' call to the frogs, they were able to show that males (who attack other males that approach them) respond only to the 'co' note, and females only to the 'qui'. Neurophysiological recordings made it clear that each sex heard only the note of relevance to itself because the neurons of the inner ear were tuned differently for males and females.

Selective responsiveness to sign stimuli has been a particularly fruitful area for showing how the behaviour of an animal can be related to underlying neurobiological mechanisms. Ewert and Traud (1979) provide a particularly good example in their study of the way in which toads respond to snakes. In response to real snakes, toads fill their lymphatic sacs with air, making their bodies appear much larger than they really are and assume a stiff-legged posture, presenting their flank to the snake. Toads will perform this stiff-legged snake posture to quite crude dummy snakes made of flexible pieces of cable. The 'sign stimulus' for this behaviour seems to be quite easy to reveal, as even a model consisting of just a horizontal piece with a little head attached produced almost as much (93%) response as a real snake.

Ewert and his colleagues have analyzed in some detail the sign stimuli that toads use for recognizing both prey and enemies. In both cases, movement and contrast with the background are important, but whereas a toad will turn towards a small, horizontal, worm-like object moving in a horizontal plane (Fig. 3.20), it will not turn towards the same object orientated vertically with the longitudinal axis perpendicular to the horizontal direction of movement. In the toad's view, worms do not walk on their heads! Large, expanding (looming) objects are sign stimuli for escape and a toad will turn away from or freeze in response to a very crude cardboard model coming towards it.

It is possible to link the toad's responsiveness to prey and enemy to cells in its retina and even specific parts of its brain. A plan of the amphibian retina is shown in Fig. 3.21 and, although it is possible to record the electrical activity of all types of retinal neurons, the most revealing have been made by placing electrodes in the ganglion cells since these turn out to be of several distinct sorts, distinguished by how they respond to different visual stimuli. In a pioneering paper entitled 'What the frog's eye tells the frog's brain', Lettvin *et al.* (1959) showed that there were six different sorts of ganglion cells, with four being most common, and this has subsequently been confirmed for toads as well. Class 1 (rare in toads) and Class 2 retinal ganglion cells show a vigorous response whenever the animal is looking at a small moving object. The retinal ganglion cells have a curious property that explains how they can 'recognize' when an object is small and moving. Each retinal ganglion cell responds to a particular section of the total

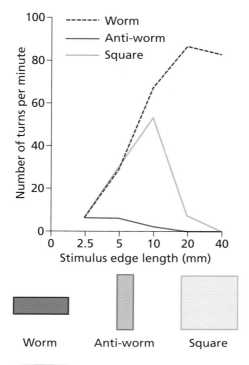

Figure 3.20 A toad's responses to three moving models. The toad turns to follow a model and its response is measured by the number of times it turns to follow it in 1 min.

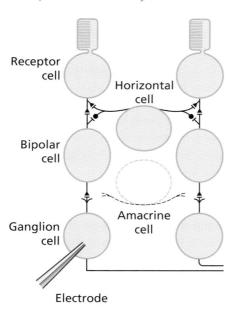

Figure 3.21 Diagram of the retina of an amphibian showing the connections between the various sorts of cells. Excitatory synapses are shown by ◄ and inhibitory ones by ─●.

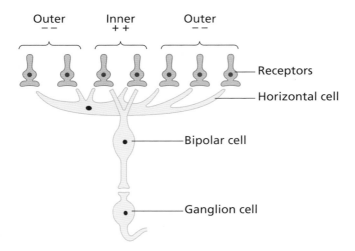

Outer   Inner   Outer
 – –   + +   – –

Receptors

Horizontal cell

Bipolar cell

Ganglion cell

Figure 3.22   Diagram of a retinal ganglion cell's on-centre receptive field. The ganglion cell receives input from a number of receptor cells through the horizontal and bipolar cells. Light falling on the receptors at the centre of the field excites the ganglion cells and evokes action potentials. Light falling on receptors on the surrounding (outer) area decreases the activity of the retinal ganglion cell and evokes fewer action potentials.

visual field. Thus, any given cell will not respond every time the toad looks at an object but only if the image of that object falls on a particular part of the retina, known as the 'receptive field' of that cell. The receptive field of each retinal ganglion cell is divided into two parts (Fig 3.22), an inner excitatory centre and an outer inhibitory ring. This means that whenever an object is picked up by the inner part of the receptive field, the retinal ganglion cell is excited and fires, but if the object falls on the outer part of the receptive field, the retinal ganglion cell's activity is inhibited. Thus inhibition, far from being a purely negative process, is crucial to the frog's ability to recognize small objects.

The different classes of retinal ganglion cells have different sizes to their excitatory receptive fields, with the Class 1 and Class 2 cells having smaller such fields (2–5°) than the Class 3 (6–8°) or Class 4 cells, which have the largest fields of all (10–15°) and do not respond at all to a small object unless its visual angle exceeds about 5°. If a square shape is moved in front of a toad's eye, the impulse frequency in the different classes of retinal ganglion cells also varies with the size of the square used. In Class 2 cells, with their small excitatory centres, only quite small shapes give maximum response because larger squares cover both the excitatory and the inhibitory areas of the receptive field and this has the effect of reducing the total activity of the cell. Class 4 cells, with their large excitatory fields, are active when the toad is viewing a much larger square because the inhibitory surround is not stimulated. What this all means is that Class 2 and 3 cells are activated mainly by prey stimuli and Class 3 and 4 cells mainly by 'enemy' stimuli. However, it would not be correct to say that there are specific prey or enemy detectors in the toad's retina and that there we had accounted for the toad's response to 'sign stimuli' in terms of the physiology of its eye. What the retinal ganglion cells do is to extract basic information about the size, angular velocity and contrast of the object the eye is looking at, leaving it to the brain to distinguish between food and enemies. Ewert (1980, 1997) describes in some detail how the toad's brain performs this task. In fact, the analysis of sign stimuli and their neurophysiological basis has been one of the most successful areas of 'neuroethology'.

Other solutions for pattern recognition: generalized feature detection

Sign stimuli – picking out a few key features – is one way of solving some of the problems of pattern recognition, but by no means the only one. We have seen that too much reliance on sign stimuli leads animals to make mistakes and much animal behaviour cannot be explained as responses to simple sign stimuli. For example, indigo buntings (*Passerina cyanea*) inherit the ability to learn complex star patterns to guide their migration routes. The young birds are not equipped with an 'innate' star map, but watch the night sky to see which star is stationary. In the northern hemisphere this would be the pole star (Polaris) and the young birds learn the pattern of stars around Polaris. By rearing young birds in a planetarium in which he could blot out certain parts of the sky, Emlen (1975) showed that the birds learnt the pattern of stars within 35° of the pole star (Fig. 3.23).

Understanding the complex pattern recognition systems that animals have evolved is beyond the scope of this book and all we can do here is to pick out a few principles and give a flavour of what we are now beginning to discover. An indication of the difference between the way a stimulus is registered by our sense organs and the way our brains perceive the world is given by Fig. 3.24. Our eyes see the same elements in the two halves of the figure, but an object jumps out at us when it can be perceived as a cube behind bars.

We know that different animals make this transition from sense organ to interpretation in different ways from us. We humans find it difficult to recognize even familiar shapes if they are presented to us upside down. (Reading upside down is difficult without practice.) Pigeons, however, experience no such difficulty and take no longer to recognize a shape when it is upside down than when it is the right way up. This suggests that the birds are processing the information in a somewhat different way to us, perhaps because pigeons are flying animals and have evolved to see objects from more different angles than we do (Hollard and Delius, 1982).

Despite the many differences, one characteristic that does seem to be very widespread amongst a variety of animals is that sensory systems are adapted to pick out 'information-rich' parts of the environment (Barlow, 1972). We all know that a complex scene can often be conveyed by a quick line drawing and this is because lines or edges

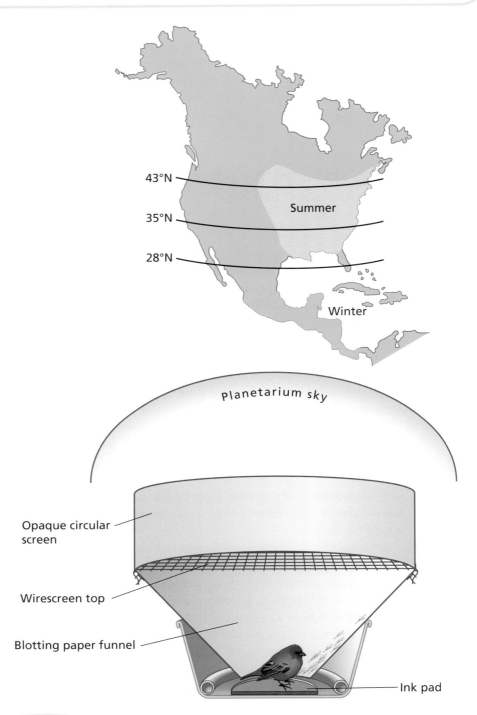

Figure 3.23   (a) The indigo bunting's summer and winter ranges. (b) The equipment used to test the buntings' response to a planetarium sky. The birds stood on an ink pad and left a record of their activity on the blotting paper funnel.

(a)   (b)

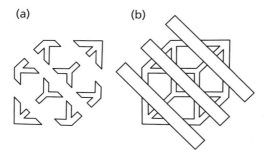

Figure 3.24   The difference between 'seeing' and 'perceiving'. Exactly the same elements of the cube are present in (a) and (b) but while the cube is not obvious in (a), it suddenly becomes so with the addition of three 'opaque' stripes.

carry a great deal of information – about where one object ends, for example, and another begins. An ability to detect lines, edges, discontinuities of light and dark and so on would therefore be a great help in analysing many different scenes, because the same general features, although in different combinations, occur in many different situations. Thus the letters 'A' and 'F', for example, are both made up of three straight lines, so line-detectors would be useful in recognizing both. To discriminate between them, on the other hand, it would be necessary to recognize the particular angles, lengths and arrangements of those three lines.

This, in a very simplified version, seems to be the basis of the way complex pattern recognition in animals occurs. At an early stage of sensory processing, generalized feature detectors are picking out spots, edges, movement or sudden changes of illumination points. These are then put together to define more complex objects at a later stage of analysis. Generalized feature detectors have now been discovered in a wide variety of animals, both invertebrates and vertebrates.

In the horseshoe crab (*Limulus*, Fig. 3.25a) it is possible to see how these general features are extracted from the overall pattern of illumination reaching the eye. Each eye is compound, i.e. it is made up of several hundred units called ommatidia, each of which contains ten or more retinula cells (Fig. 3.25b). The axons of the retinula cells come together to form an optic nerve from each ommatidium, but the ommatidia do not operate entirely separately from each other – there is interaction between them in the form of lateral inhibition. Each ommatidium inhibits the actions of its neighbours but this has the important result of accentuating contrasts in illumination.

To understand how this works, consider the case where the whole eye is viewing a uniformly lit scene. All the ommatidia will be equally active and all will be inhibiting each other with roughly equal strength. Now consider what happens when a small bright light falls on just part of the eye. The ommatidia that are stimulated by the light will respond strongly and they will also strongly inhibit their neighbours. Ommatidia just outside the patch of light will not be responding much to the light, but they will be

(a)

(b)

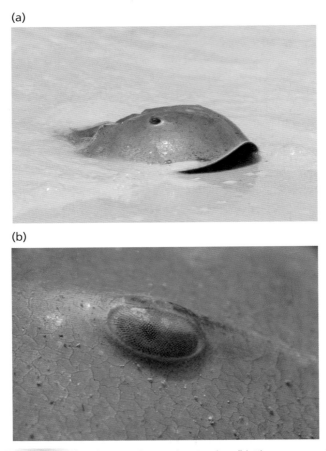

Figure 3.25   (a) Horseshoe crab, *Limulus*. (b) The compound eye of *Limulus* showing the many little facets or ommatidia.

strongly inhibited by nearby active ommatidia. They will therefore have an overall response that is less than ommatidia further away. Conversely, the ommatidia just inside the patch of light will be responding to the light and they will be less inhibited by their inactive neighbours than ommatidia in the centre of the patch of light that are all surrounded by strongly inhibiting neighbours. The result is that the boundary between the light and dark area of the scene will be the most active (Fig. 3.26), and the feature of an 'edge' will be accentuated (Hartline *et al.*, 1956).

Lateral inhibition is thus as important to the ability of this compound eye to recognize shapes as it is in the complex retina of vertebrates, where we have already seen that the retinal ganglion cells respond more strongly to a small patch of light than to a large one (p. 138). Lateral inhibition enables the retinal ganglion cells to extract the feature 'spot' from the incoming sensory information because they are excited by activity in

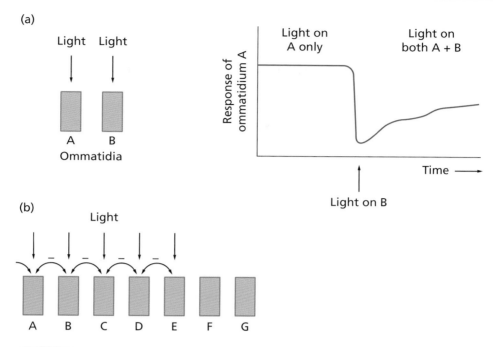

Figure 3.26 (a) The action of lateral inhibition in *Limulus* eye. The response of ommatidium A is *reduced* when light is also shone onto ommatidium B, because B actively inhibits A. (b) Shows how this lateral inhibition works to enhance the response to an edge or discontinuity. If light falls onto ommatidia A–E but not onto F or G, the largest response will be from ommadia E at the boundary between light and dark. This is because E is activated by the light but only inhibited on one side (by ommatidium D), because F is not active. However ommatidia A–D are inhibited by both their neighbours and so have lower responses.

some of the photoreceptor cells and inhibited by activity in others. By monitoring which of the lower level photoreceptor cells are active, the higher level retinal ganglion cells can begin to detect patches of light of different shapes.

There is considerable evidence that the visual system operates with such a hierarchical arrangement within, with progressive extraction of information at different points (Hubel and Wiesel, 1959), but it is not a single hierarchy. On the contrary, there are multiple hierarchies. Each area of the brain, whether concerned with vision or not, has multiple outputs, with the different areas specializing in different functions. In mammals, there are separate pathways concerned with movement and with the form of objects (Zeki, 1993). We also know that the same cells can be active in the recognition of quite different objects. At least in vertebrates, the information about what object is 'out there' is not given by the firing of one particular cell but by the patterns of activity in whole sets of interconnected neurons.

Recall the extraordinary complexity of the brain we discussed in Chapter 1. A single neuron may be connected to 10 000 others. When one neuron is active, some, none or most of the cells connected to it may also be active. These activity *patterns* – which sets of neurons within this amazing network are active and which are inactive – are what carry the information about the external world.

For example, there is a class of neurons in the temporal lobe of the brains of rhesus monkeys that fire up to 20 times more rapidly when the monkey is looking at a face (monkey or human) than they do in response to lines, gratings or complex objects that are not faces. They can therefore be referred to as 'face-detecting' neurons (Baylis *et al.*, 1985) and since many of them respond more to some faces than others, they seem to be involved in the recognition of individuals. However, a given cell can be involved in the identification of several different faces. The information about which face is seen is therefore not given by the firing of one particular cell, but is spread over a whole ensemble of neurons and it is the pattern of neuronal activity in the whole set that changes when different faces are looked at that matters (Rolls, 1994). How animals recognize patterns is an exciting and active area of research for ethologists, physiologists and psychologists. The brains of animals are regularly performing analyses that the biggest computers still cannot do. We still have a great deal to discover.

Communication

We are now in a better position to begin to understand the world of animal communication. Many of the most important external stimuli to which animals respond come from signals given by other animals of their own species – from their parents or from their own offspring, for example, or from rivals or potential mates. The discoveries that have been made about the reception and processing of sensory information help us to understand how animals respond to these signals and also how communication between them has evolved. Signals are conspicuous behaviour patterns, often combined with structures like plumes or crests that have been specially evolved to affect the behaviour of another animal. The single large claw of the male fiddler crab (*Uca*) is an example of a structure evolved for a signal function (Fig. 3.27). The claw is far bigger than is needed for feeding or other non-signal purposes. It is brightly coloured and waved rhythmically in the air at females and rival males as the crab stands by its burrow, suggesting that it has been evolved to be a conspicuous stimulus to other crabs.

In this section, we will first discuss what communication means and then we will look at the evolutionary pressures that operate on animal signals, including the importance of attracting the attention of the receiver and the question of whether animal signals are 'honest'. Finally, we will look in detail at two remarkable but very different animal communication systems – those of vervet monkeys and of honeybees.

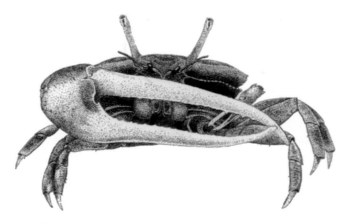

Figure 3.27 A male fiddler crab (*Uca*) whose hugely enlarged claw is purely for display. The feeding claw can be seen on the crab's left. In females both claws are the same size.

What is communication?

'Communication' is said to occur when one animal responds to the signals sent out by another animal. Although this sounds a reasonable definition, there are cases where it is not so straightforward. Burghardt (1970) cites the example of foraging ants which lay scent trails when they have found food. Its nest mates follow the scent trail, but so does a small snake (*Leptotyphlops*) which uses the scent trail to find the ants' nest, where it devours the brood. Most of us would probably refer to the scent trail as a signal, evolved by the ants for communicating with each other, but although the snake undoubtedly obtains information from the ants, we would probably not want to say that the ants communicated with the snake.

The reason we would not call this communication has nothing to do with the fact that ants and snakes are of different species because there are plenty of examples of interspecific communication. The nectar guide patterns on flowers (p. 115) communicate with bees to the mutual advantage of both and can be seen as a specially evolved signal to help the bees find the centre of the flower. Warning or startle coloration of prey animals – for example, eye-spot patterns of moths (p. 307) – are other examples of interspecific signalling.

The reason we can describe flowers and moths' eye-spot patterns as communicating but not the ants with the snake, is that the ants derive no benefit whatsoever from the response of the snake (they are of course harmed by it), so that there is no suggestion that natural selection favoured ants that attracted snakes whereas the flowers and the moths both benefit from the response they evoke. The ants have evolved to communicate with each other by their scent trails because it is of mutual advantage to recruit other workers to a food source. One of the penalties of this signalling system is that a predatory snake can use the same information but, provided snakes are relatively rare, the advantages the ants derive from communicating with each other will outweigh the occasional disadvantages.

Even this mutual advantage proviso to the definition of communication is not without its difficulties, however, and we must recognize that not everybody uses it. For example, both Altmann (1962) and Hinde and Rowell (1962) studied social communication in rhesus monkeys (*Macaca rhesus*) and attempted to describe their usual signals, but the former study lists about 50, the latter only 22. The reason for this discrepancy lies in their different uses of the term communication. Altmann's definition of social communication is a 'process by which the behaviour of an individual affects the behaviour of others'. All kinds of movements and postures may do this – the sight of a monkey feeding, for example. Hinde and Rowell, on the other hand, use a definition similar to that given above and they classified as visual signals only those that were likely to have evolved for this purpose. Even though we shall concentrate on specially evolved social signals in this brief discussion of communication, it is important to recognize that animals do receive a great deal of information in the way Altmann describes. A hierarchy is a key feature of most monkey groups and we discuss its role in Chapter 7. Here we may note that members of a rhesus monkey troop will respond to a range of diverse stimuli from the dominant male. His general body posture and manner of moving will convey information quite apart from any signals, such as threat movements, that he may make. Almost every feature of his body when moving or at rest is in strong contrast to that of a subordinate male (see Fig. 3.28). In such cases it may be difficult to decide whether a 'specially evolved' signal is involved or not.

We should also be aware that even movements or postures that are scarcely distinguishable to us may have considerable significance for the animals themselves. In blue tits (*Parus caeruleus*), for example, quite different messages are conveyed to other birds depending on whether a bird raises the feathers on the top of its head or the ones just a little way down the nape of the neck (Fig. 3.29).

A further complication is that not all communication takes the form of a clear signal followed reliably by a clear response. In many bird species, dominance is associated with special patterns of feather coloration known as 'status badges'. For example, the most dominant male house sparrows (*Passer domesticus*) have the biggest black bibs (Møller, 1987; Fig. 3.30) and in Harris' sparrows (*Zonotrichia querula*) the most dominant have the blackest heads. But whilst subordinate birds may sometimes respond to a status badge by giving way over a food source or avoiding the dominant altogether, they do not always do so. Sometimes the subordinate bird appears to take no notice of the status badge, raising doubts as to whether it is really a signal at all. However, further analysis suggests that status badges are best regarded as signals but whether or not other birds respond to them depends on factors such as where the encounter takes place (Wilson, 1992) and how hungry the two animals are. Møller (1987) showed that when food was plentiful, subordinate house sparrows responded to the status badge of

(a)

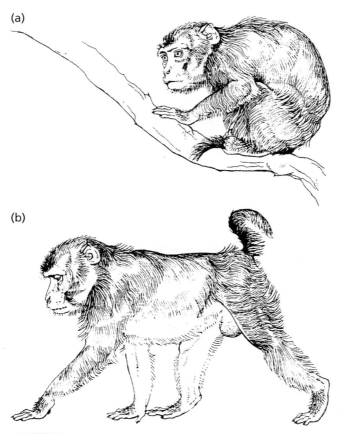

(b)

Figure 3.28   Typical body postures adopted by (a) subordinate and (b) dominant rhesus monkeys.

Figure 3.29   Two blue tits dispute over a food item on the ground. The bird on the left is giving way, about to retreat, and shows the raised crest feathers indicative of fearfulness. The winning bird on the right has a flattened crest, but the nape feathers are conspicuously elevated, giving its neck a thickened appearance. Nape raising indicates high levels of aggression.

(a)

(b)

Figure 3.30   In house sparrows, a male's dominance correlates with the size of his bib. A dominant bird (a) has a much larger bib of black feathers than a subordinate (b).

dominants by giving way to them, but when food was scarce, the high status of a dominant counted for less and fighting was much more likely to occur. This, of course, makes evolutionary sense since the cost of giving up a piece of food when there is not much available elsewhere can be very great and even fighting a dominant animal for possession of it may become worthwhile.

Well-established dominance relationships in a group of animals that know each other may cause yet more problems for an observer trying to decide whether they are communicating or not. Bossema and Burger (1980) and van Rhijn (1980) studied a small group of captive jays in which the younger birds tended to give way to the more

Figure 3.31 Metacommunication in lions. The male's posture with lowered forequarters is only seen as a preliminary to play. He invites a cub to play and cuffs it gently as the cub joins him.

powerful older members of the group. But once the rank order was established, the older birds gave only a minimal 'signal'. The older birds could sometimes displace younger ones from, say, a piece of food simply by looking at them. Only if this glance was ineffective would they escalate and perform a recognizable threat display. Similarly, most people who have studied primate groups over a long period can recount episodes where a severe fight between two individuals has had repercussions that last for months or years. This may result in the loser fleeing or showing responses to the animal that defeated it while other animals with different past experiences show little or no response. An observer unfamiliar with the history of interactions between the animals might be quite unable to detect that any communication was occurring.

We should not conclude this discussion about definitions of communication without pointing out a rather curious type of signal whose function seems not to communicate information itself, but to qualify other signals that follow it. This phenomenon has been called metacommunication and is best known from play situations in carnivores and monkeys. Figure 3.31 shows an adult male lion inviting a cub to play. His posture with forequarters lowered – a 'bow' – is not seen in any other context and its message is that all aggressive movements that follow are play. Dogs, wolves and coyotes use almost exactly the same posture and may also wag their tails during play fights (as Darwin, 1872, noted in his book, *The Expression of the Emotions in Man and Animals*). Bekoff (1995) showed that the bow is used to convey the message 'I want to play despite what I am going to do or have just done' and occurs when the animal is using actions borrowed from another context such as biting and shaking movements of the head, which are particularly likely to be misunderstood. Monkeys adopt a 'play face' in similar situations. As we saw in Chapter 2, playful aggression is a conspicuous part of behavioural development in carnivores and it has obviously been necessary to evolve some convention that allows stalking and attacking to take place without the damage of real fights.

Animal signals as effective stimuli

Since the function of signals is to elicit a response from another animal, their evolution has involved producing more and more effective ways of getting other animals to respond. But what makes an effective signal? How have signals evolved over time? The two most obvious selection pressures on signals are travelling from sender to receiver through a medium such as air or water and stimulating the sense organs of the receiver. It would be a useless signal that was not loud enough or conspicuous enough for the receiver to hear or see or was in a form that the intended receiver could not detect at all. In fact, looking across the animal kingdom as a whole, it is clear that much of the diversity of animal signals comes from the parallel diversity of their capacities to detect stimuli that we have already discussed. Animals that rely on vision for finding their way about and locating prey will also tend to communicate through visual signals. Animals that never use vision, such as the blind workers of some termite colonies which never leave their subterranean tunnels, rely on tactile communication often enhanced by scents that are detected by chemoreceptors. Water striders (*Gerris*; Fig. 3.32) use the ripples on the surface of a pond to detect both predators and prey and they also use ripples to communicate with each other. Both males and females produce surface ripples by vertical oscillations of their legs, but whereas females produce only low frequency (3–10 Hz) waves, males produce both low and high frequency (80–90 Hz) waves, to which other males respond (Wilcox, 1979). Sometimes, of course, animals use several different modalities depending on how close they are to each other. Mammals use hearing, sight and smell at a distance but for many of the more social ones that spend a good deal of time in physical contact, tactile communication is also important (Fig. 3.33). Chimpanzees touch and kiss each other's hands in gestures of reconciliation after a fight (de Waal, 2000).

Within a sensory modality, we can often explain the diversity of signals by understanding the constraints imposed by the medium through which the signal must travel. Most visual signals depend on natural daylight, available only to diurnal animals.

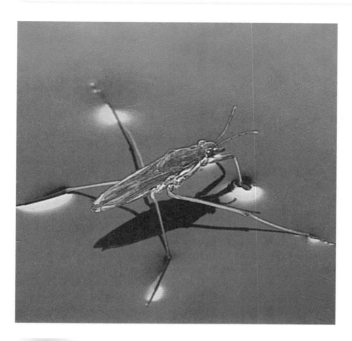

Figure 3.32 A pond skater or water strider (*Gerris*) resting on the water surface. Its feet (tarsi) repel water and so allow its tiny weight to be supported by the surface tension. These same tarsi are the means by which it responds to the ripples set up in the surface film by struggling insects that are potential prey. It also uses ripples to communicate with other water striders.

Visual signalling at night is obviously limited as even bright moonlight is only one-tenth as strong as daylight. Some nocturnal animals such as fireflies (Lampyridae – a family of beetles) overcome this limitation by manufacturing their own light and signalling at night (Fig. 3.34). In fish living in deep water, where light is scarce, we also find animals making their own light (Fig. 3.35).

Visual signals are influenced not just by the amount of light available but also by the fact that some wavelengths are scattered or absorbed more than others in their journey from signaller to receiver. This effect is particularly striking in seawater, where blue light of about 475 nm travels particularly well, but red light (longer wavelength) or ultraviolet (shorter) are simply filtered out by the water. This is the reason why so many reef fish are blue or yellow, or have striking black and white stripes (Fig. 3.36) that are also visible in seawater (Lythgoe, 1979).

Even on land, some wavelengths of light travel better than others and so some colours and patterns may be more conspicuous at long distances. In warblers of the genus *Phylloscopus*, there is particularly clear correlation with habitat and the brighter species live in the darker habitats (Marchetti, 1993). Birds such as the cock-of-the-rock (*Rupicola rupicola*) actually choose to display in places in the forest and at times of day when their signals maximize their colour and brightness (Endler, 1993). It is

Figure 3.33   Tail twining – a form of tactile communication – by adult dusky titi monkeys (*Callicebus moloch*).

Figure 3.34   Fireflies produce their own light for conspicuous night-time signalling.

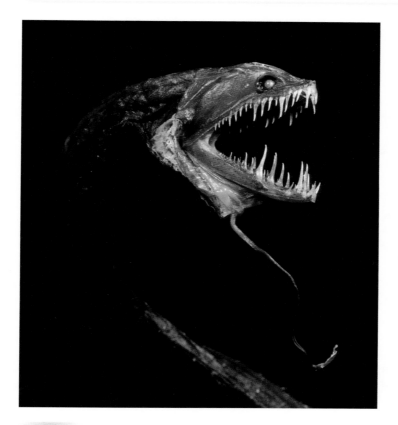

Figure 3.35 In the depths of the ocean where there is no light, this fish, *Idiacanthus antrostomus*, makes its own light.

Figure 3.36 The intense black and white markings in the blue-headed wrasse (*Thalassoma bifasciatum*) make it clearly visible through seawater.

Figure 3.37 A white-tailed deer (*Odocoiles virginianus*) shows alarm by raising its tail to reveal its white underside. The white fur of the rump is flared out to give a white flash as the deer leaps off. It may be as much a signal to the predator as to other deer.

also striking that visual signals that operate over long distances, such as the white tails of rabbits and some deer (Fig. 3.37) or the eyebrows of mangabey (*Cercocebus*) monkeys, have a broad range of frequencies. Against most backgrounds, white is very conspicuous.

Sound has one big advantage in that it can go round corners and through some barriers, although it degrades in the process. Sound attenuates (degrades) in all types of habitats but high frequencies attenuate and become scattered by obstacles much more than low frequencies. The most extreme case of this is the very high frequency (50–100 kHz) sound used by bats for echolocation (p. 117). This attenuates so much with distance that a 100 kHz sound bat call drops to 1/50 000th of its initial intensity within 4–5 cm. These calls function well as the bat comes in close to an insect before seizing it but the ultra-high frequencies would be quite useless for communicating over any greater distance. Lower frequencies travel much better over longer distances. As mentioned on p. 136, Gerhardt (1974) showed that the male green tree frog uses a call with two different frequencies. A female at a distance will pick up the lower frequency and head towards a male. As she gets closer, and the sound gets louder, she responds to both the low frequencies and also the higher ones that she can now hear.

Animals can also help sound to travel further over longer distances by elevating themselves above ground with all its obstacles. Crickets that sing from trees or shrubs can spread their signal over 14 times the area of those that sing from the ground, and consequently attract more females (Paul and Walker, 1979). The territorial songs of birds are usually delivered from a raised song post which also increases their effective range, while grassland birds such as meadowlarks (*Sturnella*) and pipits (*Anthus*) sing while they fly.

Because there is less attenuation of sound in water than in air, sound travels further and so many aquatic animals use sound extensively for communication. Since light may be dim except at the surface and at close quarters, sound is effectively all that is available. It is ironic to recall that Jacques Cousteau's pioneering book on free-swimming scuba diving was entitled *The Silent World*. The development of the underwater microphone has now allowed us to discover the amazing range of sounds produced by crustaceans, fish and whales. Payne and McVay's (1971) remarkable study of humpback whales (*Megaptera novaeangliae*) suggests that their 'song' could be picked up by other whales several hundred kilometres away! This is certainly a long-distance record for animal communication.

Chemical signals are particularly well developed in insects and mammals. (There are good reviews by Payne *et al.*, 1986, for insects; Brown and Macdonald, 1985, for mammals; and Wyatt, 2003, for vertebrates generally.) Unlike visual or sound signals, chemical signals cannot change very rapidly – there is no chemical equivalent of a flash of colour or a staccato alarm call. Some chemical signals are 'designed' to be short-lived and this requires them to be low molecular weight and be volatile. They come and go and do not linger. The chemicals used by some ants to signal alarm fall into this category of being and are very short-lived – if they were not, then it would be impossible for the ants to signal the precise localization of a new source of danger. Ant alarm pheromones disperse well over a short range of 3–5 m, but they have usually faded below detection levels within a minute or even less. It is even possible to make use of the high volatility of scents to achieve some patterning in time. For example, some moths release their sex-attractant pheromones in pulses at approximately 1-s intervals (Dusenberry, 1992).

On the other hand, some chemical signals are designed to last for a long time, allowing a message to persist in the absence of the signaller, something not possible with vision or sound. Territory markers, for example, need to be persistent and therefore their constituents must have a fairly high molecular weight (Wilson, 1965). The molecular weight cannot be too high or the substance will be difficult to secrete and may not disperse well. It may also be important for scent marks to be complex and variable for this allows the presence of particular individuals to be identified by a chemical 'signature mixture' (Wyatt, 2010). Spotted hyaenas (*Crocuta crocuta*) mark their clan territories both by smearing grass stems with paste from their sub-caudal scent glands and by depositing faeces at latrines (Fig. 3.38). Where the territory is relatively small, scent is placed strictly along the territorial boundary, but where a small clan occupies a large territory and it is impossible to keep the boundary fully scented, both scents and faeces are deposited at strategic places within the territory (Mills and

Figure 3.38   A spotted hyaena scent marks the clan territory. It carefully locates a suitable grass stem in the cleft of its extruded scent gland so that the paste-like exudate is placed conspicuously and at the right height above the ground to be smelled by other hyaenas.

Gorman, 1987). Either way, other hyaenas are informed that a territory is occupied even when the owners are not around. Brown hyaenas (*Hyaena brunnaea*) have two separate anal glands for different messages. They leave secretions as two separate pastes (one brown, one white) next to each other on stalks of grass. Moth pheromones strike a compromise between persistence and good dispersal. In favourable wind conditions, males can detect the scent of females from 4–5 km downwind.

As yet another example of an animal using for communication a system it also uses in another context, the electric fish, *Eigenmannia virescens*, that we saw earlier detects object from distortions in its electrical field (Fig. 3.12) has adapted this for an unusual short-range signal. The male discharges its electric organ in a series of 'chirps' which Hagedorn and Heiligenberg (1985), having transposed them into sound for our ears, describe as 'short and abrupt during aggressive encounters' and with a 'softer and more raspy quality during courtship'.

Once a signal has travelled through the medium, it still has to be detected by the receiver and discriminated by it from other signals. A major selection pressure on signals is that they must each be maximally effective at stimulating the receiver to which they are aimed. Here we find that a knowledge of sensory processing in receivers is particularly helpful in understanding why signals have evolved in the way they have, since it is often possible to see that signals evolve out of responses to quite different situations.

For example, female jumping spiders (*Maevia*) choose which male to mate with not by what he looks like but, at least initially, by whether he moves in a way that resembles food. Using videotape sequences of male courtship so that they had precise control of the stimuli presented to females, Clark and Uetz (1992) showed that a female spider orientated to the movement of the male's legs by turning towards him in just the same

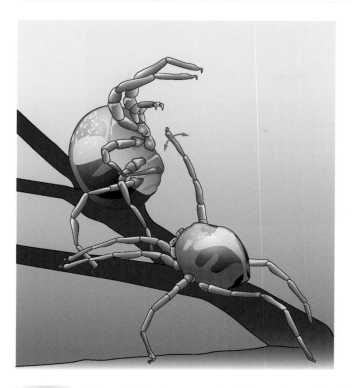

Figure 3.39 Courtship in the water mite (*Neumania papillator*). The males vibrates his foreleg in front of the female, mimicking prey. The female responds as if it were real prey by raising her first two pairs of legs into a 'net stance' and may even clutch at the male's leg. This does not deter the male, who continues with his courtship and eventually deposits spermatophores.

way that she would turn towards a moving prey item. It thus seems that successful males (provided they can turn off the female's prey-catching behaviour in time) can attract females by mimicking prey, thereby tapping into a pre-existing feeding response in the females.

Female water mites (*Neumania papillator*) also initially respond to males as if they were food (Proctor, 1991). Water mites are ambush predators and both sexes hunt prey by lying in wait with their front legs in the air and then grabbing passing crustaceans (Fig. 3.39). A male approaches a female vibrating his legs at a frequency of about 15 Hz, which is within the range of vibrations made by copepod crustaceans. The female grabs him with her front legs just as she would prey, but fortunately this does not injure the male as he is so much bigger than the female's normal prey, nor does it even deter him from further courtship. He starts to deposit spermatophores (packets of sperm) near the female and it is only at this point that she behaves any differently from the way she responds to food. She switches to sexual behaviour and picks up a spermatophore. Further evidence that the male courtship is effective because the female at first confuses

Figure 3.40   A male *Anolis* lizard in full assertion display. He bobs his head repeatedly with the brightly coloured dewlap extended.

him with food comes from the fact that, when female water mites are deprived of food, they grab courting males far more often than when they are well fed.

The male oriental fruit moth (*Grapholita nolecta*) also uses the female's response to a food stimulus to entice her to mate, but here a chemical signal is involved. The male initially finds the female by the scent she emits, but when he is very close, he wafts his own scent at her. The most effective part of his pheromone is a chemical, ethyl trans-cinnamate, which is also found in fermented fruit juices, a favourite food of both sexes (Lofstedt *et al.*, 1989).

The effectiveness of the signal in each of these cases depends on mechanisms that are already present in the receiver for non-signal functions. The signaller 'exploits' the responsiveness to the receiver by mimicking food in a quite specific way and so increases his chances of mating. In other cases, the signal may gain its effectiveness by attracting the receiver's attention in a more general way. We have already seen that the visual systems of many animals have evolved to pick up 'information-rich' features of the environment such as movement, edges or points of high contrast because these are important in the recognition of many different patterns, such as predators, prey, mates or simply for orientation through the environment. These general features are also used in signals. Male lizards (*Anolis*), for instance, display by bobbing their heads up and down and extending their dewlaps (flaps of skin under the chin) (Fig. 3.40). The 'assertion display' which is given both to attract females and to deter males starts with a series of very distinct head-bobs, which are larger and faster than the rest of the assertion display. Fleishman (1992) measured the typical distance at which the assertion display is shown and found that at this distance, the initial head-bobbing maximally stimulates the lizard's motion sensitivity in the peripheral retinal field. The result is to trigger a 'visual grasp response', mediated by the optic tectum, which causes the receiver to shift its gaze and bring the image onto the high acuity fovea. Once the receiver's attention has been grasped in this way, the rest of the display follows with lower amplitude bobs. Here we have an example of a signal having its effect by tapping in to a very general property of visual systems – sensitivity to motion in the peripheral part of the retina – rather than by mimicking any specific object.

To be effective, then, a signal has to be stimulating, either through mimicking something a receiver already responds to in another context, or through being 'conspicuous' in a broader sense by possessing those properties of movement, loudness, colour and so on that sensory systems in general have been evolved to pick up. In our discussion of the way in which animals respond to stimuli in their environment, we saw that they sometimes respond even more strongly to exaggerated or supernormal versions of natural stimuli than they do to natural stimuli themselves and it is therefore not surprising to find that many signals seem to create their effect through being supernormal in just this way. The male widow-bird with his extra-long glued on tail (Andersson, 1982) seems to be a good example of this (Fig. 3.15). Although we can understand the evolutionary origins of this female response to supernormal stimuli in the structure of sense organs and the tendency to respond to exaggerated versions of sign stimuli, we still have to explain why (in an evolutionary sense) the females continue to respond in this way. If it were not to the net advantage of females to respond to supernormal males, we would expect selection to have led to the evolution of more discriminating females. We will pick up this point again when we discuss sexual selection and female choice in Chapter 6.

Honesty and deception in animal signalling

Important though the effective transmission and reception of information undoubtedly is in explaining the evolution of signals, we need to look to other selective forces to explain the great diversity of different ways animals communicate with each other. If simply transmitting a message through the environment and making sure the female saw him was all that mattered, why should the male peacock have such a massive train, with iridescent eyes in his feathers and a dramatic display to show them off? Peahens have good eyesight and are usually not very far away from the male so something else must explain the evolution of such extraordinarily exaggerated signals.

Dawkins and Krebs (1978) identified conflict between two animals as a major determinant of signals evolving to be conspicuous over and above the need for effective transmission between signaller and receiver. If two animals are in conflict, each would benefit if the other fled without a fight, so potentially each would gain if it could convince the other that it was the biggest and strongest and that fighting would not be worthwhile. At the same time, there will also be selection on each animal not to be taken in by the bluff of the other, so that the conflict gives rise to a complex scenario of potential cheating on one side and resistance to being cheated on the other. Conflict between signaller and receiver clearly leads to selection pressure for signals to be

Figure 3.41   A red deer stag, *Cervus elaphus*, roaring. The ability to keep roaring at a high rate is an honest signal of the male's likely ability to win a fight because roaring itself is an exhausting signal. Only stags strong enough to fight long and hard will be strong enough to roar long and hard.

'honest' – that is, the signal must genuinely reflect some quality such as stamina or fighting ability and not be open against cheating or bluffing. But how can signals be made honest? What guarantee could a receiver have that its opponent is not cheating?

Red deer (*Cervus elaphus*) provide us with an excellent example. During the breeding season, known as 'the rut', the stags spend several weeks either defending a group of females or, if they have none to defend, challenging one of the owners for possession of theirs, initially through roaring matches (Fig. 3.41). These roaring matches take the form of one stag giving a series of roars or bellows and the other responding by roaring more (specifically, at a higher rate). The winner is usually the stag that can roar most vigorously, which means that, potentially, a stag could gain if he just out-roared his opponent and chased him away without a fight. But the challenger would only gain from such a retreat if the roaring itself conveyed honest information about real fighting ability. How might it do so?

The answer lies in the very fact that the roars are so exhausting, involving repeated and prolonged powerful contractions of the whole chest and throat. Only a very strong stag can keep this up for a long time and so is likely to be a strong fighter too. Clutton-Brock and Albon (1979) argued that the roaring signal is one of the ways the stags have

Figure 3.42 Thomson's gazelles (*Gazella thomsoni*) flee from a group of wild dogs that they have just detected. Some individuals do not run directly but intersperse their flight by leaping conspicuously into the air with all four feet. This is called 'stotting'.

of assessing each other's true fighting ability before letting themselves in for a damaging fight because the ability to roar at a high rate will be a reliable indicator of being a good fighter too. The point is that information about fighting ability cannot be reliably conveyed by signals that are so small and easy to produce that they can be given by all stags, regardless of their fighting ability. Receivers that respond to signals that even weak stags can give will lose out compared with receivers that only retreat when confronted with an honest, hard-to-give signal of fighting ability.

Another constant conflict in the natural world is that between predators and their prey, and here too we find the evolution of very conspicuous signals. For example, some prey animals use 'honest' signals of their capacity to escape. The conspicuous 'stotting' or leaping in the air shown by gazelles and springbok when chased is probably just such an honest signal (Fig. 3.42). Fitzgibbon and Fanshawe (1988) found that stotting is done most by animals that are in good physical condition and so are likely to be able to outrun wild dogs in a long chase. Stotting is thus an honest indicator of ability to escape. Moreover, the dogs discriminate between gazelles stotting at different rates, concentrating their efforts on the animals that stott least and thus are most likely to be weaker and ignoring those stotting at a high rate. In fact, unlikely as it may seem at first, both sides benefit from such signals. The prey benefit from giving the signal because they are less likely to be chased and the predators benefit from not pursuing prey they are unlikely to catch anyway.

Singing by skylarks (*Alauda arvensis*) when chased by merlins (*Falco columbarius*) appears to be another example of a potential prey animal signalling honestly to its predator. Cresswell (1994) found that merlins were much less likely to catch a skylark that gave a continuous full territorial song when it was chased than a skylark that did not sing or one that just gave just a brief song. Singing loudly and continuously when there is a falcon in active pursuit would seem to be a risky strategy. However, such singing can probably be performed only by male skylarks in good condition and so by learning to avoid singers, the merlin avoids prolonged chases that are likely to be unsuccessful.

But although selection for honesty is a major factor in signal evolution, it is clear that it is not always effective. There are numerous examples of successful cheating in which receivers are duped into responding against their own interests. Small fish that snap at the worm-like lure of an angler fish and promptly get eaten, or the reed warbler that feeds a baby cuckoo, are being deceived into performing actions that are not to their own benefit. There are also a number of insects, including beetles and butterfly larvae (so-called myrmecophiles), that live parasitically in the nests of ants and which trick the ants into carrying them into the nest by mimicking the signals given by ant larvae (Hölldobler, 1971). Again female fireflies of the genus *Photurus* can flash in two ways. They can either flash in the pattern of their own species to attract males, or they can mimic the flash pattern of another smaller species of the genus *Photinus*. When *Photinus* males approach in response, the *Photurus* females kill and eat them. Lloyd (1965, 1975), who discovered this extraordinary piece of deception, aptly called the *Photurus* females 'firefly femmes fatales'!

Such examples are sufficiently numerous to show that signals are certainly not always honest in the information they convey. Sometimes, as in the case of cuckoos that deceive their hosts into caring for them, such behaviour is part of an ongoing co-evolutionary arms race in which the deceived organisms are in process of evolving the ability not to be deceived. In other cases, the receiver is only rarely deceived, so that on average it is beneficial to respond to the signal. Tiny prey items are much commoner than angler fish and as long as the deception is relatively uncommon, it can persist. Many cases of mimicry seem to fall into this category. The predator that avoids black and yellow wasps will sometimes lose out through being deceived into avoiding a perfectly palatable hoverfly, but on average it gains from avoiding black and yellow prey because so many of them are dangerous or distasteful. A predator that avoids all black and yellow prey will in the long run be better off than a predator that tries them all and constantly ends up with a mouthful of distasteful insect. Conversely, the mimics gain an advantage only for as long as they are rare. Once they become common, predators that ignore the signal and attempt to eat them will be at an advantage.

None of the examples of deception we have discussed so far necessarily implies that the signallers are consciously setting out to deceive other animals. Most of the cases we have discussed are plausibly seen as evolutionary deceit, i.e. natural selection having favoured animals that behaved in certain ways and elicited particular responses from other animals with no conscious thought about what they were doing. In describing palatable insects as deceiving their predators through having body coloration that resembles distasteful ones, there is no implication that the insects think about what colour they

Figure 3.43 The 'broken wing' trick. A killdeer plover (*Charadrius vociferus*) drags its wing and hobbles away from a predator approaching its nest, which is on the ground. By remaining on the ground and stimulating the predator to chase it, the plover draws the predator away from the vulnerable nest.

should be. But in other cases, it is less easy to be so sure. Ristau (1991) has argued that plovers giving a broken wing display (Fig. 3.43) monitor the behaviour of the predator, such as a fox, that is threatening their nest and, for example, adjust their behaviour according to whether the fox is following them or still heading for the nest. Munn (1986) described two species of insect-eating birds that appeared to deceive their flock-mates by 'crying wolf'. By giving alarm calls when there were no predators present, they induced other birds to take off and so gained unrestricted access to food themselves.

Amongst primates, baboons have been reported to behave in ways that could suggest they are deliberately deceiving others. Byrne and Whiten (1988) describe a range of such incidents which are evaluated further by Whiten and Byrne (1997). For instance, a subordinate baboon which was being displaced from some food by a dominant animal gave every appearance of alarm as having spotted a predator in the distance. When the dominant animal responded by looking in the same direction, the subordinate took the opportunity of this distraction, grabbed the food and escaped! The subordinate could, of course, have learnt in the past that such behaviour resulted in the dominant being distracted and simply learnt to repeat a behaviour that gave him the opportunity to grab food. Alternatively, he might have worked out in his head that if he pretended to see a predator in the distance, the dominant would believe there really was danger would look up and this would give him a chance to steal the food. These two hypotheses – learning by trial and error and deliberate conscious deception – could both explain the observed behaviour. Deciding which is the best explanation is very difficult and takes us into the controversial issue of whether animals are conscious and have a 'Theory of Mind' about other animals and their point of view. We return to these fascinating questions in more detail in Chapter 5.

The existence of mimicry and deception do not, however, mean that 'honesty' is a useless concept in the study of animal communication. Rather, they show that the dangers of responding to dishonest signals are ever-present. Selection for honesty is on-going and dynamic. Whether the deception is evolutionary or deliberate, conscious or unconscious, signaller and receiver are engaged in a deadly arms race, signallers evolving to cheat and receivers evolving not to be cheated. This will be particularly obvious when the interests of the signaller and the receiver are in conflict, such as predators and prey or the signals of aggressive rivals. On the other hand, where the interests of signaller and receiver coincide – for example, in groups of birds or primates in which all individuals benefit from keeping together – the selection pressures to cheat and to avoid being cheated on are relaxed and the signals can become subtle and inconspicuous. Krebs and Dawkins (1984) have labelled such signals as 'conspiratorial whispers'. Examples would be the signals made by honeybees in their hive and the barely audible 'grunts' made by vervet monkeys to other members of their troop.

We draw this discussion of different aspects of communication together by looking in detail at these two remarkable examples of signalling from two very different groups of animals. Both have highly developed social systems and both have turned out to have far greater complexity in their communication than had previously been expected for any non-human animal.

The honeybee dance

Not only have honeybees evolved one of the most remarkable of all systems of animal communication, their dances also present intriguing problems of measurement and interpretation of the type which we have been discussing.

It has been known for centuries that honeybees must pass information about flower crops to one to another, but this phenomenon will now always be associated with the name of the great Austrian zoologist, Karl von Frisch (Fig. 3.44) who, together with Konrad Lorenz and Niko Tinbergen, received the Nobel Prize (for 'medicine' which is the label under which biologists are recognized!) in 1973. He was the first to unravel the nature of their communication and his major book, *The Dance Language and Orientation of Bees* (1967), provides a full survey of the whole dance system. Like others before him, von Frisch had noted that if a source of sugar solution was put out to attract bees it was often many hours before the first one found it, alighted and drank. However, once a single bee had located the source it was usually only a matter of minutes before many other foragers arrived – somehow the information had been passed on. It took von Frisch some 20 years of painstaking observation and experiment before he had worked out the bee's communication system to his own satisfaction. The conclusions that he came to were so extraordinary and unparalleled that he himself declared that no good scientist should accept them without confirmation. Following World War II, other zoologists did confirm von Frisch's results and even worked with him on some final experiments. There is now no doubt about the nature of the bee dance itself.

Von Frisch marked foragers on their thorax with a tiny spot of paint as they drank at a dish of sugar syrup and then watched their behaviour when they returned to the

Figure 3.44 Karl von Frisch in his garden watching honeybees come to the sugar solution he has put out for them. By marking individual bees he was able to show that bees inform each other of the location of food sources.

hive, using glass-sided observation hives for this purpose. The forager usually contacts a number of other bees on the vertical surface of the comb and gives up her cropful of sugar solution to them. Then, not invariably but very commonly, she begins to dance and we may first consider the case where the food dish which she has just visited is close to the hive, within 50 m. Her dance then is rapid in tempo and forms a roughly circular path just over her body's length in diameter. The bee moves in circles alternately to the left and to the right. She stays approximately in the same place on the comb and may dance for up to 30 s before moving on. Other foragers face the dancer, often with their antennae in contact with her body, and follow her movements closely, being themselves carried through her circular path (Fig. 3.45). This 'round dance' as it is called,

Figure 3.45   The 'round dance' of the honeybee worker on the vertical face of the comb. Her path is indicated by the lines. Note how she is closely attended by the other workers, who follow her round the path of the dance.

stimulates other workers to leave the hive and search nearby. It appears to convey the information, 'search within 50 m'. It may also convey some olfactory cues, because if the food source is scented the dancer will carry this scent on her body and perhaps in the sugar solution itself. If the sugar dish is not scented the forager may mark it to some degree by opening the Nasanoff scent gland on her abdomen as she drinks.

Thus far the bee dance is not very exceptional, because many ants and termites have similar alerting displays and pheromones which help to organize foraging activity when a new food source has been found near the nest (Wilson, 1965). The extent of the honeybees' communication system is not revealed until the food source discovered by a forager is further from the hive, beyond 100 m.

Von Frisch observed that as his food dishes were moved beyond 50 m the forager's round dances gradually changed in form. A short straight run became incorporated between the turns and on this run the dancer wagged its abdomen rapidly from side to side. At about 100 m distant the dance had become the typical 'waggle dance' illustrated in Fig. 3.46 and this form remained the same as the dish was moved further, to 5 km or even beyond. Subsequent work revealed that during the waggle run the bee produces bursts of high-pitched sound (Esch *et al.*, 1965). It is this waggle dance which, von Frisch claimed, transmits so much more information and is 'read back' by the dance followers as they follow every move the dancer makes.

The waggle dance certainly does contain information about both the distance and the direction of the food source. Distance is correlated with several features of the dance. Von Frisch concentrated on measuring its tempo, which falls off with distance, steeply at first and then more gradually. He marked off time at 15 s intervals. There are

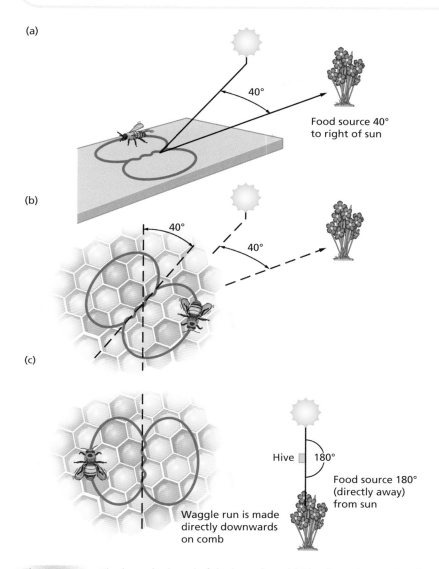

(a)

40°

Food source 40°
to right of sun

(b)

40°

40°

(c)

Hive 180°

Food source 180°
(directly away)
from sun

Waggle run is made
directly downwards
on comb

Figure 3.46 The 'waggle dance' of the honeybee. (a) The dance is occasionally performed outside the hive on the horizontal entrance board; if so, the 'waggle' run of the dance points directly towards the food source. (b) When the dance is performed on the vertical comb inside the hive, the angle of the waggle run to the vertical is equal to the angle the sun makes with the food source. (c) Shows the honeybee's convention that directly downwards on the comb represents directly away from the sun. In the same way, upwards represents towards the sun.

9–10 complete cycles per 15 s with the food at 100 m, but only 2 when the food is 6 km away. The duration of the waggle run, the number of waggles it contains and the duration of the sound pulses also correlate with distance, all three features *increasing* with it. von Frisch had no evidence that enabled him to identify which of these distance cues was important for the dance followers. Bees fly at a very constant speed in still air and they interpret distance information in terms of flight time and effort. von Frisch found that dancers indicated a greater distance for food sources upwind than for those downwind. (It is always the outward flight path which is indicated.)

It is its relation to the direction of the food source which is perhaps the most remarkable feature of the waggle dance. The little south Asian bee, *Apis florea*, is a close relative of the honeybee. This species builds a single vertical comb on a tree branch in the open with a flattish platform on top and the bees perform their waggle dances on this platform. The waggle run is made towards the direction of the food source, i.e. it operates like a pointer (Lindauer, 1961; Gould *et al.*, 1985). Now, occasionally honeybees will dance on the flat landing board provided at the entrance to most types of hive. If they do so, then like *A. florea*, their dances also point directly to the food source (Fig. 3.46a). However, this observation did not help von Frisch because he watched the dances at their normal site on the vertical face of the combs inside the hive. There he noted that the average direction of the waggle run was consistent within a dance and that it was the same for all the foragers who danced after feeding at the same dish. Given that other bees foraging at different dishes made waggle runs orientated at other angles even when the distances were comparable, this was strong circumstantial evidence that the angle related to direction in some way. Now came the crucial observation. von Frisch recorded dance after dance throughout the day as foragers repeatedly returned from the same food source and he found that the direction of the waggle run gradually changed. Its mean direction shifted by about 15° per hour and this could mean only one thing: that it relates to the apparent movement of the sun. Figure 3.46b and c show the way it does so. The foraging bee, like many other insects, uses the sun as a compass and records the position of the food source with respect to it. To get to the food it steers, say, 40° to the right of the sun. When dancing on the vertical comb the sun is not visible but the bee transposes the angle to the sun into the same angle with respect to gravity. The honeybees' 'convention' takes vertically upwards to represent the present position of the sun. Thus the forager dances with her waggle run 40° to the right of vertical. As her course with respect to the sun will have to change as the day progresses, so she changes the angle with respect to gravity on the comb to match the sun's apparent movement across the sky. von Frisch had at last understood the honeybees' dance language and the world was forced to accept that another animal apart from ourselves – and a humble insect at that – could convey information in a symbolic fashion.

Von Frisch and his co-workers had no doubt that the dance did communicate. Other foragers picked up the dance's rhythm and orientation as they followed through the dancing bee's movements on the comb and they then transcribed back from an angle with respect to gravity to an angle with respect to the sun. This assertion was based on numerous experiments in which an array of food dishes was offered, so arranged as to test the accuracy with which foragers recruited by the dance interpreted its information on distance and direction.

Figure 3.47 shows the two commonest types of configuration: a line of dishes at different distances on the same bearing from the hive tested distance communication; a fan pattern of dishes in an arc equidistant from the hive tested for direction. The figures

(a)

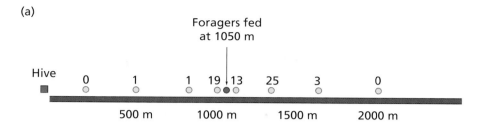

(b)

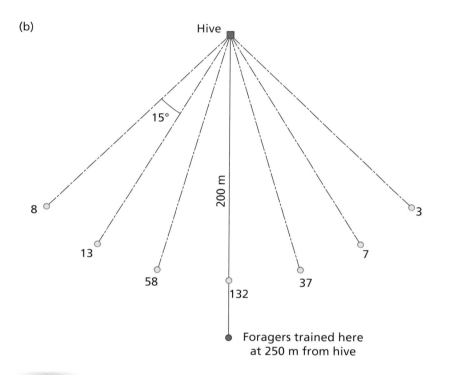

Figure 3.47 Experiments carried out by von Frisch and his co-workers to test the communication of distance and direction by the honey-bee waggle dance. (a) Distance test. Foragers were trained to a dish of dilute scented food 1050m east from the hive. Then a series of scented plates without food were put out at distances varying in the same direction from 100m to 2000m. Dancing was induced by suddenly increasing the sugar concentration at the feeding dish. Recruits were counted, but not captured, as they approached the different scent plates. The numbers above them record the number of visits made to each scent plate during the test. Most recruits appeared at plates close to the feeding station. (b) Direction test. The procedure is much as for the distance test, but the scent plates are put in an array at the same distance but different directions from the hive. The majority of visits were made to dishes close to the bearing of the training station.

also indicate the results of typical experiments showing that the number of recruits is highest at the dish closest to that at which the original forager fed and recruiting falls off to each side. (Further details of the experiments are given in the captions.)

This type of evidence was very generally accepted as convincing proof that the dance was the effective communication system, although it was clearly not perfect and there is some spread of recruits around the direction and distance indicated by the dancer. This 'noise' is not surprising in view of the crowded, jostling bees on the comb. Commonly, dancers do not have adequate space to keep a consistent line or rhythm. Nor, with natural food sources, do minor inaccuracies matter at all. For the most part the dancers will have been foraging at large sources of food – a grove of lime trees in blossom or a field of clover. Provided the recruits get some idea of distance and direction this will suffice, the more so because the foragers are very responsive to the scent which the dancers carry into the hive on their body. In any case, the scent of a food crop is a powerful stimulus for bees to approach and alight. It was doubts over the importance of scent which led to the later controversy over the interpretation of von Frisch's results. Scent and wind direction tended to be ignored in some of the original tests and it could just be possible that the dance merely alerted and random search did the rest. There is a review strongly arguing that this is actually the case by Wenner (2002).

Nobody can deny that there is a clear relationship between the form of the dances and the position of food sources which a forager has just visited. Therefore, if one is to argue, as Wenner does, that after being alerted, random search and scent are the means by which foragers find new food sources this must imply that the information contained in the waggle dance exists but is not communicated. The most obvious riposte is to ask, why then has such a remarkable relationship evolved? However intuitively reasonable this reply seems, it does not really supply a secure argument for it assumes that every biological phenomenon must be functional. We do indeed have examples of behaviour which certainly contains information that could be useful to conspecifics but which just as certainly is not used. Dethier (1957) has described the searching movements made by flies after they have located and then exhausted a small source of food, such as a drop of sugar solution. Their 'dance' can convey information to a human observer about the 'shape' and concentration of the source, but other flies do not respond. Certainly other bees do respond to the waggle dance, but the 'olfactory' hypothesis also requires that it arouses them to go out and search, just as ant foragers do. Von Frisch had already shown that recruits pick up olfactory cues about the food source from the dancer – its scent will adhere to her body.

As is often the case, contradiction and alternative hypotheses are a spur to good science. Gould (1975) and Dyer (2002) review all the recent experiments on how the bee dance functions. Gould himself did some elegant and better controlled experiments which effectively proved that the information in the dance was communicated as von Frisch originally described. It proved possible to induce bees to 'misinterpret' dances and search for food at a site that had never been visited before. Later, after many unsuccessful attempts, Michelson *et al.* (1992) succeeded in constructing a

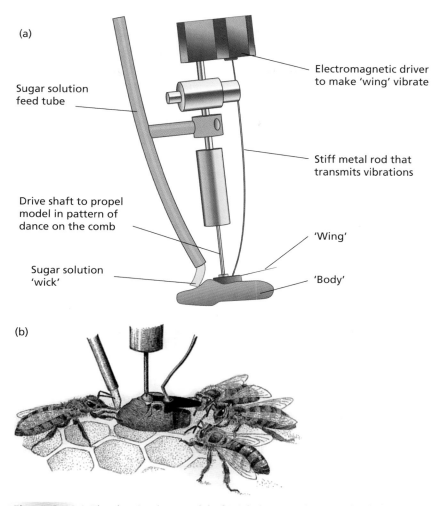

(a)

Sugar solution
feed tube

Electromagnetic driver
to make 'wing' vibrate

Stiff metal rod that
transmits vibrations

Drive shaft to propel
model in pattern of
dance on the comb

Sugar solution
'wick'

'Wing'

'Body'

(b)

Figure 3.48   The dancing bee model of Michelson, Lindauer and colleagues. Its construction is shown in (a). The body can be made to move on the comb in the form of a dance, 'waggling' during the straight part of the run, whilst the wing (a single piece of razor blade) is vibrated to produce sounds mimicking those of a dancing bee. (b) the model is shown on the comb. Workers fan out to its rear and follow its dance. One is accepting sugar solution which the model offers at its head end.

relatively crude model which can be moved through a waggle dance path on the comb. It has an artificial 'wing' attached to it which can be vibrated electrically so as to produce an acoustic field around the model similar to that of a real dancer (Fig. 3.48).

The model works! It recruits foragers to visit food dishes which have not been visited at all previously, and the proportion of bees turning up at different directions and distances follows that of the dance pattern. Sound turned out to be crucial for without it, no bees were recruited and this may be the reason for earlier failures with silent models. Not only does the fact that a model works at all provide proof of true communication in the bee dance, it may also eventually give us the opportunity to investigate the communication process in detail. To complement work with models, remarkable advances in technology allow minute transponders to be attached to free-flying bees, enabling their flights to be followed by radar! Riley *et al.* (2005) record the flights or recruits responding to waggle dances and find that they fly very close to the direction indicated by the waggle run for about 200 m, after which they begin to search about more widely. There can be no doubt that the initial direction is a response to their hive mate's dance, for if recruits are displaced laterally before they are allowed to fly out, their flight path is also displaced exactly as would be predicted by the dance.

There have been a number of beautiful studies which reveal how the whole dance system fits in to the life of honeybees. It seems likely that such a remarkable communication system could not have evolved unless it conferred a big selective advantage, which has enabled the genus *Apis* to develop large and complex colonies. Experimental work has amply confirmed this conclusion. Seeley and Visscher (1988) could not find evidence that foragers located food more quickly by following dances but showed that it enabled them to concentrate their efforts on the richest food sources. The foraging behaviour of tens of thousands of workers can be so coordinated that they spend the maximum amount of time working the most productive flower crops. As one crop fades, dancing dies away and foragers tend to stay in the hive until they are attracted by the dances of workers who have found other crops which are now more productive. The duration and frequency of dancing by a returning forager is directly related to the concentration or abundance of the nectar or pollen she has found and, accordingly, the number of recruits will come to match the available supplies. By some ingenious measurements, Sherman and Visscher (2002) were able to show just how effectively the waggle dance operates at the colony level. As mentioned above, if they have no alternative, honeybees will dance on horizontal combs, like the wild bee *A. florea*, angling their waggle run like a pointer with respect to the sun. If a horizontal comb is enclosed and dark, the workers will treat a bright single light as if it were the sun and orientate their dances accordingly. Under diffuse, non-directional light, the impulse to communicate the location of a food source is just as strong and the bees dance just as persistently, but there is no point of reference for them and their waggle runs are directed at random. Sherman and Visscher used rich food supplies at a suitable distance from two such combs and simply measured how their total weight increased as foragers brought back food. That of the combs whose dances were directed grew much faster.

When they return to the hive, the foragers give up their nectar to house workers, who are engaged in brood-rearing and food storage. These latter also play a key role in matching supplies and demand. When the colony is short of food, a returning forager is instantly solicited to regurgitate and her response is to set out again, probably dancing first to attract more recruits if her food source is a good one. Conversely, there are times when there has been a great flow of nectar or pollen and supplies in the hive are superabundant. If so, the returning forager will have great difficulty in finding a house bee to accept the nectar she proffers or a pollen storage cell with space enough in it to receive her load. Foragers can be seen running rapidly around the combs, those with nectar offering food to a succession of their sisters with no success. At such a point there is commonly a shift in their behaviour and they can be seen moving about, often in the brood-rearing regions of the combs, in a characteristic jerky fashion, shaking their bodies as they do so, with frequent brief pauses followed by an abrupt change in direction. This 'tremble dance', as it is called, may continue for a long time – 30 minutes in some cases. It has been recognized for many years, but only recently has it become understood. It is a kind of anti-recruitment signal. Seeley (1992) has shown that it is induced by the delay in having food accepted and it reduces quite sharply the amount of dancing by foragers who encounter a tremble dancing bee. It may also cause bees to switch their behaviour towards brood care. The immediate effect is to reduce the incoming surplus of food.

Seeley's group (see Seeley and Visscher, 2004) have also worked on the manner in which the waggle dance may be used in a completely different context. If a colony has grown so much that there is no longer adequate room for workers and brood, as often happens during a really good summer rich in flower crops, the colony splits. The old queen leaves with a swarm of many thousands of workers, which soon alight to form a dense cluster, often on a branch near the old hive. Then scout bees leave the swarm and search for new nest sites. If successful they return and perform waggle dances on the vertical sides of the cluster to indicate its position. Others are induced to visit and when enough have done so, the swarm departs, amazingly, guided visually by what Seeley's group calls 'streakers' who lead the way (Beekman *et al.*, 2006). (In the next chapter, p. 190, we discuss this behaviour of the honeybee swarm in the context of decision-making.) Tremble dancing and guiding to nest sites are just further examples of the honeybee's remarkable communication repertoire.

The calls of vervet monkeys

A very different communication system, but one that has also been referred to as a 'language' is found in the vervet monkeys of Africa (*Cercopithecus aethiops*). Much of our knowledge of the meaning of the vocalizations of these elegant, group-living primates (Fig. 3.49) comes from detailed field studies carried out by Dorothy Cheney and Robert Seyfarth in Amboseli National Park in Kenya (1990). They were able to get very close to the monkeys, document the interactions between different members of the group and, most importantly of all, experimentally manipulate their communication system by playing sound recordings back to them from loudspeakers hidden in the grass. It was already known that the monkeys gave different alarm calls when they saw different predators. While this is not unique in the animal kingdom, as both chickens and

Figure 3.49 Vervet monkeys are highly sociable animals and spend a lot of time grooming each other.

ground squirrels also have different alarm calls for aerial and ground predators, the predator-specific alarm calls are particularly well developed in vervets.

When a vervet monkey sees a leopard (*Panthera pardus*) or other large cat, it gives a loud barking alarm call (Fig. 3.50). When the other vervets hear this, they run up into trees, a manoeuvre that appears to give them some protection from the leopards because, being so small and light, they can escape to branches where the leopard cannot catch them.

On the other hand, if the vervets catch sight of either of the two large species of eagle that prey on them (the martial eagle, *Polelaetus bellicosus*, or the crowned eagle, *Stephanoaetus coronatus*), they give a completely different alarm call – a sort of double-syllable cough (Fig. 3.50). Since both these eagles are highly skilled at taking monkeys both from the ground and from the trees, running up into tree-tops, which is highly adaptive against leopards, would be disastrous against the high speed aerial attack of the eagles. A monkey on the ground hearing the eagle alarm call immediately looks up into the air and then runs into the thickest bush available. A monkey in a tree will actually come down from the tree and hide in a bush. Yet another type of alarm call is given when the monkeys encounter snakes such as pythons, mambas or cobras, which hunt for monkeys by hiding in the grass. The snake alarm call is a sort of 'chutter' (Fig. 3.50). When a vervet hears a snake alarm call, it stands up on its hind legs and looks down into the grass. Once several monkeys have spotted a snake, they will often approach and mob it.

Vervets thus have three different alarm calls for three different sorts of predators. Using the technique of hiding loudspeakers in the grass near where the monkeys were feeding and then playing different calls to them, Cheney and Seyfarth were able to show that the monkeys responded quite differently to the three alarm calls. In each

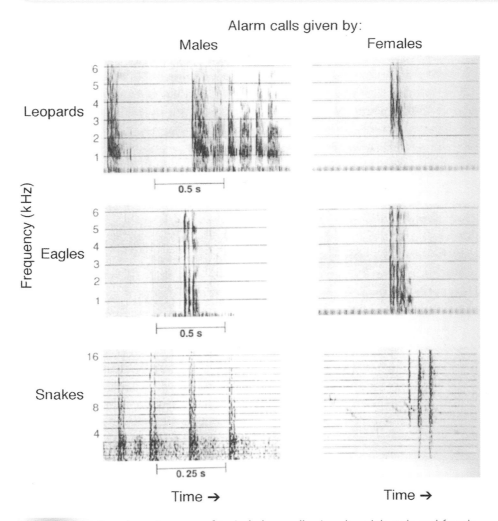

Figure 3.50 Sound spectrograms of typical alarm calls given by adult male and female vervet monkeys to leopards, martial eagles and pythons.

case, their initial response was to look towards the loudspeaker and scan the surrounding area. But then their behaviour was quite different. If they were on the ground, the leopard alarm call would cause many of them to run into trees. The eagle alarm call resulted in them looking up into the air and often hiding in bushes, whereas the snake alarm call caused them to stand upright on their hind legs (Seyfarth *et al.*, 1980a, b). The interesting thing about these playback experiments is that, while they do not rule out the possibility that during the approach of a real predator, the monkeys could simply be responding to the predator itself, they do show that the alarm calls themselves carry information about which predator is around. Even without seeing the predator, the vervets 'know' what action to take.

Cheney and Seyfarth (1980, 1986) were also able to show that young vervets are less specific than adults in what they give alarm calls to. Infants often make 'mistakes' and give alarm calls to warthogs, pigeons and other animals that are not dangerous to them. Their calls are not, however, given at random. An infant may give an eagle alarm call to a harmless bird or even a falling leaf, but at least it attaches the eagle alarm call to 'something overhead' and restricts the leopard alarm call to mammals, even harmless ones, on the ground.

There is no evidence that the adult monkeys actively teach the younger ones to be more accurate and discriminating in making their alarm calls. Rather, what seems to happen is that, on hearing an alarm call from a juvenile, adult monkeys will look up, just as they would to a call from an adult. If they see no danger, they will return to what they were doing before. If, however, they see a real predator nearby, the adults will give an alarm call themselves, followed by the appropriate behaviour. In this way, by watching the response and hearing the calls of adults, the young vervets learn what the various alarm calls should be attached to (Seyfarth and Cheney, 1986).

As well as their dramatic and obvious alarm calls, vervet monkeys have other, more subtle, ways of communicating with each other using quieter calls. One such is a series of grunts which are harsh, raspy noises that vervets make in a variety of social situations, usually when the animals are alert but relaxed. For example, vervets often grunt when they approach a socially dominant individual, when they approach a social inferior, when the group is about to cross an open space and when they have just seen another group of monkeys. To the human ear, all these grunts sound much the same. Cheney and Seyfarth (1982) played grunts recorded in these various situations (approaching a superior, approaching a subordinate, starting to move into an open space and having seen another group of monkeys) through loudspeakers hidden in the grass. They found a very consistent difference between the responses to at least two of these context-related calls. Grunts that had been recorded when the monkeys were about to cross an open space resulted in the monkeys looking away from the speaker towards the horizon, regardless of what they happened to be doing at the time. By contrast, grunts that had been recorded when a monkey had been approaching a dominant animal caused the monkeys to look towards the loudspeaker.

Although humans could detect no differences between the grunts, the monkeys can do so. When the calls were analysed with a sound spectrograph, physical differences between them could be seen and it was apparently these that the monkeys were picking up. Such subtle distinctions in animal communication are, of course, the opposite of the situation we described earlier, where signals become large and distinctive under the selection pressure of being clearly and quickly understood by the receiver. However, we also saw that when animals are cooperating and their interests coincide, they may adopt 'conspiratorial whispers' or very inconspicuous signals. Vervet grunts appear to be a very good example of this. The monkeys all have a common interest in staying together as a group, partly because they may be related, but also because they derive mutual benefits such as protection from predators. It is as much in the interest of one monkey to understand and respond correctly to the grunt of another as it is to the giver of the grunt to be responded to: if the group is to cross an open space or to respond to the presence of a strange group of monkeys, they all benefit if they all respond together. As a result, even the subtlest nuances in the vocalizations of other animals are responded to. The fact that it takes sophisticated sound analysis to demonstrate to humans that there are physical differences in the monkey sounds, which the monkeys are already sensitive to, should alert us to the intriguing possibility that we still have much to learn about how animals communicate with each other.

SUMMARY

The study of how behaviour 'works' (Tinbergen's causation question) starts with understanding how animals detect what is going on in their environments. Stimuli are features of the environment to which an animal responds. Stimuli may have an immediate 'trigger' effect or they may have longer lasting 'arousing' effects.

Different animals sense the world in many different ways, many of them very different from the ways we do. In addition to vision, hearing and chemical senses, animals may use magnetism, distortions of an electrical field and other senses beyond our direct knowledge and we may need to use special instruments to understand their worlds.

Sensing the world, however, is only the first step and we then have to discover how animals make sense of all the information their sense organs are giving them. Many animals use sign stimuli, picking out key features to respond to and ignoring the rest. Sometimes we even find them responding more to exaggerated versions or supernormal stimuli than they respond to natural stimuli. Although we are beginning to understand the neuroethological basis of how animals respond to stimuli, we are still a long way from fully understanding how animals (including ourselves) perceive the world.

Many of the stimuli animals encounter come from other animals and these form the basis of communication. Animals communicate with each other using signals, which often closely match the sensitivities of the sense organs of the animal receiving them. Signals also vary depending on the medium through which they are travelling. Not all signals are honest and deception and mimicry are widespread.

Two remarkable communication systems seen in very different animals are the dance of the honeybee, where the insects 'tell' each other where food is to be found, and the diverse calls of vervet monkeys that carry information about which predator a monkey has seen.

Decision-making and motivation

Just before dawn, a red junglefowl (*Gallus gallus*) cock crows from his roost in the bushes. Then, as it begins to get light, he makes his way to the ground with his group of three females and starts his daily search for food, scratching in the leaf litter for seeds and, if he is lucky, worms and grubs. In the heat of the day, he will return to the bushes for a siesta, but later he will resume his feeding on the ground, perhaps breaking off to dustbathe, mate with the females or search for water. All the time, he keeps a wary eye out for predators and rival males, ready to disappear into the undergrowth or see off an intruder at a moments' notice. As it gets dark, the group leave the ground in preparation for spending the next night in the relative safety of the bushes.

This simple description of the day in the life of one animal illustrates perfectly just how complex animal behaviour is. At every moment of every day an animal can be said to be making 'decisions' about what to do next – choosing to do one thing rather than another, or choosing to stop what it is doing and start something else. And those decisions can be studied at every level – from the level of the whole day, in which the decisions are between which parts of the day to be active and which parts to hide or sleep, right down to the moment-to-moment decisions about whether at this second to peck down at a food item or look up to see if there is a predator around. Indeed, we can look at even longer time scales, such as how behaviour changes over a whole year or even a whole lifetime.

The first thing to be said about such decision-making processes in animals is that they are disconcertingly complex. For any behaviour to occur, nerves, muscles and hormones all have to operate in particular ways. The brain has to direct and coordinate the behaviour in ways we do not fully understand and, most significantly, it has to weigh up the competing demands from different kinds of stimuli both inside and outside the animal's body. Only rarely are animals stimulated to do just one thing at a time. More often, they are confronted with a whole barrage of stimuli – some goading them to feed (the sight and sound of ripe fruit), some goading them to hide (the smell and sound of a predator) and they have to decide which one or ones to respond to. Woe betide the animal that – like Buridan's ass[1] – is unable to make such decisions, and remains transfixed and motionless half way between feeding and escaping, achieving neither. Such conflicts occur almost all the time in an animal's life. The junglefowl cock going to drink at a waterhole in the forest is in a state of conflict: he is in need of water, but the waterhole is also where predators lie in wait. He has got to drink sometime, but he also risks death at the waterhole. This chapter is about how animals make such life and death decisions, and balance the various external stimuli they encounter with the various internal factors that give rise to behaviour such as those arising from food deprivation ('hunger'), which we will here refer to by the term 'motivation'.

Tinbergen's 'four questions' help us make some sense of this complexity of decision-making and motivation in animals. Particularly helpful here is the distinction between

---

1 Buridan, a fourteenth century philosopher, posed the paradox of a hungry and thirsty ass who, standing between a pail of water and a bundle of hay, died of inanition because it could not decide which to approach first!

questions about adaptation and questions about mechanism. For example, in asking how a junglefowl decides when to go to roost, we can first of all view this as an adaptive question: why is it that birds which decide to roost at night and during the middle part of the day survive better than those that have a different pattern of activity? Such adaptive questions are often phrased in terms of the consequences to the animal if it did *not* perform the behaviour (Houston and McNamara, 1988). If it did not go to roost or did not go to drink at a particular time, how likely would it be to die or fail to reproduce? To put it in its most basic terms, which behaviour can the animal least afford to miss out?

The problem is that there is no fixed answer to this question. The behaviour that has the worst consequences if it is not done will shift with time and circumstances. For an animal that has drunk recently, death from dehydration is a long way off, so it can afford not to drink for a while, but for one that has not drunk all day, going without water could be much riskier. The balance of risks may shift throughout the day depending on how many predators there are around or how much cover there is available. From an adaptive point of view, the animal that is able to weigh up all these and other factors will survive better than one that cannot make such complex decisions.

But how, you might reasonably ask, can animals possibly weigh up all these complicated factors? How can they know when predators are likely to be about and how close they are to death by dehydration or starvation? How can they possibly balance short-term benefits of gaining food or water against their long-term chances of surviving long enough to reproduce successfully?

This is where we have to ask another of Tinbergen's questions: that about mechanism. Given that animals behave 'as if' they were in possession of all this knowledge and 'as if' they were making adaptively sensible decisions about what to do next, what is the mechanism by which they achieve these remarkable feats? As we will see, they do not have to 'know' in any sense we understand the word, that it is safer up a tree at night. They just need proximate mechanisms for getting them up trees at the right time of day and down again at other times – a preprogrammed daily rhythm, for example. Of course, if all they had was a daily rhythm, that would not be as successful for survival as more detailed decision-making mechanisms that could operate on an hour to hour or moment to moment time scale and allowed the animal to cope with unexpected and unpredictable events. This chapter is about how animals manage to behave adaptively 'as if' they were making complex conscious decisions involving extensive knowledge not just about their present environments but about what is going to happen in the future as well. It is about the mechanisms by which decisions are actually made.

As we shall see throughout this chapter, there are many different opinions as to how best to go about trying to study decision-making in animals. Some people believe that since the aim is to study the actual mechanisms at work, progress can be made only through the techniques of physiology – by looking inside the animal and recording nerve impulses, muscle movements and hormone levels. Such views have recently gained additional plausibility through the development of new techniques of brain imaging that enable us to study the activity of the living brain, and so to 'probe' inside without damaging it at all.

Others, however, believe that complete physiological explanations are still so far off that we do best to study the behaviour of animals and to use words like 'motivation' when we do not yet understand how it is produced. We will look at these different ways of explaining behaviour, how the concept of 'motivation' is used and why it still serves a useful purpose. We will do this in the way we stress throughout this book – by looking at different levels of explanation as well as constantly keeping our eyes on the interplay between them. We thus start by looking at decision-making on a behavioural level and then go on to discuss what is known of its basis in physiology.

 ## Decision-making on different time scales

Time budgets and daily routines

Almost all animals, like the cock junglefowl in our opening example, show different behaviour patterns depending on the time of day and many show great regularity in their habits. To some extent, these daily changes in behaviour depend on animals having an 'internal clock', which times their behaviour more or less independently of events in the external world. Figure 4.1 records the activity of a flying squirrel (*Glaucomys volans*) living in a rotating cage. Each row represents 24 hours and the dark bands show when the animal was actively moving around. The squirrel is a nocturnal animal and it was normally active from just after 'dusk' until the lights came on again at 'dawn'. At the beginning of the records shown in Fig. 4.1 the lights were switched off permanently. Now, even though it was living in complete darkness, the squirrel nevertheless maintained fixed periods of activity. These lasted almost exactly the same time as the original dark periods and occurred at extremely regular intervals. Interestingly enough, these intervals are not 24 hours, but just over 23 hours, so that in real time, the squirrel begins its activity a little earlier in each 24-hour period. A rhythm of activity of this type is called circadian from circa diem 'about a day'. Circadian rhythms are widespread amongst plants and animals and affect not just behaviour but all aspects of metabolism (Aschoff, 1981, 1989).

Wild animals spend a surprisingly high proportion of the 'active' part of their days resting and apparently doing nothing, but this too can be seen as adaptive. Periods of inactivity may be essential for digestion (Diamond *et al.*, 1986), energy conservation (Capellini *et al.*, 2008) or simply keeping out of the way of predators. Indeed, spending certain parts of each day sleeping in a hidden place may be one way that animals have of removing themselves from potentially dangerous situations (Meddis, 1983; Lesku *et al.*, 2006). The internal clock is not, however, the only factor determining what animals do at different times of the day, because animals can also adjust their time budgets according to circumstances. We can see quite major changes in how animals fill their days, depending on what demands are being made on them. Female baboons spend more and more time feeding as their infants get older and make bigger lactational

Original 12 hour light/12 hour dark cycle

Activity records below are from continuous darkness

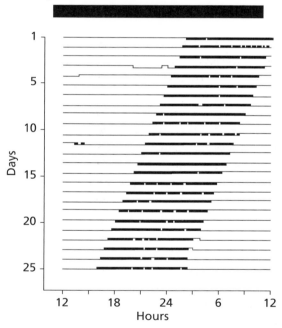

Figure 4.1 The spontaneous wheel running activity of a flying squirrel in total darkness. Active periods appear as broad dark lines. Before these records were taken, the squirrel had been kept on a regular cycle of 12 h dark and 12 h light, as indicated at the top. Even in total darkness, the squirrel retains its rhythm and is active only in the period that had previously been dark. In the absence of external cues, however, its natural circadian rhythm, which is naturally slightly less than 24 h, asserts itself so that the period of activity 'slips' and occurs slightly earlier each night.

demands on them (Dunbar and Dunbar, 1988) and they give up some of their rest periods in order to feed more. Then when most of that available rest time is used up, they will feed at times that would otherwise be devoted to social interactions. Ruddy turnstones (*Arenaria interpres*) are migratory shorebirds, many of which overwinter on rocky shores in Western Europe and then fly up to 3500 km to breed in the Arctic tundra of northeastern Canada and Greenland. In the 3 weeks just prior to migrating to their northern breeding areas, the birds increase body weight by over 40% and do so by a major adjustment of their time budgets: they spend a greatly increased amount of time feeding, even though as a result they allocate less time to looking out for predators and become markedly less vigilant (Metcalfe and Furness, 1984). In this case, feeding (to give them enough fuel for their long journeys) is given a higher priority than usual. Like many other migratory birds, turnstones eat much more than they normally do just before migration and lay down extensive body fat reserves. Hormones such as prolactin appear to influence this fat deposition, but the physiological basis for the seasonal changes in feeding behaviour involves a complex interaction between different hormones, changing day length and endogenous factors (Follett *et al.*, 1974; Wingfield *et al.*, 1990; Berthold, 1993).

Decision-making from minute to minute

Most studies of animals have involved studying their activities over periods rather shorter than that of a whole day. For example, if we watch a bird foraging through the trees or an animal building a nest, the behaviour may change from one minute to the next. If we look closely, we can see that the animal is not just feeding, in the sense that it is continuously taking in food, but that its feeding is divided into searching, capturing food, preparing the food, eating it, flying to another tree and so on. The animal is constantly making decisions about the next stage in its sequence of behaviour – when to stop doing one thing and go on to the next.

An important idea for understanding such sequences is to explore the possibility that animals make decisions that 'optimize' some factor such as the net rate of energy intake. So, as an animal eats up the food available in one area, it will be getting less and less food for more and more effort if it continues to search there. However, if it moves to another area where the food has not been depleted it may gain energy at a higher rate, despite the fact that it would have to use up some energy in flying or walking to the new place. Optimal Foraging Theory (OFT), which in its original form assumed that the well-adapted animal would optimize its energy intake, predicted that the decision to stop feeding in one area and move to the next could be calculated by knowing how much energy was involved in both staying and moving away. The point at which a bird, say, would gain more by flying to another tree than by staying to look for food in an already depleted one would be the point at which it should 'decide' to move. Clearly, flying to very distant trees, which would use up a lot of energy en route, will become the optimal behaviour only when the present tree is severely depleted (Fig. 4.2). Optimal Foraging

Theory predicts that an animal should stay in one area, even though it is gradually depleting the food there, until the net rate of energy intake drops to the average for the environment. At this point, the animal should decide to move on to seek out something different.

Optimal Foraging Theory is essentially an adaptive, not a mechanistic theory. In other words, it answers Tinbergen's adaptive question 'Why, in survival and reproductive terms, does the animal behave in this way?' by seeing how closely real animal behaviour follows the biologists' predictions of how animals should behave on the assumption that they are perfectly adapted to optimizing energy intake. The fact that, at least in some cases, OFT does describe behaviour remarkably well, suggests that, at least to some extent, that assumption is justified. The adaptive reason 'why' a bird spends a certain amount of time foraging in one place and then 'decides' to move to the next patch is indeed because it is making the best possible use of energy. Cowie (1977) studied great tits foraging for pieces of mealworm hidden in 'patches' consisting of sawdust-filled pots in different parts of an aviary. He made moving on to the next patch more or less difficult and energy-consuming by putting lids on the pots that were easy or hard to remove. So a difficult lid, by taking more time, effort and energy to remove, was taken to be the equivalent of a long distance flight between patches. The fit between the time the birds spent in a patch depending on 'distance' to the next one and the time predicted by OFT was remarkably good (Fig. 4.2). The birds' behaviour could be explained in terms of the optimum use of energy – that is, they balanced the energy that might be gained by staying in one food patch against the energy obtainable in another patch, and allowing for the energy expended in travelling between the two.

Optimal Foraging Theory is not, however, a theory of mechanism. In other words, it does not answer Tinbergen's question about how (causally) such optimal behaviour is achieved. It leaves us with the conclusion that that the animal behaves 'as if' it could estimate what conditions are like elsewhere in the environment and that it has some way of comparing the net rate of energy gain it is experiencing now with what it would experience if it flew to another patch and started eating there. But what it does not do is to explain how 'behaving as if' is translated into real behaviour. Does the animal really know all the things about its environment that OFT would appear to attribute to it? Perhaps all that is happening is that a bird moves to another patch whenever it has failed to find a food item for a certain length of time, that length of time varying with its past experience of hunting in that environment. It may be following such a relatively simple 'rule of thumb' and not even unconsciously working out 'travel times' or 'energy gains' at all (and certainly not understanding the equations of OFT!). We do not consciously calculate the trajectories of balls thrown into the air, but we often catch them all the same. An automatic process takes over and we behave 'as if' we knew all about the physics of flight and the aerodynamics of moving objects. Some possible 'rules of thumb' that foraging birds follow might be 'leave after catching n prey', 'leave after x seconds', 'leave after y seconds of unsuccessful search', etc., depending on the exact conditions a bird finds itself in (Stephens and Krebs, 1986).

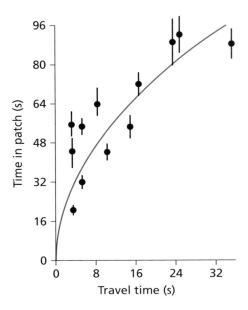

Figure 4.2   Optimal Foraging Theory (OFT) in blue tits. OFT is the idea that animals forage for food as efficiently as possible – that is, they always obtain more energy from the food they find than they expend in obtaining it. When an animal first arrives at a patch of food, there will be plenty of it and so the net energy gain will be high. However, as it depletes the food, it will have to spend more and more effort finding food until eventually it is more energy efficient to fly on to the next patch where there is more food. The animal is thus balancing the amount of food it gains by staying in one patch against the energy it would have to expend moving to the next patch where there might be more food. The graph shows how blue tits respond to experimental variation in their travel times. The longer the travel time (and so the greater the energy expended in getting to the next patch), the longer they stayed in their present patch, closely following the predicted optimal use of energy predicted by OFT (solid line).

Not all cases of animal decision-making fit the predictions of OFT as well as did Cowie's great tits. A possible and very interesting reason for this is the one we identified right at the beginning of this chapter – namely, that animals have to tread a careful balance between the different threats to their survival. In winter, when food is scarce and the survival of a small bird may depend critically on its obtaining enough energy and not expending too much in the process, it might be reasonable to assume that energy gains are paramount and that the bird that did not do everything possible to optimize energy intake would simply not survive at all. But when food is plentiful or when an even greater danger is posed by, say, being eaten by a predator, we might expect natural selection to have favoured animals with wider horizons and more flexible 'rules of thumb'. In other words, we might expect that their decisions about what to do next would be influenced by factors other than just those to do with feeding.

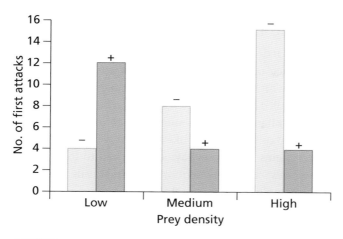

Figure 4.3 Number of first attacks at food by sticklebacks when presented with prey (water fleas) at different densities either with (+) or without (−) a model kingfisher overhead.

Although OFT helps us to understand some important evolutionary reasons why animals change their behaviour, we should positively expect animal decision-making to be adapted to more than just energy intake. As a consequence, the mechanisms of most decisions should reflect a wider range of stimuli and internal motivations than just those relating to feeding.

It is not difficult to find examples. Even great tits which, as we have seen, have been observed to optimize the time spent in a patch and travelling between patches, do so most obviously when they are very hungry. When their hunger is reduced, great tits become less than optimal foragers (Ydenberg and Houston, 1986). They compromise their feeding and spend more time on the look out for other birds intruding into their territory. Fish, too, balance the demands of feeding with other activities. When sticklebacks are very hungry, they prefer to feed in high density swarms of water fleas because in this way they get the maximum amount of food with the least effort (Milinski and Heller, 1978). However, when a model kingfisher was flown over a tank containing hungry fish, the sticklebacks preferred to feed in a much lower density swarm (Fig. 4.3). A plausible explanation of this effect is that kingfishers are predators of sticklebacks and if a stickleback feeds in a high density swarm, it may get a lot of food, but finds the many, jerking water fleas so distracting that it is less able to keep a look out for predators. The low density swarm of fleas, while giving less food, allows a stickleback to keep an eye out for danger so the decision to feed on the swarm of lower density paves the way, in turn, for a series of even shorter term decisions, to snap at a water flea or to look upwards to see if a predator is about to strike (Milinski, 1984). In all of these cases, the internal state of the animal – such as its state of hunger or its state caused by the presence of a predator – has to be taken into account.

Decision-making from second to second

We have so far looked at decision-making at the level of gross categories of behaviour such as 'feeding' or 'vigilance' or 'flying to a tree', each of which takes at least a few seconds to perform and each of which involves very large numbers of muscles. But we can be even more specific and detailed than this. When an animal is feeding, what movements does it actually make? Does it hammer open a mussel like an oystercatcher (*Haematopus ostralegus*), graze like a sheep or stalk its prey like a lion? If we are ever to understand how the action of muscle systems leads to behaviour, we need to look in more detail at these movements themselves – the movement of limbs and heads and bodies. We should not underestimate the complexity of the machinery underlying animal movement. Even an apparently simple behaviour such as a water snail opening its mouth and scraping its radula along the surface of a plant – an action that takes only 1–2 s – involves the contraction, in the correct sequence, of 25 different pairs of muscles controlling the mouth (Kater and Rowell, 1973). There are, of course, many other muscles controlling the head and the snail's foot as it moves along. Every time a dog swallows, there is a highly coordinated sequence of activity in the muscles of its throat (Fig. 4.4). There is a veritable 'symphony beneath the skin' involving many different muscles, each coming into action at a particular time. And if this symphony is complex for water snails, or for the single act of swallowing it is even more so for the complex activity of a male mallard courting a female, or the oystercatcher, the sheep and the lion, all feeding in their idiosyncratic ways. We should remember that no man-made robot comes anywhere near the complexity, the finely tuned adjustment and the flexibility of the behaviour of an animal.

When confronted with all the different possible movements an animal can make, it is often useful to focus on easily recognized and distinct sequences of behaviour lasting only a few seconds or so. These short sequences are the behaviour patterns such as the courtship strut display of the sage grouse that we discussed in Chapter 1. They are given descriptive names like 'pecking', 'drinking', 'wing-flapping' and together they constitute the ethogram or behavioural repertoire of a species.

Early ethologists laid great stress on these short sequences of behaviour, which they called 'Fixed Action Patterns', implying that they are immutable and always performed in exactly the same way. Certainly, the courtship patterns of some ducks studied by Lorenz (1941) and the 'strut display' of the male sage grouse we discussed in Chapter 1 do have this rigid 'clockwork' quality (Dane *et al.*, 1959). However, not all behaviour patterns are so stereotyped. There is enough similarity between different instances of 'drinking' in chicks or tail-wagging in dogs that we can recognize these behaviour patterns on different occasions, but they are not exactly the same from individual to individual or even from occasion to occasion in the same individual. For this reason, the term Fixed Action Pattern gives an inaccurate impression and a more neutral description – such as simply 'behaviour patterns' – is preferable.

Sometimes behaviour patterns that themselves last only a few seconds can be strung together into sequences with a high degree of predictability that last much longer. The

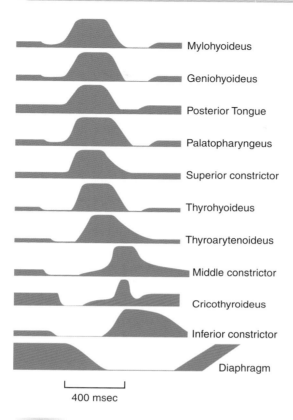

Mylohyoideus

Geniohyoideus

Posterior Tongue

Palatopharyngeus

Superior constrictor

Thyrohyoideus

Thyroarytenoideus

Middle constrictor

Cricothyroideus

Inferior constrictor

Diaphragm

400 msec

Figure 4.4 Diagram of the activity in various muscles during the act of swallowing in an anaesthetized dog.

'song' of the humpback whale, mentioned in Chapter 3, is a long series of highly diverse sounds, linked into phrases. All is then repeated in *precisely* the same way. Each repetition lasts about 40 minutes – about the same time as a performance of Beethoven's Fifth Symphony! The elements here are often common to all humpbacks but males are constantly changing and combining the elements in different ways. Often a group of males will change songs to match each other during the course of a breeding season (Payne and Payne, 1985).

When Lorenz originally used the term Fixed Action Pattern, he was not just referring to the fixed or stereotyped nature of the behaviour. He also believed that such distinct-ive, species-specific behaviour patterns were developmentally fixed or innate and that they develop in the absence of learning. We now know that fixity of performance and fixity of development are not necessarily the same thing. Stereotyped behaviour can develop in many different ways, including complex interactions with the environment as we discussed in Chapter 2, or as the result of keeping animals in conditions of poor welfare, as we will see later in this chapter.

Mechanisms of decision-making

The theory of natural selection leads us to expect that animals should make their various decisions – on all the different time scales we have discussed – in such a way that they make the best or optimal choices of what to do at any one time, which means choices that maximize their chances of successful reproduction. As we have seen, animals may not always 'optimize' feeding efficiency because of other conflicting demands, such as having to look out for predators or care for young. But in a well-adapted animal, the choice of what to do next should be an optimal compromise between all the different options open to it. This should be as true whether we are considering daily time budgets or second-to-second switches in behaviour. We now consider the mechanisms by which this is achieved, at all levels from groups of animals to groups of neurons because, although these might seem very different, there are some surprising similarities between them.

Let us begin by looking at how whole groups of individual animals collectively make decisions about what the flock or herd should do next. The way such group decisions are made is not only fascinating in its own right, it also shows how a single collective outcome can emerge from apparently independent units, whether those units are birds in a flock, fish in a school, people in a football crowd or even neurons in a brain.

Uniquely among social insects, honeybees reproduce by 'swarming' (Fig. 4.5). This happens when numbers in the hive increase beyond a threshold point which is probably signalled by the 'queen substance' pheromone becoming too diluted as it is shared among the workers inside the hive (see Chapter 7 on the honeybees' life cycle). At this point, the old queen, together with about half the workers, leaves the hive in search of a new home. The remaining workers begin at once to rear a new queen.

The swarm usually settles in a tight cluster quite close to the old colony and now the swarm with their queen have to find a new nest site that is suitable for building combs and rearing young. They achieve this in a most remarkable way. A small number of scout bees will venture out from the cluster and search for suitable nest holes, which have to be of a suitable size and shape, not too far off the ground, not too damp, facing in the right direction and so on (Seeley, 1982). When a bee discovers a potentially suitable nest hole, she returns to the swarm and dances on its surface to indicate the hole's location. They use the same conventions for distance and direction as for feeding sites. Other workers, having discovered other nest holes, also 'dance' about their locations, so there can be more than one dance going on at the same time, each indicating a different potential new home. The deadlock is broken, however, because dancing bees sometimes break off their own dance and visit the place indicated by another worker's dance. If they find that the alternative location is better than the one they themselves discovered, they will switch allegiance and start dancing about the new location, dancing more vigorously (more dancing circuits) for high

Figure 4.5 Honeybees swarming in a tree. Having left the hive, the swarm forms a temporary home while the bees decide where to go next.

value sites (Seeley and Visscher, 2008). This has the effect of recruiting more and more scouts to the site until eventually all the scouts end up dancing to the 'best' location (Fig. 4.6). When this happens, all the scout bees that have visited the new nest site emit 'piping' signals that have the effect of stimulating the rest of the bees in the swarm to warm up their wing muscles in preparation for flight. In the 60 minutes before the whole swarm takes off, the workers produce more and more piping signals that reach a crescendo just before the whole swarm departs, coordinating their departure and making sure that the swarm does not get split up (Rangel and Seeley, 2008). The decision about which nest hole to choose is thus determined by the behaviour of the scouts, each acting according to relatively simple rules (e.g. dance more if returning from more favourable nest site) and each responding to the behaviour of other scouts so that the result is what looks like a democratic 'vote' on which site best fulfils the criteria for a desirable nest. The decision about when to take off is also made by use of the relatively simple signal of 'piping' that the scouts increase to stimulate the simultaneous take-off of the whole swarm.

But even 'relatively simple' is more complex than we can explain at present. An insect such as a bee has about a million nerve cells and we certainly do not understand how

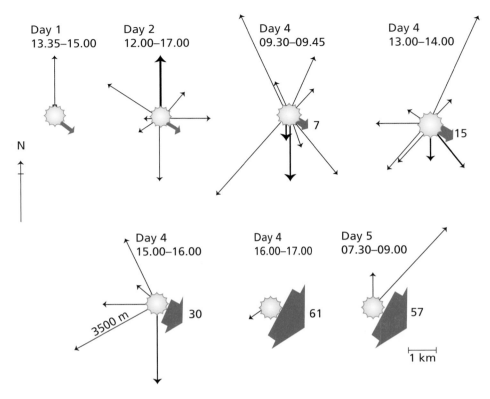

Figure 4.6 Decision-making by bees over a period of 4 days. Scout bees fly out from the swarm in search of a new nest site. When they return they dance with a vigour in proportion to the suitability of the site. They follow each other's dances and visit the alternative sites. When a consensus is reached, the swarm departs for its new home. In the diagram, the length of the arrows represents the distance of the site and the width of the arrow shows the number of bees departing in a particular direction. Note how the red arrow gradually becomes thicker as the decision about which nest site to go to is reached.

these work to produce bee behaviour. A mammal has far more nerve cells. A cat's brain, for example, has about 10^9. Sense organs, muscle cells and hormones add to the huge number of elements that may be involved in explaining how animal decisions are made. Many workers have dealt with this complexity problem by studying mechanisms of behaviour without referring directly to neurons or hormones at all. Instead, they invoke concepts such as 'motivation' or 'state' to explain the as yet unknown causal factors that are at work inside the animal. Of course, these 'unknown causal factors' include neurons, hormones and other physiological elements and the ultimate hope would be to explain motivation in these terms, but some people believe that such explanations are so far in the future that much useful groundwork can be done by not invoking physiology directly. We will therefore start our discussion of the causal basis of decision-making by looking at what such an approach can tell us.

Decision-making and 'motivation'

If an animal changes its behaviour over a time scale of minutes or hours when everything in its external environment has remained constant, then we can deduce that something inside the animal must have changed. It is this something – however little we yet know of its nature – which we call 'motivation'. For example, if we gave an animal some food and initially it took no notice but some hours later it began to feed, this would suggest that the animal's internal state had now changed. This can be described as the animal having increased its feeding 'motivation' even though we might not know anything at this point about what had actually changed inside the animal.

Of course, not all changes in behaviour are attributable to changes in motivation, even when external factors stay relatively constant. A particularly important distinction is between the long-term, near-permanent changes due to learning or maturation, where an animal changes its behaviour as it grows from infancy to adulthood on the one hand, and short-term reversible changes on the other. It is these latter – for example, animals changing from feeding to drinking and then back again all within the course of a few hours – that we refer to as being due to changes in motivation.

Motivation, despite its lack of physiological detail, has been, and still is, a very important concept in the understanding of how animals make decisions (Berridge, 2004). Our previous discussion of the adaptiveness of behaviour sequences showed how important it is to consider the internal state of the animal (its motivation) at all times. For example, if an animal is deprived of food and its food reserves are very low, it is more likely to optimize its net rate of energy intake (i.e. to forage optimally) than if it has recently fed and there are predators or females around. We may not know the exact physiological differences between a hungry and a less hungry animal, but we certainly know that whatever the key feeding motivation differences are, they are having a major effect on what the animal decides to do.

Measuring motivation

If the concept is to be useful we must be able to measure motivation quantitatively. Otherwise, we are in danger of talking about levels of mysterious causal factors without having any independent measure of how high or low these are. The whole idea of an animal being 'highly motivated' would thus become at best speculative and at worst circular. Our so-called explanations would turn out to be little more than re-descriptions of what is observed. Fortunately, there are several different ways in which we can independently measure motivation using an animal's behaviour.

Amount of behaviour performed

Perhaps the most straightforward way of measuring an animal's motivation is to give it the opportunity to perform a response and then see how much or for how long the behaviour is performed. In the case of feeding behaviour, we might measure the amount of food eaten, and it is usually easier to weigh the amount of food eaten than to count the number of feeding movements. Drinking motivation can be measured by the amount of fluid drunk, and even sexual and aggressive behaviour can be measured with a little ingenuity. Sevenster (1961) put an 'object' stickleback into a glass tube and held it inside the territory of a male, counting the number of bites directed at the tube by a test male as a measure of its aggression. Similarly, the number of 'zig-zag' courtship movements directed by a male towards a tube containing a female was used as the measure of sexual motivation. Vestergaard (1980) measured the motivation of domestic hens to dustbathe (Fig. 4.7) by how long they did so when moved from wire floors (where they could not dustbathe) to litter floors (where they could). Interestingly, he found that the hens showed a big surge in dustbathing when they were finally allowed access to somewhere where proper dustbathing was possible, as if they were making up for lost time. Such 'compensatory rebound' is also shown in people deprived of sleep (Borbely and Achermann, 1999) and suggests that at least in some cases, levels of motivation build up during a period of deprivation (Olsson and Keeling, 2005). We have already mentioned the phenomenon of 'reflex rebound' in Chapter 1 when discussing Sherrington's work with dogs and the scratch reflex. Now we can see that the same phenomenon operates at the behavioural level as well.

Figure 4.7   Chicken dustbathing. If given the opportunity, chickens 'dustbathe' regularly every 2–3 days, using dry earth, sand or litter. They use a complex sequence of movements that result in all their feathers becoming covered with particles.

How aversive a stimulus can be made before it is avoided

The aim here is to attempt to make it difficult for an animal to perform a behaviour pattern and see how far it will persist in spite of this. For example, quinine is an intensely bitter substance to us, and other mammals seem to find it equally so. If quinine is put into food pellets or drops of condensed milk at increasing concentrations, a rat will eventually reject usually attractive food as too bitter. The concentration of quinine it will tolerate can be used as a measure of feeding or drinking motivation.

An alternative version of this method is to place a stimulus, such as food, in full view of an animal and then contrive that, in order to reach it, the animal has to overcome some obstacles or 'run the gauntlet' of something it would normally avoid, such as an electric shock or a blast of air. By varying the intensity of the shock and finding how much an animal will accept in order to reach the food, we have a measure of motivation. In the heyday of experimental psychology with rats, these techniques were used to examine changes in both sexual and maternal motivation (Munn, 1950). For example, female rats will cross an electrified grid to reach a male and the strength of shock needed to stop them reveals cyclical changes corresponding to their oestrous cycle; it is highest during their 'oestrus' or 'heat'.

Rate of bar-pressing or key-pecking

The 'Skinner Box' is a useful piece of apparatus for studying both learning and motivation (Fig. 4.8). An animal is put into the box when hungry and will readily learn that it will receive a small food or water reward when it presses a bar that protrudes into the box or, in the case of a bird, when it pecks at a key. The apparatus is so arranged that rewards do not follow every bar-press or key-peck but come at irregular intervals, averaging out at, say, one reward every 30 s. In psychological jargon, this is called a 'variable interval reinforcement schedule' and it means that the animal never knows whether or not a given bar-press will give it food. Somewhat surprisingly perhaps, animals will often press their bar (or peck their key) much more consistently when they get their rewards only irregularly than if a reward comes after every response. One-armed bandits in our casinos operate on the same principle!

The rate at which the animal presses the bar under a variable interval schedule is so predictable that it can be used as a measure of motivation. The rate at which water-deprived rats will bar-press under these circumstances is reliably related to the length of time for which they have been deprived of water. In some cases, an animal's motivation appears so strong that it will show 'compensation' or 'resilience': thus, if the number of bar-presses required for a given amount of reward is gradually increased, hungry rats may be prepared to work harder and harder to obtain it. Hogan (1967) showed that Siamese fighting fish (*Betta splendens*) can be taught in an analogous way, to swim through a loop to obtain a food reward, and they will similarly show 'compensation' if they have to swim the loop many times to get the same amount of food. However,

Figure 4.8   B.F. Skinner with a rat in a 'Skinner box'. Rats can easily be trained to press a lever to obtain a food or other reward. By making the rat press the lever not once but many times, it is possible to see if the rat is sufficiently motivated that it will still continue to press the lever even when it has to pay a higher 'price'.

aggressive behaviour does not show the same effect. Although fish will readily learn to swim through a loop for the reward of gaining access to a rival male fighting fish, which they then display to, they will not show much compensation. They quickly stop swimming through the loop when the number of responses they have to make for each sight of the rival rises.

Extremely ingenious ways have been devised to measure animal motivation. All that is needed is some way in which the animal can perform a behaviour to bring about a change in its environment. For example, the animal can be given a key to peck, a lever to press or a computer screen to touch to give it food, water, or access to social companions. We can then ask how hard the animal will work to obtain these things and whether it will still work if, say, it has to press the lever hundreds of times for each piece of food. Such studies give us a quantitative measurement of the motivation (Dawkins, 1990).

Sows kept commercially often have to give birth to their piglets in a bare cubicle with a concrete floor. They have no straw to build nests with even though domestic pigs will build quite elaborate nests if given the right materials. Arey (1992) showed that

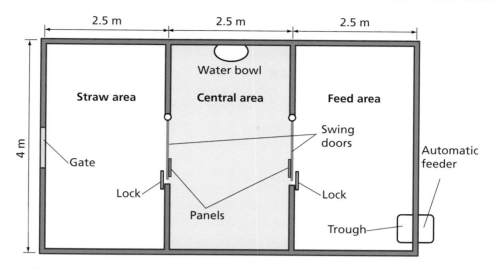

Figure 4.9 Apparatus used by Arey (1992) to test the motivation of pregnant sows for nest material and food. The sow could unlock the swing doors to either the straw compartment or the feed area by pushing the panel on the appropriate side with her snout. The 'price' of each commodity could be altered by changing the number of times the sow had to push the panel to unlock it. One push is a low price; many presses is a high price. The price the sows are prepared to pay to obtain straw for nest-building rises steeply just before they give birth.

pregnant sows would, if given the opportunity, work to obtain straw for nest-building. He gave the sows access to two chambers, one of which contained straw and one of which contained food (Fig. 4.9). The pig could enter either chamber by pushing with its nose a panel that unlocked the door. All the sows quickly learnt the trick of opening the doors and would choose to go into the straw area almost as often (mean of 17 times per day) as they chose the food chamber (21 times a day). But a somewhat different result was obtained when the 'cost' was increased. Instead of being able to unlock the doors with one push of the panel, the pigs suddenly found that they had to push it 10, 20 or even 300 times before the door opened. In other words, they had to work much harder to get what they wanted. Sows that were 2 days away from giving birth kept pushing the panel to get at the food so that they were now pushing up to 300 times for one entry into the food compartment and would still choose to enter the food chamber 11.4 times a day. On the other hand, they did not seem as willing to work for access to the straw chamber since when they had to push the panel 50–300 times for each entry, they only rarely entered the straw chamber (2.6 times a day). However, the day before they were due to give birth to their piglets, their motivation for access to straw appeared to increase dramatically. They started pressing the panel leading to the straw chamber even though they had to work hard to get in. Even with 300 panel pushes, they entered the straw chamber nearly as often (16.4 times a day) as they entered the food chamber (17 times a day). This suggests that motivation for access to straw is as high as motivation for access to food in the 24 hours before the birth of piglets and therefore, at this time, deprivation of straw would be as serious for the sow as deprivation of food.

Using a different way of imposing a 'cost', Duncan and Kite (1987) gave cockerels access to hens, but only if they pushed through doors with weights. The weight of the door that a bird was prepared to push was used as a measure of its motivation.

'Vacuum' activities

When very highly motivated, animals sometimes carry out behaviour even when the appropriate stimuli are not present. Lorenz (quoted by Tinbergen 1951) describes the case of a starling which went through all the movements of catching and eating an insect even though no insect was present – so the behaviour was performed 'in a vacuum'. Hens kept in wire-floored cages sometimes go through all the movements of dustbathing, even though there is nothing for them to dustbathe in. Vestergaard (1980) calls this 'vacuum' dustbathing and suggests that when behaviour is performed in the absence of suitable stimuli, or at least with very minimal stimuli, this must imply a very high state of motivation (Fig. 4.10).

Figure 4.10 A chick dustbathing on glass. Even though it does not receive the usual feedback from dustbathing (such as particles in the feathers), it continues to go through the whole sequence of dustbathing movements.

Is motivation specific or general?

These, then, are some of the ways in which levels of motivation have been measured using purely behavioural indications of motivational state. Superficially, they all appear to be measuring the same thing and we would expect that when, say, an animal is deprived of food, all these measures would increase in the same way. However, Miller (1957) describes a number of experiments which show that at least three of the measures of feeding motivation – amount eaten, quinine accepted and rate of bar-pressing – do not all rise together. Over the range of 0–54 hours of food deprivation, the amount of quinine that rats will accept steadily rises and so does their rate of bar-pressing, but their food intake reaches a maximum after only 30 hours and actually declines slightly thereafter. Thus a rat appears to be becoming 'hungrier' in that it will accept food that is more and more bitter and yet if offered normal food, it eats less. There are also some situations in which rats will eat abnormally large amounts of food (hyperphagia) and yet do not show any other signs of hunger such as being prepared to overcome a barrier or press a bar for food, or eat quinine-treated food. Hyperphagia can be produced by certain brain lesions and it would seem that what we commonly lump together under the term 'hunger' may in fact be a combination of factors; brain damage elevates some but depresses others.

There is evidence for a similar multiplicity of factors under the heading of 'thirst'. Very much in parallel with Miller's experiments on feeding mentioned above, experiments by Choy (quoted by Miller, 1956) show that different measures of drinking motivation do not correspond with each other very well, either (see Fig. 4.11). Choy had rats with tubes implanted directly into their stomachs so that they could be given liquids without having to drink. He measured drinking in rats previously satiated with water, after 5 ml of concentrated salt solution was put into their stomachs, recording three measures of thirst over time. Up to 15 minutes after giving salt, bar-pressing does not increase, even though the amount of water drunk does. Yet the amount of water drunk levels off 3 hours after the salt is given, even though bar-pressing continues to rise and so does tolerance of quinine in the water. It is obvious that we need a combination of measures to get a reasonable picture of the effect of salt on drinking motivation, which is clearly complex and not a simple, single entity. 'Motivation', as we have seen, covers a range of causal factors inside the animal (hormones, activation of specific parts of the nervous system, etc.) and so it is hardly surprising that its component parts do not always change exactly in step.

On the other hand, a characteristic feature of motivational changes – and the one that justifies grouping behaviour into major motivational systems – is that typically a whole range of responses that are functionally related to one another are affected by many of the same factors. An animal's thresholds of response to all stimuli connected with food and feeding behaviour will not be identical but they will tend to rise and fall together. Similarly, during the breeding season the threshold of responses to many

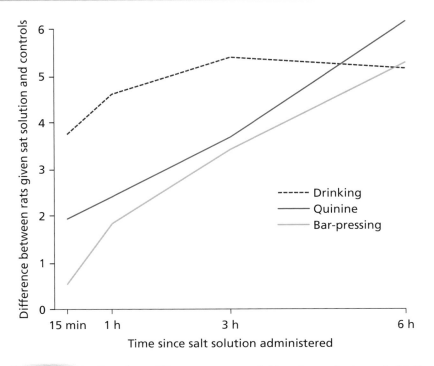

Figure 4.11   How three different measures of thirst change in the period following the placing of 5 ml of a strong salt solution directly into the stomach of a previously water-satiated rat. The units on the *y*-axis are arbitrary and are simply the difference between control and experimental rats on three measures of thirst.

factors relating to sex and reproduction will fall together. It is because of this association of effects with different sets of stimuli that ethologists and other workers have talked about 'specific motivational states', implying the existence of a set of internal causal factors that affect not just one behaviour but a whole functional group. Instead of invoking a different motivational state for every single element of its behavioural repertoire, such as a 'chasing motivation' and a 'swallowing motivation' the animal is said to have 'feeding motivation', implying that most or all of the behaviours functionally related to obtaining food (stalking, chasing, biting, etc.) are affected by many of the same causal factors.

Evidence for the operation of common causal factors comes from work such as that of Baerends and his collaborators (Baerends, 1976), who studied the changes in various behaviour patterns that occurred when herring gulls were disturbed at the nest while incubating an egg. In the 90-minute period following the interruption, three behaviours (building, preening and re-settling) all decreased at first and then increased at very much the same time, suggesting that all three shared at least one common causal factor, the level of which was falling and then rising again. Baerends (1976) also used an analysis of what behaviour precedes or follows another one to identify which ones might have common causal factors. He found that there were two sets of behaviour patterns. One set involved re-settling, pecking, picking up material and sideways building. The other set was composed of turning, head-shaking, scratching, shaking, yawning and looking around or looking down. If a gull had just performed one of the behaviour patterns in the first set, its next action was most likely to be one from the same set, not one from the second set. If it had just performed a second set behaviour, the converse was true. This suggested that, within one set, the various behaviour patterns shared common causal factors and that this was different (or at least different in effect) from the other set. Possibly the two causal sets of factors acted antagonistically.

With some motivational systems we have to be careful that categories which have particular associations in human terms, such as 'aggression', do properly correspond to those we observe in the behaviour of animals. Aggression is certainly an interesting case in point here for in animals what looks like fighting can be seen in several different contexts. Aggression towards members of the same species – often called social aggression – is reasonably distinguished from the interspecific fighting shown in catching and killing prey and also from the defensive behaviour shown by an animal faced by a predator. Such a distinction can be made both on the basis that the behaviour patterns used in these latter situations are often different (although both may show some overlap with social aggressive behaviour) and also on the grounds that the external and internal factors giving rise to them seem to be different. For instance, in mice, electric shock-induced fighting often takes the form of upright postures with bites directed at the opponent's nose, which more resembles the behaviour of a defensive 'cornered' mouse than that of a mouse attacking an opponent (Blanchard and Blanchard, 1981). However, we should not assume that the motivational systems of attack and defence are entirely unrelated. Huntingford (1976) has shown that in sticklebacks there must be some common causal factors between the aggression shown to rival males and the aggression – or perhaps we should call it defensive behaviour – shown towards predators such as small pike (*Esox lucius*). Fish which behaved in a 'bolder' fashion towards pike were also those which showed the highest levels of fighting in territorial disputes. Bakker (1986) used artificial selection to examine the genetic basis underlying the association of these two types of aggression; his work is discussed further in Chapter 6, p. 311.

Animal behaviour researchers are now less chary of using the term 'personality' with reference to general motivational 'biases' which animals may show in different contexts. Thus consistent differences between individuals on how bold or shy they are can have major effects on their fitness. Such personality variation has been found in animals as diverse as birds, mice and primates. Individual wild caught great tits that show the most exploratory behaviour in the laboratory also reveal this same characteristic when released back into the wild. They disperse further and subsequently their offspring are more likely to set up territories away from where they were born than those whose parents were shyer, less exploratory individuals (Dingemanse *et al.*, 2003). Adult survival was correlated with personality, which affected how individuals responded to novel environments, seeking them or avoiding them. The best strategy, conservatism versus enterprise, depended on the exact circumstances which – unsurprisingly – varied from year to year. It may be a case of 'who dares wins' – but only sometimes!

Note that we are already beginning to be able to say something about what appears to be going on inside an animal, using the correlations between its different behaviour patterns, even though the exact nature of the various causal factors has not been determined. It would, however, be misleading to think of animals as though they had fluctuating sets of causal factors for 'feeding', 'drinking', 'sex', etc. all operating independently of one another. Many stimuli produce non-specific 'arousing' effects which render animals more responsive to a wide range of stimuli and this could be described as a rise in 'general motivation' or, as it is called by some psychologists, 'general drive'. Although it might seem a straightforward question, it is in fact very difficult to collect really conclusive evidence as to whether motivation is general or specific; see good discussions of this point in Toates (1986) and Berridge (2004). In what follows, we shall assume that there is a considerable degree of specificity, although it would certainly be wrong to think of motivation as rigidly compartmentalized. Indeed, we have already seen that different motivational systems interact with one another and that one of the key issues in animal decision-making is how these interactions are brought about to produce the adaptive sequences of behaviour that we see. Why does an animal decide to stop doing one behaviour and start doing a different one?

Goals as decision points

An important idea in the study of decision-making has been that of a 'goal' which can be defined objectively as that situation which brings a whole sequence of behaviour to an end and the animal then moves onto a new set of behaviours. A dog shown his bowl of food on the other side of a fence will go through all sorts of different behaviour – running, jumping, scratching the fence and so on – and these would all cease once the dog had achieved the goal of getting its food. In fact, it would be difficult to describe the dog's behaviour concisely without calling it goal-directed. The implication would be that all of these different behaviours shared common causal factors such as the sight of a food dish and food deprivation, and would all be activated together and then fall together when the goal had been achieved.

The nest-building behaviour of the female great golden digger wasp (*Sphex*) fits the idea of goals beautifully. The wasp digs a burrow in the soil and then provisions it with katydids (large grasshoppers) for her offspring to eat when they hatch. Once she has achieved the goal of one completed burrow, she moves on to build another one, but she also has secondary goals in the excavation of each individual burrow. The wasp first digs a downward-sloping main burrow and then a side tunnel, at the end of which is the

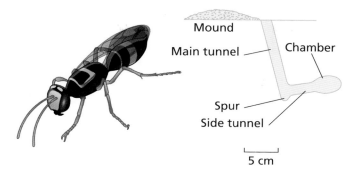

Figure 4.12 A female great golden digger wasp digs a burrow consisting of a main tunnel and a side tunnel ending in a nest chamber.

nesting chamber (Fig. 4.12). Brockmann (1980) altered the depths of main tunnels artificially and showed that the wasps adjusted their behaviour accordingly. If a wasp was confronted with an artificially lengthened main tunnel, she switched to making the side tunnel much sooner than normal, apparently satisfied with the construction even though she had not made it herself. The wasp evidently has a goal of a main tunnel of a particular length and moves on to the next part of the building sequence when this is present.

A similar cessation of behaviour once a goal has been achieved is shown by the reduction in sexual behaviour of the male three-spined stickleback when he has achieved the usual result of his courtship of a female – a new clutch of eggs in his nest. Male sticklebacks of various species build quite elaborate nests out of plant material (Fig. 4.13). A male will then court passing females and then, having persuaded one to enter his nest and lay eggs, he follows into the nest after her and fertilizes them. After he has done this, his sexual behaviour declines for a while, not because he is exhausted from his activities, but because the sight of eggs in the nest reduces his sexual motivation. Sevenster-Bol (1962) placed eggs in the nests of male sticklebacks and found that their sexual behaviour was reduced whether or not the males had been through the act of fertilization.

In the case of both the stickleback and the digger wasp, the goal that brings the behaviour to an end and causes the animal to start doing something else consists of the usual result of the behaviour, demonstrated in each case by an experimenter artificially mimicking the normal outcome and showing that this was what switched off the behaviour. If there had been no interference, the normal result of the behaviour would have acted as a negative feedback mechanism (Chapter 1) to end the sequence. The essential feature of a negative feedback mechanism is that some of the results of the behaviour are monitored and these results are fed back into the control system of the behaviour, reducing or sometimes increasing the animal's motivation to keep doing the same thing, depending on whether the goal has been achieved or not. It would be a mistake, however, to think that all animal behaviour can be neatly divided into separate

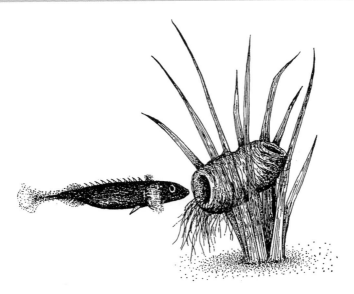

Figure 4.13 A male 10-spined stickleback 'fanning' at the entrance to his nest to ventilate the eggs. He drives a current of water through the nest by vigorous beating of his pectoral fins. This pushes him backwards so he compensates by driving forward with this tail so that he stays in the same place.

sequences each brought to an end by the achievement of a separate goal. In most cases, the situation is a great deal more complex. For example, domestic hens go through a sequence of nest-building activities that involves scraping a shallow depression in the ground and placing nest material around the edges. It might be thought that the goal would be a completed nest. However, hens will continue to show nest-building behaviour even when they are given ready-made nests (Hughes *et al.*, 1989). Unlike the digger wasp and the stickleback, therefore, the hens are still motivated to perform the behaviour even though their apparent goal has been achieved.

We gain a fascinating glimpse into the way animals are motivated with the discovery that often they seem to prefer to 'work' for something such as food rather than just take it 'for free'. For example, pigs prefer to expend time and energy foraging in straw for food even when the same food is offered freely from a trough where they can help themselves (De Jonge *et al.*, 2008). The animal's goal is thus not simply the food itself but, in addition, the animal appears to want to perform the behaviour normally associated with feeding. In Chapter 5, we describe some other examples of animals persisting with behaviour which delays their achieving a goal directly.

Thus goals are not always just simple stimuli like 'food' or 'a completed nest'; there may be other things that can act as goals or sub-goals along the way, such as a nest or a social response from another animal. Negative feedback means that something about the result of performing the first behaviour feeds back and begins to reduce it. As the first sub-goal is reached, the animal becomes less likely to persist and goes on to the next stage of the sequence. Drinking, nest-building and court-ship are just some of the behaviours that are also drastically affected by what the animal itself has done earlier, through the animal's own action in changing its environment. Feedback effects have now been incorporated into a number of motivational models under the general heading of 'homeostatic models' (McFarland, 1971; Toates, 1986).

Homeostasis and negative feedback

'Homeostasis' was the term coined by W.B. Cannon in his pioneering book, *The Wisdom of the Body* (1932, republished 1974) to describe the ways in which the body achieves relative stability despite the changes that go on in the outside world. Our internal body temperature does not depart much from 37°C even though the external temperature may be much hotter or much colder. In this sense, the goal of the body is to maintain a constant body temperature. We constantly lose water by evaporation, urination and so on, and yet the body's fluid volume remains roughly constant. Homeostasis implies that the body has some means of correcting deviation, so that if temperature or fluid volume falls, steps are taken to restore the balance. The starting and stopping of drinking behaviour can be seen as part of this homeostatic mechanism. Body fluids decrease, this stimulates the animal to drink and this in turn helps to correct fluid loss, returning the body fluids to some 'ideal' or normal value. Figure 4.14 shows a simple homeostatic model of motivation in which this process is represented diagrammatically.

Homeostatic models assume that there is an ideal state or set-point for the animal (Fig. 4.14). If there is a difference between this set-point, say, for body fluids and the actual state in the body, this so-called 'error' or 'discrepancy' is said to provide the motivation for drinking. With 'negative feedback' error is 'fed back' into the system and stimulates the behaviour or physiological response to operate until the error itself is reduced. The system is thus self-correcting.

In many instances, such a negative feedback model of motivation seems to provide a reasonably good analogy to the behaviour of a real animal. Fitzsimmons (1972) showed that rats injected with salt, which dehydrates them, drink just enough water to restore their fluid balance. The effects of placing food directly into the stomach are also understandable on a homeostatic model: loading the stomach with food is enough to 'turn off' eating because a full stomach is normally part of the negative feedback loop of the homeostatic feeding system.

As we have already stressed, it would be quite wrong to think of animals as nothing more than simple thermostats, switching behaviour on and off whenever they are in particular states. Homeostasis is much more complicated than this in practice

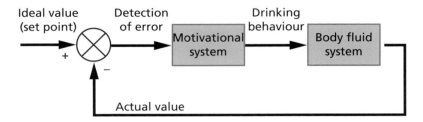

Figure 4.14   Simple homeostatic model of drinking behaviour. The actual state of the body fluids is compared to an ideal 'set-point' and any difference activates drinking. The water drunk changes the level of body fluids and corrects the deficit until the set-point and the actual level are the same. This then switches off drinking.

(see Rolls and Rolls, 1982), and for good functional reasons. Consider the problem of an animal having to maintain a constant fluid volume and composition in its body despite losses of water due to urination, perspiration, etc. and despite variations in the external temperature, the composition of its food and the availability of water in its environment. Its first difficulty is the time-lag which exists between behaviour (drinking) and the eventual effect on fluids in the body. When rats and other mammals are deprived of water for a long time, the cells of their bodies become dehydrated and their extracellular fluid (such as plasma) also diminishes in volume. One of the main reasons why we become thirsty when we take in salt is that the salt stays outside cells (since it cannot get through the cell membranes) and draws water out of the cells by osmosis. Water loss from cells is then detected by specialized osmoreceptors in the brain, which are spread out over quite a wide area of the lateral hypothalamus. (For a diagram of the brain, see Fig. 4.15.) Blass and Epstein (1971) injected small amounts of saline or sucrose solutions into the lateral hypothalamic areas of rats and found that the rats started drinking, but they did not do so following injection with urea. Since urea is, unlike salt or sucrose, able to cross cell boundaries, it does not draw out water from the cells and the stimulation of drinking therefore seems specifically related to such cellular dehydration.

Now, in the very simplest sort of homeostatic model, drinking would occur when cells become dehydrated and cease when they become rehydrated again. But since most animals drink much more rapidly than the fluid can be restored to their cells, the animal would drink far too much if it drank until its tissues and plasma were fully rehydrated. If human beings are deprived of water for 24 hours and then allowed to drink, they will drink almost all that is needed to restore fluid balance within 2.5 minutes even though changes in plasma dilution cannot be detected for 7.5 minutes and are not back to normal for about 12.5 minutes (Rolls and Rolls, 1982).

There must, therefore, be a means of detecting that water has been taken into the body before the full physiological consequences have made themselves felt. Miller *et al.* (1957) showed that at least some of these water detectors are in the mouth and oesophagus and that activation of these causes termination of drinking even before

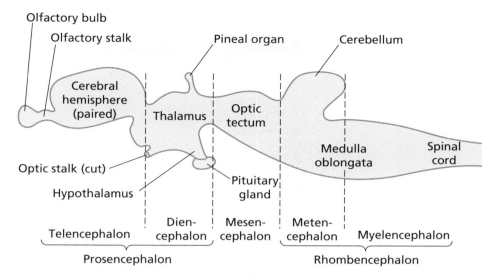

Figure 4.15 The basic divisions of the vertebrate brain. The brains of all vertebrates pass through a stage like this during development and the brains of adult sharks and rays are little different from this plan. However, in mammals, the adult brain is dominated by the growth of the enormous cerebral hemispheres and cerebellum, which lie on top of the rest and obscure the original plan.

fluid balance is restored. They allowed three groups of previously water-deprived rats to drink for 18 minutes. The first group just had 14 ml of water loaded directly into their stomachs (i.e. by-passing the mouth and oesophagus altogether), the second were allowed to drink 14 ml in the normal way and the third group of rats had no water at all. Putting water into the stomach had relatively little effect on the amount drunk immediately after: the stomach-loaded rats drank almost as much (average 16 ml) as the completely deprived rats (average 21 ml), whereas the rats that had been allowed to drink for themselves drank only 6.7 ml during the 18 minutes of the test (Fig. 4.16).

Water is thus detected in the mouth and throat during the course of normal drinking and reduces the subsequent tendency to drink. Distension of the stomach and stimulation of the intestine (together with other factors in the blood and liver) may also play a part in terminating drinking, the effects being different in different species. In dogs, for example, gastric distension is relatively unimportant for drinking, but it is very important in monkeys. Furthermore, some drinking occurs not because of a fluid deficit at the present moment, but because of an anticipated fluid deficit in the future. Fitzsimmons and Le Magnen (1969) showed that when eating dry food, rats drink in anticipation of a water deficit (the food will make them thirsty), a phenomenon which McFarland (1971) called 'feed-forward'. The state of the body fluids is certainly detected but it constitutes only one among many other factors in determining whether an animal will drink at a given moment. But whatever the precise mechanism, it is quite clear that a simple

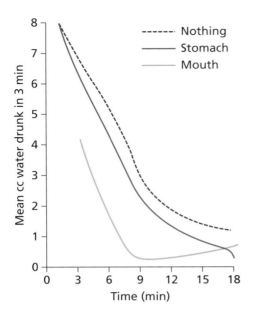

Figure 4.16 The amount of water drunk by thirsty rats under three conditions: 14 ml of water placed in the mouth, 14 ml placed directly into the stomach or nothing. This shows that preloading with water in the mouth or the stomach leads the rats to drink less than drinking for themselves.

homeostatic model, involving only the detection of fluid imbalance to initiate and terminate drinking, would be a poor predictor of the drinking behaviour of real animals.

Simple homeostatic models similarly fail to explain all aspects of feeding behaviour. Animals fed to satiety on one kind of food will often resume feeding if given a wider range of foods. Wirtshafter and Davis (1977) showed that rats used to a normal laboratory 'rat food' diet and then given a 'supermarket' diet (a variety of sweet and palatable foods) will eat a great deal more than usual and may even become grossly obese. Such observations have led researchers away from a simple view of a particular brain site as the place where feeding is switched on or off (the so-called feeding centre) towards a view of the control of feeding in which many different parts of the brain are involved. Another important idea is that of 'sensory specific satiety' (Rolls *et al.*, 1983). Animals do not satiate to food in general, but to the taste of specific foods, which explains why new foods can induce further eating in animals (or people) already apparently satiated on another. Furthermore, as we have already seen, feeding behaviour is also affected by the availability of food, the presence of predators, whether other conspecifics are present or not, and so on.

All models are simplifications. Nevertheless, the attempt to identify certain principles, such as 'negative feedback' can be an important step in understanding some aspects of animal behaviour, even though we may be fully aware that it is not a complete explanation. However, while it is often the case that functionally related groups of behaviour – for example, those related to feeding or sexual behaviour – rise and fall together, the reason for the changes in the level of causal factors for these behaviours may not relate in any simple way to the achievement of a goal or to the maintenance of a homeostatic state. In any case, when we look for explanations of behavioural sequences on a somewhat larger time scale, involving interactions between different motivational systems, we need to invoke some different principles. An animal may have a number of different goals simultaneously and somehow it has to decide which one of them to go for.

 ## Competition between motivations

At several points in this book, we have stressed the idea that animals are often stimulated to do different things simultaneously and are thus in a state of conflict about what to do. Their different motivational systems compete for control of what McFarland and Sibly (1975) called the 'behavioural final common path'. This is really an extension of Sherrington's idea (Chapter 1) that rival reflex arcs compete for control over the muscles and, therefore, what the body actually does. Given this idea of constant stimulation to do different things, an obvious way to look at sequences of behaviour is in terms of changes in which motivational system has control over the behavioural final common path at any one time. We can expect that this will shift over time as animals are exposed to different hazards, use up their food supplies, and so on. One way in which sequences of behaviour could result from such shifting motivational priorities would be if an animal always started by doing the behaviour for which it had the highest level of causal factors (the highest motivation) and then, when the causal factors for that behaviour declined, to switch to the behaviour with the next highest motivation, and so on. The sequence of different behaviours would thus come about by competition between the different motivational systems (McFarland, 1971). For example, suppose an animal was very hungry but only moderately thirsty. It would start by feeding but then as it took in food and approached satiation, its motivation to feed would decline until it dropped below the level of causal factors for drinking, at which point it would change its behaviour and start drinking. This might be one way of explaining the sequence of feeding and drinking in doves we discussed in Chapter 1 (Fig. 1.12).

A very dramatic example of competition between motivational systems comes from the incubation behaviour of junglefowl. Junglefowl hens undergo a major change in behaviour before and after incubation starts. Normally, the hens will spend some 60–70% of the daylight hours scratching in leaf litter and feeding. When they are incubating their eggs, however, they spend almost all their time on the nest, only leaving it for 5–10 minutes a day to feed and drink. As a result, they may lose up to

17% of their body weight during the 21 days it takes for the eggs to hatch (Sherry *et al.*, 1980). Clearly their motivation to sit on the eggs is very strong (a broody hen does little else all day), but what has happened to the motivation to do other behaviour such as feeding? Is it still high and inhibited by the strong motivation to brood? Or has it fallen and is simply outcompeted? The evidence suggests that during incubation, motivation to feed drops dramatically. Even if broody junglefowl hens are offered food on the nest, they refuse to eat and still lose weight. Their whole metabolism changes so that they are effectively anorexic and their motivation to incubate easily outcompetes their motivation to feed for most of the day. They are certainly not desperately hungry, despite losing body weight.

However, the causal factors for incubation also seem to fluctuate during the day. An incubating hen is much more likely to leave her eggs and eat at the beginning or middle of the day than she is in the afternoon, even though her state of hunger presumably increases throughout the day (Hogan, 1989). One hen in Hogan's study chose to go without food for 5 days running when food was only available after 1700 h, but the same hen ate normally when food was available in the morning. These results suggest that the competition between incubation and feeding is complex and constantly changing. The causal factors for incubation fluctuate during the day and are lowest in the late morning, whilst the causal factors for feeding rise steadily throughout the day. The change from incubation to feeding only occurs if feeding temporarily wins the competition.

While the shifting balance of competition between motivational systems undoubtedly accounts for some sequences of behaviour, competition is unlikely to be the only mechanism involved. The trouble with 'pure' competition is that it could lead to maladaptive 'dithering' between different behaviours, as the motivation for first one and then the other gets the upper hand. It could even lead to prolonged conflict in which neither emerges as the winner and – like Buridan's ass – the animal is literally immobilized by its motivational conflict. Suppose an animal's food and water are located in different places some distance away from each other, so that there is a cost (in energy, time, conspicuousness to predators, etc.) to changing from feeding to drinking. If there was a straight competition between the motivation to feed and the motivation to drink, the animal would dither between the two – taking a drink and then, when the motivation to drink had dropped slightly, immediately run to feed until feeding motivation had dropped and then go back to the water, and so on. Such rapid alternation would carry a high cost of changing from one behaviour to the other, with the animal constantly exposing itself to the risk of predation and not properly satisfying either its hunger or its thirst. A much more adaptive strategy for deciding what to do when there is a conflict of motivation is to have a degree of 'authority' with inhibition of competing responses, at least for short periods of time before the next behaviour occurs, just as with reflex organization at the spinal cord level (Chapter 1). In fact, the patterns of feeding and drinking in doves (Fig. 1.12) do indeed show a bout of one behaviour followed by a bout of the other, rather than constant dithering between the two.

Inhibition/disinhibition

Inhibition is of critical importance in animal behaviour, at all levels. In Chapter 1, we saw how the most basic movements of limbs depend on inhibition. Sherrington recognized that flexor muscles must inhibit extensors and vice versa, or their contractions would cancel each other out and no movement would occur. In Chapter 3, we saw that lateral inhibition is very important in enhancing edges and points of contrast. Inhibition at a neuronal level means that the firing of one nerve cell is actually suppressed because of the action of another. Inhibition at a behavioural level means that behaviour that would otherwise have occurred is prevented from occurring by the action of another motivational system.

As an example, we can look at the courtship of the smooth newt (*Lissotriton vulgaris*; Figs 4.17, 4.18). Newts breathe air but perform their courtship under water and this poses a particular problem for the males, since their courtship consists of vigorous side-to-side movements of the tail which demand a lot of oxygen, while the female's role is much more passive. During courtship, the male rapidly finds himself in a motivational conflict: he is stimulated both to go up to the surface to breathe and to continue with a sequence of behaviour that may lead him to a successful fertilization. Resolving such a conflict by competition would be fatal to his reproductive chances since if he took time off to breathe when the causal factors for breathing reached the level that normally stimulated him to go to the surface, the female might lose interest or even pick up the spermatophore of another male. Temporarily, therefore, breathing is inhibited by courtship. The male stays near the female despite rapidly depleting oxygen levels. Halliday and Sweatman (1976) showed how powerful this inhibition can be by altering the behaviour of a female in such a way that courtship was artificially prolonged. They

Figure 4.17 Female (above) and male smooth newts (*Triturus vulgaris*).

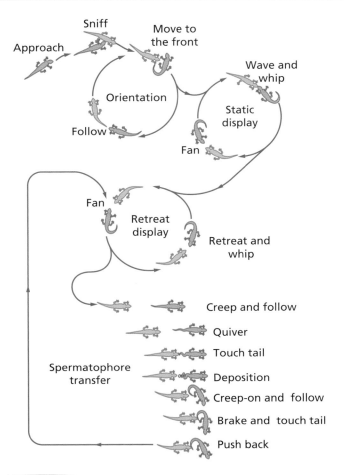

Figure 4.18   The courtship sequences of smooth newts. The male is shown in green and the female in blue. The male goes through a series of phases (orientation – static display – retreat display), each phase is affected by whether the female responds. After the retreat display, the female follows the male and when she touches his tail, he deposits a spermatophore. He moves forward just enough so that her cloaca is over the spermatophore. He then stops her so that her cloaca picks it up.

encased a female newt in a 'strait-jacket' (Fig. 4.19). This immobilized female provided enough stimulation for the male to go through the initial parts of his courtship display, but since the female could not move towards him, he did not receive the stimuli necessary to move on to the next stages. He was thus trapped, repeatedly displaying to a female who did not respond. Under these circumstances, the male held his breath for much longer than he would normally do under non-sexual (i.e. no female present) circumstances. Only when the female was removed would he go up for air. Sex inhibited breathing very effectively and breathing occurred only when 'disinhibited' by the removal of the sexual stimulus.

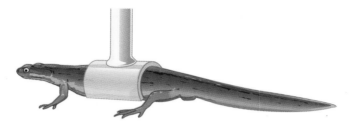

Figure 4.19 Female smooth newt in a 'strait-jacket'. By controlling the female's behaviour so that she does not go on to the next stage in the courtship sequence (Fig. 4.18), the male's courtship is prolonged.

It will be obvious that this could not continue indefinitely. Clearly, the need to take in air imposes an upper limit on the amount of time that a male can devote to sexual behaviour and so eventually the inhibition of breathing by sexual behaviour breaks down, and he goes up to breathe, leaving the female behind. The sequence of courtship and breathing is thus a complex mixture of inhibition and disinhibition of breathing by sexual behaviour but with an element of competition if the need for oxygen becomes too great.

These two basic elements of interaction between motivational systems – competition and inhibition – are thus used in different circumstances to give adaptive outcomes to the motivational conflicts that occur when animals are stimulated to perform more than one behaviour at the same time. A resolution of motivational conflicts based entirely on competition would result in dithering or even in complete immobility. A resolution based entirely on inhibition would result in an animal never doing second-in-priority behaviour until top priority behaviour had no causal factors left. A male newt that never left a female however long courtship took would die of lack of oxygen. An adaptive mixture of shifting motivational priorities with a degree of 'authority' between them is what we find in most animals most of the time.

Decision-making with incomplete information: the role of signals

We have so far emphasized the idea that animals have decision-making mechanisms that enable them to resolve their motivational conflicts fairly rapidly and immediately choose one option over another. But sometimes the situation which an animal finds itself in makes a quick decision impossible and it remains in a state of prolonged conflict.

Tinbergen (1952) argued that prolonged conflict played a part in to the evolution of some signals. For example, when a lesser back-backed gull (*Larus fuscus*) is in a conflict between attacking a rival and escaping (Fig. 4.20), it performs signals which reveal

(a)   (b)

(c)   (d)

Figure 4.20 Threat and appeasement postures. (a) The 'upright threat' of the lesser black-backed gull and (b) the hunched appeasement posture of the same species. This latter represents an almost perfect opposite to threat. The head is held low on a shortened neck and the bill points upwards, whilst the wings are pressed close into the flanks so that the wrist joints – so conspicuous in the threat display – are completely hidden. Now compare the gull postures with those of the domestic dog – an illustration taken from Charles Darwin's *The Expression of the Emotions in Man and Animals* of 1872. He entitled them (c) 'Dog approaching another dog with hostile intentions' and (d) 'The same in a humble and affectionate frame of mind'. The parallels with the gull postures are remarkable and both animals exhibit clearly what Darwin called the principle of antithesis in their communicatory behaviour.

elements of both attack (raised wings and down pointed bill) and escape (turning sideways). These were interpreted as 'threat' signals, with the function of intimidating the rival. More recently, the idea that motivational conflict can give rise to signals has taken on a new lease of life, but now with a different interpretation. Instead of seeing animals in a conflict as threatening their rivals or signalling that they are in a conflict, they are now seen as using signals to obtain further information that will help them to decide one way or another. The signals thus become part of the mechanism by which motivational conflicts are resolved. The new information tips the balance between one option and another.

The animal in a conflict between attacking a rival and fleeing, for example, may initially not have sufficient information to enable it to make a decision straight away. If the rival is likely to win the fight, then the optimal decision would be to give up immediately and not risk getting injured. But if the rival is weak and easily defeatable, then there could be considerable benefit in going ahead and obtaining the territory, females, food or whatever is at stake. By taking a little extra time to collect information about the opponent's strength, endurance and therefore likelihood of winning, the animal is more likely to reach a decision that maximizes its chances of winning than if it takes a decision without such information. Many signals are now seen as having this information gathering or 'assessment' function, directly contributing to the mechanism of the decision-making process by supplying vital information about the likely outcomes of the various options (Dall *et al.*, 2005).

As we saw in Chapter 3, red deer stags may assess each other's fighting ability through roaring at each other before deciding whether to attack or flee (Clutton-Brock and Albon, 1979). Sometimes roaring matches between well-matched stags fail to give a clear indication of which one is the stronger and the stags then move on to a further stage of assessment called the parallel walk. They walk up and down, sideways on to each other, keeping the same distance apart and inspecting each other's bodies. In this way, the stags obtain even more information about their relative size, physical condition and likely fighting ability. Either stag may give up at this point, but if neither does, the encounter escalates into a full-blown fight (Fig. 4.21). The whole sequence of roaring and parallel walking means that the animals remain in a state of prolonged motivational conflict for a considerable length of time while they gather the relevant information about each other's fighting ability. This extra information adds to the causal factors for either fighting or fleeing until the motivation for one of them clearly is the higher. Thus, initially, the causal factors for attacking and fleeing might be more or less equal so the challenging stag is in a motivational conflict. But if, during the course of the roaring match, he is 'outroared' by the other stag, or decides on close inspection that his rival is bigger and stronger than he is, the motivation to flee becomes stronger than the motivation to attack and the conflict is resolved in favour of retreat.

Information gathering is important in other types of difficult decision-making situations too. Prey animals are constantly having to decide whether to continue feeding or to flee from a predator. Obviously if the predator is close and heading straight for them,

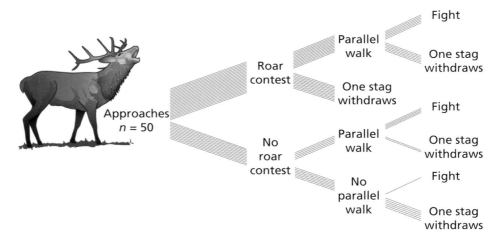

Figure 4.21 Sequence of events in roaring contests between red deer stags. As the stags approach each other, they may challenge each other by roaring as an indication of fighting ability. The roaring contest may lead to one stag withdrawing or to 'parallel walking', which may or may not then lead to actual fighting. In this diagram, each line represents one encounter.

it is adaptive to escape immediately. But if the predator is a long way off, the animal may gain more by staying where it is and not losing valuable feeding time. Prey animals do not flee every time a predator appears but take different decisions depending on circumstances (Ydenberg and Dill, 1986). Zebras watch hyaenas carefully and are much more likely to flee on some days than others, apparently sensitive to small differences in the hyaenas' behaviour and the size of the hunting group that indicates what prey the hyaenas are after on any one occasion (Kruuk, 1972). Such assessments are highly adaptive but take time to make. The motivational conflict is resolved only when, eventually, the new information received tips the balance of causal factors in the direction of one behaviour or another.

Maynard Smith and Riechert (1984) used these ideas of information gathering leading to shifts in motivation and ultimately to decision-making to develop a detailed model of the fighting behaviour of the desert spider, *Aegelopsis*. This spider is very aggressive and fights over access to web sites. In Maynard Smith and Riechert's model, there are two motivational variables, fear and aggression, which fluctuate during the course of an encounter between two spiders according to such factors as who owns the disputed web site, relative body weight of the two, territory quality, and so on. There are also some genetic factors involved for it is known that some spider populations are more aggressive than others, even when the external situation is kept constant. On their model, if fear greatly exceeds aggression, the animal simply withdraws. If fear is equal to or less than aggression, then the contest continues, but what behaviour is performed (locate–signal–threat–contact) depends on the absolute levels of the two variables.

Figure 4.22   'Grass-pulling' in the herring gull. This occurs when a male is facing his rival at the boundary of his nesting territory. Interspersed with bouts of threat or actual fighting, males will seize a firmly fixed clump of grass or turf and tug at it vigorously.

The model accounts well for many observed facts about spider fighting behaviour, such as the fact that owners of web sites tend to win, larger spiders tend to win, and so on. Maynard Smith and Riechert argue that a model with two variables is the simplest one that will account for all the observed data. It does appear, then, that the traditional idea of agonistic encounters (including threat) as extended motivational conflict has stood the test of time although there has been a definite shift of emphasis. Animals may display when they are in a state of conflict, not necessarily to signal that they are in a conflict, but, by displaying and seeing the displays of others, to get themselves out of the conflict. They make their decision on whether to attack or to flee from a rival or even a predator on the basis of new information received during the course of the encounter.

Conflict and 'abnormal behaviour'

Tinbergen (1951) described a very curious category of behaviour, typically seen in animals in situations of conflict that are called *displacement activities*. This behaviour seemed to arise from the animal being in a conflict or being frustrated in some way, but to be apparently irrelevant to the resolution of the conflict. For example, in the middle of a fight, two cockerels will each turn aside briefly and peck at the ground, sometimes picking up stones or grains which they then drop again. Fighting gulls may show 'grass-pulling' in the middle of a fight (Fig. 4.22). A tern that is incubating eggs on its nest makes preening movements just before it takes off at the approach of an intruder. A thirsty dove that is prevented from getting to its water bowl by a sheet of glass may preen itself.

In all these examples, the animals seem either to be thwarted or in conflict between two opposing motivations. The appearance of, say, feeding behaviour in the middle of a fight would seem to be 'irrelevant' to the conflict (between attack and escape). Similarly

Figure 4.23 A polar bear continuously pacing around its cage in a zoo. The bear may follow exactly the same path over and over again in a stereotyped pattern.

when a tern is faced with a decision of whether to escape from a predator or stay and incubate its eggs, preening does not seem relevant to doing either.

Although the term 'displacement activity' is now rarely used, the idea that conflict or frustration may give rise to unexpected or unusual behaviour in animals remains important, particularly in the context of animal welfare. Animals confined over a period of time may develop behaviour that is quite different from anything seen in wild animals. For example, polar bears in zoo enclosures, even if they are quite large, may pace around a set route so often that they wear away the ground (Fig. 4.23), or stand constantly swaying their head and neck from side to side. Sows confined in stalls may repeatedly rub their mouths backwards and forwards over the bars, even making themselves bleed (Fig. 4.24). Such repeated, functionless and often damaging behaviours are called *stereotypies* and are widely seen as examples of unnatural behaviour indicating poor welfare, either currently or in the animal's past life (Mason and Turner, 1993). Domestic hens in small cages may develop stereotypic pacing backwards and forwards, apparently out of frustrated attempts to escape (Duncan and Wood-Gush, 1972). Even when let out of their cages,

Figure 4.24 A confined sow bites repeatedly at the bars of her pen. She may repeat this action so often that her mouth bleeds.

however, they may continue to show the same stereotyped behaviour. So the performance of a stereotypy may not necessarily indicate that the animal is stressed at the moment. Rather, it may indicate that at some time in the past, its welfare was compromised and that it still carries the behavioural scars of its past experience.

The use of the word 'stereotypy' to imply that something is abnormal or even pathological is perhaps unfortunate in view of the fact that, as we have already seen, many quite normal behaviour patterns may be fixed or stereotyped (Chapter 1). Perfectly natural behaviour such as breathing, walking or pecking and whole sequences of bird or whale song may be repeated in the same way over and over again and are in this sense 'stereotyped'. Despite this, 'stereotypy' is now commonly used to imply that the animal's welfare is or has been compromised.

A major problem with equating stereotypies with poor welfare is that, at least in some cases, performing the stereotypy may actually help the individual doing them to 'cope' with its environment, in the same way that, say, pacing back and forth sometimes relieves tension for a human (Mason and Latham, 2004). Young calves taken from their mothers and fed on milk from a bucket often develop stereotyped sucking, which can be directed at other calves or at objects in the environment. If given a nipple which gives them no milk whatever, the calves suck vigorously even though they are obtaining no milk. This would appear to be a clear example of a functionless stereotypy. However, this stereotyped non-nutritive sucking actually stimulates the calves' digestion. Calves that suck on a dry teat after drinking milk show greater secretion of the hormones insulin and cholecystokinin, which help with them take in nutrients (de Passillé *et al.*, 1993). The 'functionless' sucking benefits the animal after all.

It is perhaps not surprising that the behaviour of animals when they are in a motivational conflict is still something of a puzzle to us. The mechanisms that animals have evolved to decide between different courses of action come under particular strain when two motivations are finely balanced or when what an animal has decided to do is blocked for some reason, such as the presence of another animal or lack of appropriate stimuli. Given the diversity of the mechanisms controlling each behaviour, diversity in how conflicts between them are resolved is only to be expected.

The physiology of decision-making

We have now seen that careful analysis of behaviour can get us quite a long way in understanding the mechanisms of animal decision-making, including the range of factors that go into each decision of what to do next. Next we turn to decision-making inside the body, at the physiological level. Encouragingly, we find that many of the same ideas that we used for behaviour also have their counterparts in neurophysiological explanations of behaviour. Positive and negative feedback, competition and inhibition are all essential to the way nervous and hormonal systems operate.

One particularly good example of a physiological approach that even enables us to locate the neurons responsible for decision-making is the sequence of movements made by an escaping crayfish (Fig. 4.25). When a crayfish is touched, it has a sudden and dramatic way of escaping: it flexes its abdomen so that the body is drawn into a tight curve. This has the effect of propelling the crayfish rapidly backwards. This behaviour is brought about initially by the activation of receptor cells that are sensitive to high frequency water movements and to touch. These are in turn connected to sensory interneurons (SIs) which in turn connect to a lateral giant interneuron (LGI) (Fig. 4.25). There are over 1000 tactile sensory cells and over 20 sensory interneurons. The LGI is a very distinctive nerve cell and is unusually large – about 100 μm in diameter. There is one LGI per abdominal segment and each one connects to the LGI in the next segment in front, all the way up the nerve cord. In each segment, there is also a very large motor neuron (motor giant or MoG) which innervates a fast flexor muscle (FFM). The contraction by FFMs of all the segments together gives rise to the escape movement of the whole crayfish.

The LGIs in each segment obviously have a key role in making the escape response. Many years ago, Wiersma (1947) showed that electrical stimulation of any single LGI gives rise to a fast abdominal flexion very similar to the complete escape response of the whole animal. But without being prompted by the LGIs, the motor nerves do not stimulate the muscles to contract.

It also seems that it is the LGIs which make the decision whether or not to show the escape behaviour at all. Olson and Krasne (1981) showed that when shocks of low voltage were applied to the sensory nerve, these were not large enough to stimulate the

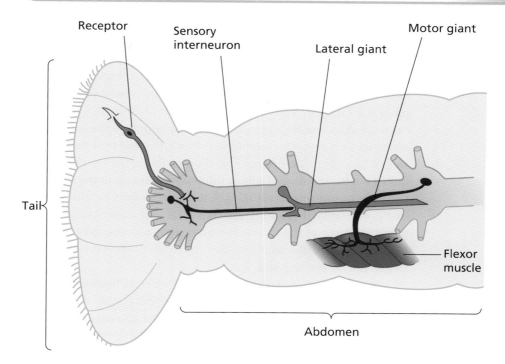

Figure 4.25 Diagrammatic ventral view of the abdominal nervous system of the crayfish showing the excitatory pathway from receptor cell to the flexor muscles. The cells are not drawn to scale and only one segment of the lateral giant is included.

LGI to produce an action potential (see Chapter 1). No escape behaviour occurred. Then, as the voltage applied to this nerve was increased, the excitation of the LGI grew until suddenly it 'decided' to respond with an action potential and a full abdominal flexion occurred. An all-or-nothing response by the LGI (strictly, by any LGI in any of the abdominal segments) decides whether the crayfish shows the escape response or not. In this sense, we can see that with a rising level of causal factors (here identified as a voltage), the probability that the response will occur suddenly rises as a threshold is exceeded and the escape behaviour outcompetes everything else.

But we can also see the importance of inhibition. A single action potential in an LGI also inhibits other behaviour that might interfere with escape. In the crayfish, the circuits that extend the abdomen are inhibited when the escape response (which, of course, involves flexion) is in progress. A single action potential in an LGI is enough to bring about this inhibition. This evokes inhibitory postsynaptic potentials (IPSP) in the motor neurons of the extension and prevents any of the muscles acting. It also inhibits slow postural movements (both flexion and extension) which would also get in the way of rapid escape and which have their own slow extensor and flexor muscles. Each slow flexor and extensor muscle has five excitatory motor neurons and one inhibitory one.

The LGI uses the single inhibitory neuron to suppress postural movements and make sure that the crayfish's whole body gives top priority to escaping as rapidly as possible.

In the giant sea slug, *Pleurobranchaea*, we find another dramatic example of inhibition, this time mediated not through neurons but through a hormone. When the sea slug is laying its eggs, feeding behaviour is strongly inhibited by egg-laying. *Pleurobranchaea* is a carnivore and normally eats whatever animal matter it can find, including other sea slugs and their eggs. Inhibition of feeding during egg-laying prevents it from eating its own eggs. The inhibition is brought about by a hormone released when the animal lays eggs. The hormone works directly on the buccal ganglia that control the muscles of the mouth and simply inhibits movements of the mouth (Davis *et al.*, 1977). Even egg-laying, important though it is, can in turn be inhibited by escape. The sea slug's motivational priorities are thus clear and understandable: escape is given top priority because of the serious and immediate consequences of not getting away from immediate danger but second comes egg-laying because it is not advantageous for a sea slug to eat its own eggs (Fig. 4.26).

Even sea slugs, however, have a greater degree of flexibility in their behaviour than might appear from this 'inhibition-only' priority system of escape and egg-laying. When the oral veil of the sea slug is touched, the animal usually withdraws defensively. If the animal is feeding, however, this withdrawal response is not shown and the animal continues feeding. This occurs because neurons in the buccal ganglia that are active during feeding inhibit output to the oral veil (Davis *et al.* 1977). While this looks like a straightforward case of inhibition, the inhibitory action can be two-way. If the animal is satiated with food, the inhibition of withdrawal normally brought about by the presence of food does not happen. When its head is touched, the animal now does withdraw, but this time gives priority to defence over feeding. The low level of causal factors for feeding in a satiated animal means that withdrawal outcompetes feeding, but inhibition still characterizes the interaction between the two systems.

In vertebrates, particularly mammals, decisions tend to be made by whole networks of neurons and often involve two distinct stages. Firstly, specialized cortical areas process particular inputs. For instance, they recognize food items, integrating memory of the food's taste and its energy content. The output from this specialized area is then modulated to produce a suitable level of the appropriate motivational state, in this case hunger. Then a more global decision takes place, in which the competing claims of the different cortical areas (e.g. to eat the food item or to flee from a predator) are evaluated and structures such as the basal ganglia 'decide' which one should have control over the body (Rolls, 2005).

Although we are far from understanding how the brain makes and integrates its multitude of decisions, we nevertheless find that many of the principles we encountered when we looked at decisions at the behavioural level also apply when we look at what is happening at the physiological level. Competition and inhibition are fundamental to the way the brain works and we also find that the concepts of feedback and homeostasis are just as applicable at this level. This is shown particularly well through the study of one area of the vertebrate brain that has proved to be of major importance in motivation, the

(a)

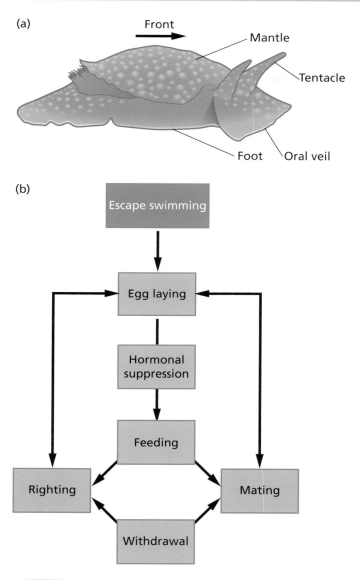

Figure 4.26 (a) The carnivorous sea slug, *Pleurobranchaea*. (b) The behavioural hierarchy of *Pleurobranchaea*.

hypothalamus. The hypothalamus is a relatively minute piece of brain tissue – in the human brain it is smaller than the last joint of the little finger – but is of primary importance in a whole host of reactions. Its position in the brain is shown in Fig. 4.15, where it can be seen that it has a very intimate connection with the pituitary gland. This key endocrine gland controls the whole hormonal system of the body and is connected to the hypothalamus by the pituitary stalk, which is rich in both nerves and blood vessels.

From its connections with other parts of the brain, its rich blood supply and its links with the pituitary gland, the hypothalamus is well adapted both to measure changes in the metabolism of the body and to set in motion activities which will rectify them. It is, in other words, well suited to serve as part of the homeostatic control systems we have already discussed. We saw (p. 206) that the hypothalamus has 'osmoreceptors', cells that detect water loss by responding to increased concentration of circulating body fluids. They can set in action two compensating systems. The first, acting via links with the pituitary gland, causes the secretion of anti-diuretic hormone (ADH) which increases the resorption of water by the kidneys. The second system, as we have already seen, causes the animal to seek out and drink water. If the link between the hypothalamus and the posterior lobe of the pituitary gland is damaged, ADH may never be secreted. In such a situation, the kidney continues to excrete copious quantities of urine and, to compensate, the animal continually drinks large quantities of water – a condition known as *diabetes insipidus*.

Another homeostatic system that involves the hypothalamus is the control of body temperature. There are areas of the hypothalamus which are highly sensitive to changes in the temperature of the blood. If these areas are heated artificially by implanted wires, a rat will start sweating and panting. Sweating is controlled peripherally by the autonomic nervous system and the hypothalamus can initiate its activity. The reverse effect is produced when temperature-sensitive areas are cooled; now the animal shivers, another autonomic response. If a rat is cooled in this way for an even longer time, shivering alone is inadequate and, if given the right material, the rat begins to build a nest, or enlarge the one it has, in order to insulate itself. Both the reflex response of shivering and the more complex behavioural one of nest-building are initiated by the hypothalamus.

The hypothalamus is also implicated in the control of feeding. Mammals and birds normally keep their weight very constant and adjust the amount of food they eat accordingly. Rats with damage to the central areas of the hypothalamus (the ventromedial nucleus) were known to lose this sensitive control of their eating and develop the hyperphagia we mentioned earlier. Figure 4.27 shows the food intake of rats whose ventromedial nucleus has been damaged compared with that of sham-operated control animals. After a few days' post-operative depression of food intake, the brain-damaged rat begins to eat huge amounts of food – at least four times the normal amount. This so-called 'dynamic phase' of hyperphagia lasts for 3 weeks or so. Beyond this, food intake declines and eventually settles down, with the animal eating about double the normal amount. Needless to say, hyperphagic rats become grossly fat and are very inactive. (The same effect is seen in mice and it is mice shown here in Fig. 4.27.) By contrast, rats with damage to a different part of the hypothalamus, the lateral area, show completely opposite symptoms. Initially, they stop eating and drinking altogether and would die unless some intervention were made to stop them. Eventually these rats would start taking in food and water again and slowly regain normal body weight. However, they failed to respond appropriately to food or water deprivation – in that

(a)

(b)

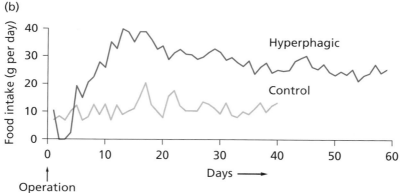

Figure 4.27 (a) A normal mouse and one with a lesion in the ventromedial hypothalamus. (b) The daily intake of normal rats and those with bilateral lesions in the ventromedial nucleus of the hypothalamus. Further explanation is given in text.

they fail to eat or drink enough to make up quickly for their initial deficit as normal rats would do (Teitelbaum and Epstein, 1962).

It used to be thought that these results together constituted evidence for two feeding 'centres' in the hypothalamus, the lateral hypothalamus being the hunger centre and the ventral hypothalamus being the satiety centre. However, more recent work has shown that the role of the hypothalamus is more complex than this. New techniques enable damage to the brain to be monitored much more precisely, and these have shown that the failure to eat and drink that had been thought to be characteristic of rats with damage to the lateral hypothalamus could be induced by damaging only the fibres containing the neurotransmitter dopamine that pass through the lateral hypothalamus. Such rats would stop feeding and drinking even though there had been no damage to the lateral hypothalamus itself. Indeed, damaging the lateral hypothalamus without affecting these fibres resulted in rats that fed and drank normally (Winn *et al.*, 1984). However, such rats did behave oddly in some respects: they did not drink as much as a normal rat when injected with saline or with angiotensin, a substance that is formed naturally in the body during dehydration. In normal rats, both saline and angiotensin make them start drinking. Similarly, the rats with damage to the lateral hypothalamus did not start eating when injected with 2-deoxyglucose, which normal rats do. This suggested that, although the hypothalamus was clearly important to feeding and drinking behaviour, it does not have the key control function that would justify it being called a feeding or drinking 'centre'. Rather, several different brain regions, including the frontal cortex, would be involved in the control of ingestion (Winn, 1995). This idea in turn fits well with what we have already discussed at a behavioural level. Animals do not eat and drink only in response to food and water deprivation. On the contrary, it is remarkable how many different factors, ranging from the presence of predators and members of their own species to the palatability of the food itself and even their expectation of where it is to be found, affect what they do at any one time. The complexity of decision-making that we observe in the natural lives of so many animals inevitably means that input from many different parts of the brain will be involved as the animal 'decides' what it will do next.

We can see this very dramatically through new imaging techniques that show which brain systems are involved in different aspects of behaviour. One such technique, called functional magnetic resonance neuroimaging (fMRI), measures the increased activity of groups of brain cells by detecting increases in local blood flow when a brain region becomes more active. Active parts of the brain become deoxygenated and changes in the amount of deoxyhaemoglobin (which is paramagnetic) can be picked up by a brain scanner to give a measure of brain activity. An example is shown in Fig. 4.28, which illustrates what happened when human subjects were given some food and then asked to say either how pleasant it tasted or how strongly it tasted. In trials where they were asked to rate the intensity but not the pleasantness, the primary taste cortex in the insula was active. But when they were asked to rate the pleasantness of the taste, a different area of the brain became active – the orbitofrontal cortex where the secondary taste cortex is located (Rolls *et al.*, 2008).

The brain has separate regions for processing different aspects of a stimulus, such as what it is and how pleasant it is. For example, the taste of a sweet food becomes less pleasant after we have eaten it to satiety and this forms part of the normal control of food intake (Rolls, 2005). But we are still able to taste it and describe it as sweet, even when we no longer want to eat it. This means that we can still perceive food and learn where to find it even if we don't want to eat it at the time, perhaps because we have just eaten a meal. Such examples show how we are beginning to understand the causal basis of behaviour right down to the level of the brain mechanisms that control it.

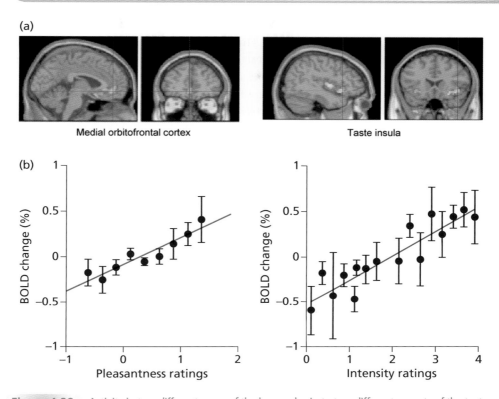

Figure 4.28   Activity in two different areas of the human brain to two different aspects of the taste of monosodium glutamate, the 'fifth taste', sometimes called umami. Activity is revealed by functional neuroimaging (fMRI) which shows the change in the BOLD (blood-oxygen level dependence) signal. Left: The median orbitofrontal cortex registers the reward value or subjective pleasantness rating of the taste (-2 = very unpleasant, 0 = neutral and $+2$ is very pleasant). The more pleasant the person says the taste is, the more active this part of their brain is. Right: The primary taste cortex in the insula, on the other hand, registers the subjective intensity (and concentration) of the taste (0 = very weak, 4 = very strong). The activity of this part of the brain is correlated with whether the person says the taste is weak or strong, not pleasant or unpleasant.

Hormones and sequences of behaviour

Hormones form a chemical message system within the body which is probably as old as the nervous system itself. The endocrine organs, which produce the hormones, and the nervous system share the common functions of communication and coordination both within the animal and between it and the outside world.

The functions of the two communication systems – endocrine and nervous – remain essentially complementary. The nervous system can pass information from one part of the body to another very rapidly, but it operates on a time scale from milliseconds to minutes and is not particularly well suited to transmitting steady, unchanging messages over long periods of time. The endocrine system cannot respond so rapidly, but its cells can maintain a prolonged steady secretion into the bloodstream, lasting for months, if necessary. For example, the rise in hormone levels that occurs in birds as day length begins to increase in spring initiates the whole breeding season. Moreover, hormones can reach every cell in the body via the bloodstream, whereas the nervous system generally controls only the muscles. During reproduction there need to be a number of persistent morphological and physiological changes as well as short-term changes to behaviour.

Circulating hormones are one of the major components of the 'motivation' of behaviour: we have already seen how feeding behaviour is suppressed by egg-laying 'motivation' through the mediation of a hormone in the sea slug *Pleurobranchaea*. In vertebrates, too, we have direct evidence of hormones being one of the major causal factors leading to the occurrence of behaviour. Sometimes hormones apparently 'force' behaviour from an animal even under the most inappropriate circumstances. Injections of the pituitary hormone prolactin will cause isolated male cichlid fish – which have not been through any of the early stages of breeding behaviour – to perform the parental fanning behaviour normally associated with keeping eggs aerated in the nest (Blüm and Fiedler, 1965). Injected males will begin to fan in a bare tank devoid of all the normal external stimuli for fanning – CO_2 in the water, fertilized eggs, nest, etc. – and the amount of time they spend fanning is dependent on the dose of prolactin they have received.

Such central action is dramatic, but hormones are highly potent substances which can affect 'target organs' in other parts of the body and thus affect behaviour indirectly. For example, they may affect the responsiveness of a sense organ. Under the influence of oestrogens (sex hormones produced by the ovaries), the sensory field of the perineal nerve supplying the genital region of the female rat enlarges and is maximal at the time of oestrus (Komisaruk *et al.*, 1972). This means that at the peak of oestrus she is more easily stimulated by the thrusts of the male's penis and can orientate her body so as to help intromission.

Correspondingly, testosterone (the sex hormone secreted by the testes) causes changes in males such as thinning of the skin covering the glans penis in rats. This means that the tactile sense organs are more exposed and, under the influence of the hormone, the male becomes more sensitive to the movements of the female during intromission. In male birds, testosterone affects a wide variety of behaviours including aggressiveness, rate of singing, and even incubation and feeding of the young (Ketterson and Nolan, 1994). In some female birds, oestrogen causes shedding of ventral feathers and increased vascularization of the skin on the lower breast to form a brood patch (Fig. 4.29a, b). Hinde and Steel (1966) found that a female canary is more responsive to stimuli from the nest cup after her brood patch has developed and that this affects behaviour during nest-building and incubation.

From the behavioural viewpoint, however, the most important target organs for hormones are regions of the central nervous system itself. Here we often find an

(b)

(a)

Figure 4.29 (a) The blood-red brood patch of an incubating bird. It has shed all its feathers and the rich capillary blood supply is just below the surface allowing maximum heat exchange between the bird and its eggs or young. To an unsophisticated eye it does look like a wound. Pelicans develop just such a conspicuous, vivid brood patch and glimpses of this as the parent tucks its small young beneath its body for brooding must be the basis of the early Christian belief that the pelican tears its own breast to feed blood to its young. (b) This image taken from the crest of Corpus Christi College, Cambridge is also very commonly seen painted, carved or sculpted in churches. It is usually taken to symbolise Christ's sacrifice of His own blood.

elaborate interaction between the way hormones act on the brain and the way the brain's response to an environmental stimulus can, in turn, affect hormones by altering the secretory activity of the endocrine organs. One of the best examples of this inter-action comes from studies of reproductive behaviour in the ring dove (*Streptopelia risoria*; Fig. 4.30) initiated by Lehrman and his co-workers (Lehrman, 1964). It was known for a long time that female pigeons and doves do not normally lay eggs if kept alone or with other females, but begin to lay soon after a male is introduced. Simple tests show that the sight of a male is enough to cause ovulation, provided he is able to court the female. Thus a male which performs the typical courtship bows, even though separated behind a glass screen, is much more effective than a castrated and therefore non-courting male in the same cage.

Under the warm and well-lit conditions of a laboratory colony, a male dove is usually ready to court the moment he is put with a female, which presumably means that his testes are active and secreting hormones. Visual stimuli from his courtship activate the female's hypothalamic centres, which control secretion by the pituitary gland of two hormones: follicle stimulation hormone (FSH) and luteinizing hormone (LH). These stimulate the growth of her ovary, which in turn secretes the female hormone oestro-gen, under whose influence her reproductive tract begins to grow. After a day or so,

Figure 4.30 A pair of ring doves. The male and female look very alike but can be distinguished by their behaviour. The male courts the female and attracts her to a nest site. They build the nest together.

both birds select a nest site and begin to build. During nest-building, courtship continues and the birds copulate. When the nest is complete, the female becomes increasingly attached to it and she soon lays her first egg. This follows the release of LH from her pituitary, and for a few days before this her ovary has been secreting another hormone, progesterone, which induces incubation behaviour. The male, too, secretes progesterone, which acts antagonistically to testosterone so that his courtship and aggression die away and he becomes willing to incubate eggs. The birds take turns to sit on the eggs and after a few days' incubation, the cells lining their crops begin to proliferate and slough off, eventually producing 'crop milk' for the young birds. In both sexes, the growth of the crop is controlled by yet another hormone from the pituitary, prolactin. Prolactin not only leads to crop growth but also inhibits the secretion of FSH and LH, which in turn shuts off the secretion of sex hormones. As a result, sexual behaviour between the pair dies down as they incubate the eggs and feed the squabs. After 10–12 days, the young birds leave the nest, by which time parental feeding and prolactin secretion have begun to decline. As prolactin levels in the blood fall, so FSH and LH are once again secreted and the male begins to court the female again, ready for the next cycle of reproduction. Both visual and auditory stimulation are important for the female's hormonal response (Barfield, 1971), but physical contact is not essential. Castrated males are less stimulating for females unless they are given an injection of testosterone and it seems to be the male's display that has the stimulating effect (Erickson, 1985).

Some of the experiments involved in working out this sequence are quite dramatic. Thus the secretion of prolactin normally follows when the doves have been sitting on the eggs for a few days, but the male dove's crop develops even if he does not actually incubate himself – provided he can see his mate incubating! Here, then, a specific visual stimulus leads to the secretion of a specific hormone, but it will only do so if the male is in a particular physiological state. He must previously have participated in nest-building, because if he is separated from his mate earlier in the cycle his crop does not develop, even if he can see her incubating.

Hormonal effects in mammals are just as dramatic. The prairie vole, *Microtus ochrogaster*, is unusual as a monogamous rodent with long-term pair bonds and biparental care. The males are highly aggressive to intruders in defence of the female and young, but only after they have mated. Before mating, naïve males are neither particularly aggressive to intruders nor do they have a preference for one female over another. The act of mating stimulates the release of arginine-vasopressin (AVP), a nine amino-acid neuropeptide, and within a few hours the male becomes highly aggressive to intruders and shows a very strong partner preference (Winslow *et al.*, 1993).

Direct effects of hormones can be demonstrated by implanting specific hormones into different parts of the brain. Controlled amounts of hormone can be placed in particular sites through a hollow needle without interfering with an animal's freedom to behave. Hutchinson (1976) and Komisaruk (1967) showed that castrated male doves would show courtship behaviour to females and aggression to other males if testosterone were implanted in the anterior hypothalamus and the preoptic nucleus (so called because it lies just anterior to where the optic chiasma enters the brain). Progesterone implanted in the same places suppresses these behaviour patterns and causes an increased tendency to incubate – a direct demonstration of antagonism between progesterone and testosterone.

The relationship between behaviour, hormones and the nervous system can be very intricate. One dramatic and, at first sight, rather puzzling example is the 'Bruce effect', named after Hilda Bruce (see Bronson, 1979). Pregnant female mice abort their litters and reabsorb the embryos if a strange male mouse (not the father of the litter) comes into contact with them. Just the smell of a strange male will suffice to induce abortion, it happens even if females come into contact with litter soiled by one. The pregnancy-blocking effect is brought about by an inhibition of prolactin secretion and consequent reduction in the progesterone which is needed to support pregnancy. The smell is the behavioural signal and females with olfactory bulbs removed remain pregnant even when in contact with strange males.

There are also close ties between hormones and the immune system, some of which are quite unexpected. Testosterone, important though it is for male behaviour and appearance, also has some costly side-effects, the most serious of which is to make the immune system less able to fight infection (Grossman, 1985). This immuno-suppressive effect of testosterone appears to be widespread among vertebrates and probably accounts for the fact that sexually mature males are particularly vulnerable to infections from parasites and diseases. The resplendent male, displaying vigorously to females is, paradoxically, least likely to be able to fight infection. This may explain why, when choosing a mate, females appear to pay particular attention to how healthy and disease-free a male is (Zuk, 1994), an idea we discuss much more fully in Chapter 6.

Conflict and physiological stress

If conflict or thwarting are prolonged for days or weeks, with the animal allowed no chance of escape, then many different bodily changes are likely to occur. These are often put together under the single term 'stress' because the body's response to a wider range of stressors (thwarted escape, overcrowding, extreme cold and burns, for example) are very similar. Moberg (1985), Broom and Johnson (1993) and Toates (1995) describe the physiological changes involved and point out that most of them are attempts to restove the delicate balance of the body's metabolism when it has been upset. If activated for a short time, they are a perfectly normal and adaptive response. For example, when an intruder enters an animal's territory, various physiological changes occur in the body of the defender. There is increased activity in the autonomic nervous system which supplies the viscera and smooth muscles and which is hormonally under voluntary control. The autonomic system also supplies the adrenal glands, endocrine organs close to the kidneys which have a double structure, an internal medulla (supplied by autonomic nerve fibres) and an external cortex. When stimulated via its autonomic nerve supply, the adrenal medulla releases the hormone adrenalin into the body. This causes changes in numerous parts of the body: the sweat glands of the skin begin to secrete, hair becomes erect, the heart beats faster, breathing becomes more rapid and deeper, and blood gets diverted to the skeletal muscles from the alimentary canal. These responses are adaptive in the sense that they prepare the animal for the strenuous action of either fighting or fleeing. The increase in heart rate, in breathing rate and the diversion of blood to the muscles from the alimentary canal all ensure that there is plenty of oxygen available for physical activity. Corticosteroid levels are also elevated during coitus and physical exercise in humans and, to judge by its physiological response, a passionate kiss is as much of a stressor as an electric shock (Toates, 1995)!

In natural territorial defence, there will be a rapid flush of adrenalin as the intruder is detected, the territory holder will become aggressive, and this will usually result in the intruder leaving. Adrenalin levels in both intruder and defender will then subside. But if the stressful situation persists – as might happen if two animals were confined in a small cage – then a further reaction begins. This involves the other part of the adrenal glands, the adrenal cortex, which is stimulated to release its hormones, not directly by nerves as is the medulla, but by another hormone, adrenocorticotrophic hormone (ACTH), produced by the pituitary gland. Here, as with the release of adrenalin, the nervous system initiates the response. Stress activates cells in the hypothalamus which itself then stimulates the pituitary to release ACTH. It is not fully clear how the adrenal cortex hormones help the animal to adapt to stress. Some of them are concerned with glucose metabolism and may serve to mobilize the body's long-term food reserves. Corticosteroids both make glucose available and slow down its rate of metabolism, thus keeping glucose reserves available in the bloodstream. Whatever their action, the release of adrenocortical hormones is most dramatic. The cortical cells become drained of their contents and, if stress persists, the adrenals enlarge, sometimes by 25%. Prolonged stress reduces immune responses and suppresses of growth and reproduction (Maier *et al.*, 1994). Barnett (1964) showed how a wild rat, unable to escape from the territory of a dominant resident male, may die after a few hours of intermittent attack, even though it has no significant wounds.

Stress responses that lead to pathology and death are clearly indicative of poor animal welfare. We now conclude this chapter by showing how some of the other aspects of motivation and decision-making we have been discussing also contribute directly to our understanding of animal welfare and indeed provide one of the most important tools we have available for improving it.

Decision-making, motivation and animal welfare

We have so far discussed the mechanisms of decision-making 'as if' animals were able to weigh up the different courses of action open to them. We have carefully described motivation and behavioural conflict in terms that are either directly understandable in physiological terms or potentially could be understandable as our knowledge of the complex mechanisms involved becomes greater. We have avoided any discussion of the emotions animals might be experiencing when they show aggression or fear and made it clear that decision-making can be studied without implying that they necessarily understand what they are doing when they weigh up the various options facing them.

There is a good reason for this. The subjective experiences of other animals (or other people for that matter) are not accessible for study in the same way that behaviour and physiology can be studied. So, for much of the twentieth century, they were not studied at all by scientists. Psychology and ethology stuck to what was observable and testable, and that eliminated any discussion of what animals might be feeling. We shall meet similar issues in the next chapter when we discuss questions concerning thought processes in animals in relation to their learning abilities.

Whilst avoiding speculation about subjective experience is, from many points of view, desirable for scientific investigations of animal behaviour, it has had some unfortunate consequences. As more and more people became concerned about animal welfare and the possibility that animals might experience pain and suffering, the study of animal behaviour was contributing less to the debates about animal welfare than it could, and indeed arguably, should have done. Added to this, one of Tinbergen's four questions, the one about adaptation, consumed far more research attention than the ones about mechanism and development. The enormous excitement generated by sociobiology, behavioural ecology and the revolution in adaptive thinking brought about through a gene-centred approach to evolution left the other questions relatively neglected.

Figure 4.31 Mink are valued for their fur. As active carnivores, wild mink travel over long distances. In small cages, many of their natural behaviours such as swimming are prevented. Fur-farming is now banned in the UK.

But, of course, it is precisely those questions about mechanism and development that are most relevant to animal welfare. Many, if not most, welfare issues arise because animals are reared and kept in conditions where it is impossible for them to perform much of their natural behaviour (Fig. 4.31). Junglefowl live in tropical forests whereas now their chicken ancestors live in cages or huge flocks of commercial broilers. Ancestral pigs, too, were forest dwellers. If we want to understand the consequences of keeping animals in unnatural environments and of preventing them from behaving in ways for which they were evolved, then it is to the study of animal behaviour that we should turn. In particular, the mechanisms animals have evolved for making decisions and their motivation to perform different behaviours are of crucial importance in understanding the welfare consequences of what humans do to animals.

For example, we saw earlier in this chapter (p. 194) that if hens are kept in barren cages where they have wire floors rather than sand or litter in which they could dustbathe, they will show behavioural 'rebound', which means that if they are then given access to a dustbathing substrate, they will compensate for their previous deprivation by doing extra dustbathing. This implies that hens in cages are strongly motivated to dustbathe but are prevented from doing so by the paucity of their environment. This conclusion is strengthened by the further finding that hens will push heavily weighted doors in order to gain access to somewhere to dustbathe. We also saw that sows that are about to give birth will 'work' extremely hard and press a lever repeatedly to obtain straw for building a nest for their piglets (Fig. 4.9). Their motivation to build a nest is very high, suggesting that their welfare is compromised if they cannot do so, as is the case in many intensive systems where there is just a bare floor with no nest-building material.

Establishing the strength of an animal's motivation to perform a behaviour is one of the most important steps we can take in identifying a potential source of suffering in animals and also provides us with a powerful means of improving their lives. The reason it is so powerful is that it effectively 'asks the animal' what is important to it (Dawkins, 1990). It gives the animal the means to express not just a preference for one environment over another, but puts a quantitative value on the strength of its motivation to obtain or get away from something. It is a very animal-centred approach to welfare, the nearest we have yet come to being able to ask the animals what they want and what they want to get away from.

Clearly, motivation, or lack of it, is not all there is to welfare. It is almost beyond dispute that an animal's welfare cannot be good if it is diseased or injured so that physical health is included in almost everyone's definition (Broom, 1991; Webster, 2005; Fraser, 2008). Hurnik extended this idea to say that longevity, particularly in captive animals, is a good net measure of welfare because if an animal lives a long time, that must mean that for most of its life, it basic needs for survival, health and comfort must have been met. Clubb *et al.* (2008) argued that the welfare of zoo elephants was poor compared to that of wild or even working elephants in India on the grounds that the median lifespan of zoo elephants was half that of their wild counterparts.

Thus it is essential to include physical health in the definition of welfare, particularly because animals (and people) may choose things that are not necessarily good for physical health in the long run, for example, they may choose to overeat and then suffer the ill effects of obesity. So, combining these two ideas of motivation and health, we reach a simple working definition of good welfare as where animals are healthy and have what they want. They are in a state of good physical health and not frustrated by being highly motivated to obtain something they do not have. We supplement information from motivation and decision-making with what keeps animals healthy.

This simple approach – asking if animals are healthy and have what they want – can also encompass a wide range of other definitions of welfare (Dawkins, 2008), such as whether the animal can perform all of its natural behaviour patterns (Kiley-Worthington, 1973; Bracke and Hopster, 2006), whether it is 'stressed' (p. 232) and how we should interpret unusual or abnormal behaviour (p. 217). In each case, asking these two apparently simple questions tells us what we want to know and whether what we humans think would be an improvement in animal welfare actually is so from the animal's point of view. For example, many people believe that 'environmental enrichment' always improves animal welfare or that keeping animals outside is necessarily better for welfare than keeping them inside. The two question approach makes it clear what we need to know: is the animal's health improved by the enrichment or being allowed outside and do the animals want the enrichment or do they want to go outside? The answers may surprise us when we get them, but at least they put us on the road to making genuine improvements in animal welfare.

Conclusions

Throughout the last two chapters on the causal basis of behaviour, we have tried to strike a balance between trying to explain what animals do in terms of known physiological entities, such as hormones and neurons, on the one hand and avoiding over-simplification on the other. Animal behaviour, with its shifting motivational priorities and its decision-making on many different time scales, will inevitably have complex explanations and we should not underestimate how difficult it is to 'explain' it. At the same time, major advances have been made in our understanding of behaviour at a physiological level and are likely to continue at an accelerating rate as new techniques such as brain scans enable us to gain access to what was previously inaccessible. We can look forward to a future in which we have a greatly increased understanding of how the brain works to control behaviour. Even with our present knowledge, it is striking that behavioural and physiological explanations have proceeded hand-in-hand and reinforced each other's findings. Physiological studies have added to and sometimes even superseded vaguer ideas of 'motivation' while behavioural studies have shown what real animals do and guarded against our viewing animals as simple control systems switching on and off when a single variable reaches a particular value. We can no longer think in terms of single 'centres' for feeding, drinking, aggression, and so on, but have to look at the integration of information from many different areas of the brain. What an animal decides to do will be affected not only by the stimuli in its present environment but also by what happened to it in the past. For this reason, we now turn to the important topic of learning and memory. Although we dealt with some aspects of development in Chapter 2, our emphasis there was on the role of innate behaviour. In the next chapter, we emphasize the role of experience. Equipped with some understanding of the causal basis of behaviour, we are now in a position to see how what animals learn affects the machinery of behaviour itself. The past affects the present state of the animal so that questions about causation and questions about development, although distinct, are intimately related.

SUMMARY

The way animal behaviour 'works' is very complex (much more complex than man-made machines) and we have only just begun to understand the causal basis of what animals do. Different behaviours on different time scales have many different explanations but we can pick out a number of important principles. Some behaviour is 'goal directed' – that is, a whole sequence is brought to an end when the animal achieves a particular goal state such as eggs in a nest. Homeostasis or negative feedback (where the animal becomes less likely to do a behaviour as a goal state is approached) often plays an important part. Key to understanding animal behaviour is the idea that animals are often motivated to do more than one behaviour at a time and have to 'decide' which to one to do. Sometimes such motivational conflicts are resolved by a form of internal competition, sometimes there will be active inhibition between the behavioural elements and sometimes positive feedback tips the balance in favour of one or the other. Animals may show specific behaviours at times of conflict and some of these have become signals for communication. Motivation and decision-making in animals have great practical importance for animal welfare because understanding what animals are motivated to do and what happens when they are prevented from doing what they have decided to do are crucial for improving their welfare.

CHAPTER FIVE

Learning and memory

CONTENTS

Learning plays a major role in the lives of many animals. We discuss different types of learning and go on to consider evidence that animals may share some human abilities.

- Learning as part of adaptation
- Sensitization and habituation
- Associative learning
- Specialized types of learning ability
- What do animals actually learn?
- Are there higher forms of learning in animals?
- The comparative study of learning
- Social learning and culture
- The nature of animal minds
- The nature of memory

Learning as part of adaptation

Throughout this book we have made frequent reference to learning as a form of behavioural development retained into adult life and as one pathway by which behaviour becomes adapted to an animal's requirements. Now attention must be focused on learning itself. Learning and memory go together because whilst the former involves changing behaviour as a result of experience, its effects cannot be put to use unless the results of the experience can be stored in some way and recalled the next time they are needed. A discussion of learning will finally lead us on to a consideration of the mental or cognitive abilities of animals and how far they share some of those abilities with ourselves. The issues raised there take us far beyond just learning abilities but it is essential first to consider what are the different types of learning and biologists will want to examine how and whether they are expressed in a range of very different animals. For this reason, Thorpe's (1963) book, although old, remains useful because it does deal with the whole animal kingdom. It covers something of the context and the phenomena associated with learning in molluscs and insects, fish and amphibians, as well as the more familiar studies with birds and mammals. Thorpe defines learning as, 'that process which manifests itself by adaptive changes in individual behaviour as a result of experience'. This definition draws attention to two important features. Firstly, learning normally results in *adaptive changes* and, as we discussed in Chapter 2, learning and instinctive behaviour are both ways for equipping an animal with a set of adaptive responses to its environment. Normally both are found in combination and logically they have much in common. In the one case, we have adaptation by genetic modification within a population of animals biasing them to respond in a particular way. If it is successful it will be passed on to following generations. With learning, it is individuals which select and retain the best responses over the course of their own lifetime. Offspring cannot inherit any specific learnt response but only the more general ability which enables them, in their turn, to benefit from experience. Another important point arising from Thorpe's definition is that, strictly speaking, learning is a process which we cannot usually observe directly; we measure what has been remembered as a result of learning.

This ability to benefit from experience is a link to our non-human fellow creatures because they so obviously share this striking behavioural capacity with us. Animal learning and memory differ from ours both quantitatively and qualitatively, but there remain many features in common and we can make useful comparisons. In fact, because we can communicate so easily with human subjects, they are in many respects better material for learning studies than animals. For instance, in the laboratory, it is possible to test our memory in two ways: 'recall', i.e. by reciting or writing down a list of nonsense syllables which we have previously learnt; and 'recognition', i.e. presented with a set of nonsense syllables which includes those we have learnt, we record which syllables we recognize. Recognition is always an easier task than recall because the situation provides stimuli which, as we say, 'jog our memory' and help the process of recall. If we have trained a rat to run through a maze, we cannot ask it to draw a map of its route on a piece of paper. The only way to test what it has learnt and retained is to put it back in the maze and observe its behaviour. If the rat makes mistakes we have no means of knowing whether it failed to learn adequately or learnt but failed to recall.

This difficulty brings us to the question of learning mechanisms. What happens in the nervous system when an animal learns something? When a young game bird crouches on the first exposure to the sound of its parent's alarm call, it must utilize pathways already present in its brain whereby a particular auditory input has easy 'access' to the motor system controlling crouching. On the other hand, when a rat learns to press the bar of a Skinner box, new pathways must become established, because prior to learning the bar evoked no special response. We know too that mere presentation of the bar to the rat is not enough; there must be some sort of reward as a result of doing so. Some part of the learning mechanism records the result of the rat's reaction and if this result is 'good' it increases the probability that the reaction will occur again the next time the situation is presented. Further, we know that somewhere in the nervous system there is stored a more-or-less permanent record of the learning that can be consulted or recalled on future occasions. The investigation of the physical basis of memory has been an active area of research for decades. Recently there have been some remarkable advances, and although much is beyond the scope of this book, we shall return to it briefly after looking at the phenomenon of learning itself.

There was a time during the first half of the twentieth century when a huge amount of effort was directed towards the study of animal learning. Many experimental psychologists were attracted by the notion that animals could serve as models for human beings in this respect and the majority of their work concentrated on two convenient laboratory species, the rat and the pigeon (Fig. 5.1). By investigating an amazing variety of learning tasks and with diverse reinforcements (rewards or punishments), they tried to generate 'laws of learning' which would enable us to predict performance and outcome no matter what subject or what situation. Munn's *Handbook of Psychological Research on the Rat* (1950) is a fascinating record of this 'heroic age' of experimental psychology, the various models and theories of learning which it produced. It has to be said that, for the most part, they attract little attention now. The classification of learning into distinct types, the proscribed learning situations in which they were studied and the rigid models to which all animals and all learning were expected to conform, have been replaced by a much more biologically based approach. This recognizes that an animal's learning abilities must have evolved to suit its own special requirements just as any other aspect of its behaviour; they are learning by instinct. We should not expect learning in pigeons to resemble that of honeybees except in a very general manner. Furthermore, learning is now seen as a way in which animals attempt to identify key aspects of a fluctuating environment: to detect its regularities and ignore the distracting 'noise' which is not important for them. They will integrate their learning with the inherent biases of perception and response that all animals bring to the world. Encouragingly, these moves towards a more truly comparative and biological approach have come both from psychologists, dissatisfied with the old learning theories, and from ethologists who observe the role of learning in their animals' natural lives. Articles by Mackintosh (1983), Roper (1983) and Rescorla (1988) form a good background to this new synthesis. Books by Dickinson (1980), Hinde and Stevenson-Hinde (1973), Seligman and Hager (1972) and Pearce (1996) provide more examples and a more detailed analysis of some of the changing attitudes towards learning.

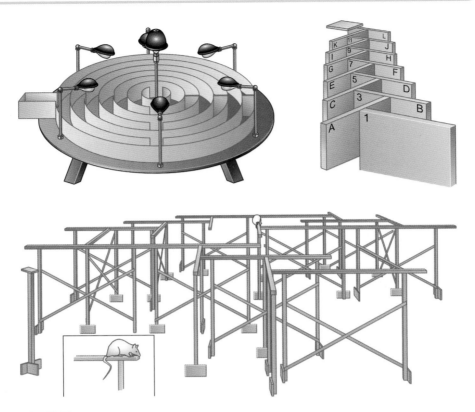

Figure 5.1   Three examples which give some idea of the elaboration of mazes used for measuring learning processes in rats during what might be called 'the heroic age' of American experimental psychology.

 ## Sensitization and habituation

In Chapter 1, we described how responses to stimuli, both at the level of neurons and of reflexes, change according to their strength and their repetition, showing facilitation ('warm-up') or fatigue, for example. In the natural world, the pattern of stimuli reaching animals will always be changing. If they are to learn to change their behaviour to meet a new situation, then clearly anything new appearing in the environment – and we do not mean just a visual stimulus – will have to be taken note of and its importance assessed. The easiest way to get some estimate of importance is to note what happens just before or just after it appears. An alert animal, for instance, one which has had its appetite aroused by the smell of food or has just fled from a predator's attack, is particularly sensitive to such stimuli. To take a laboratory example, a rat which has just

received a small electric shock to its feet jumps in alarm to any novel stimulus –
e.g. a flash of light, a tap on its box – which would normally evoke little or no response.
Evans (1968) studied marine ragworms (*Nereis*), which live in tubes they construct in
the sandy floor of the sea, emerging to stretch out the front part of their bodies to forage
on the surface nearby (Fig. 5.2a). Evans kept his *Nereis* in laboratory aquaria in dim
light. Under such conditions he found that retracted worms sometimes emerged from
their tubes following a flash of light and in one experiment 21% of worms did so.
However, if he fed a second group of unexposed worms just once in the dim light
conditions, over 60% of them emerged from their tubes to a subsequent light flash. The
arousal following feeding had made them much more sensitive.

This phenomenon is a kind of facilitation – a period of high responsiveness
following arousal by rewarding or punishing experiences. It is, reasonably enough,
called *sensitization*, and is widespread in such situations. Why then, do not animals
respond with alacrity to all stimuli, for opportunities to have become sensitized to
them must often exist? The solution lies in a second and in many ways opposite
effect of stimulation – a kind of fatigue. If repeated several times in the absence of
any significant accompaniment – no more foot shocks to the rat, no more food for
Nereis – then animals gradually cease to respond. From being initially sensitized, they
calm down and eventually ignore the stimulus. This waning of responsiveness is
called *habituation*.

If we now think back to the definition of learning – adaptive changes as a result of
experience – then perhaps we should regard both sensitization and habituation as
learning. Habituation, as we shall see, fits quite well but sensitization really does not.

Sensitization, although probably an essential preliminary to learning in many
cases, is not really to be regarded as a form of learning itself. It is too short-lived
and too indiscriminate – for a short time the animal is hyper-responsive to a variety
of events in the outside world and pays more attention to them. Soon one of two
things must follow. If the novel stimuli which initiated sensitization are not repeated,
then the effect wanes quickly. If they are repeated enough times, then the animal
does go on to learn. What it learns depends on the results of the situation. Do the
stimuli signal pleasant or unpleasant events – food, water, a predator, pain? If so
then the stimuli which give the best prediction of these rewards or punishments
(together we can conveniently call them reinforcements) will be picked out for
continuing attention. We say that the stimulus and the reinforcement have become
associated – and association learning is a major category for discussion later. There is
an alternative result of repeated stimulation – that nothing happens, there is no
reinforcement. The stimulus then captures less and less of the animal's attention.
Birds soon come to ignore the scarecrow which put them to flight when it was first
placed in a field. We say that the birds habituate, and habituation can be regarded as
a simple form of learning. It is relatively long-lasting and it is 'stimulus-specific', i.e.
only the stimuli which are repeated without reinforcement are affected – the animal
remains alert to others.

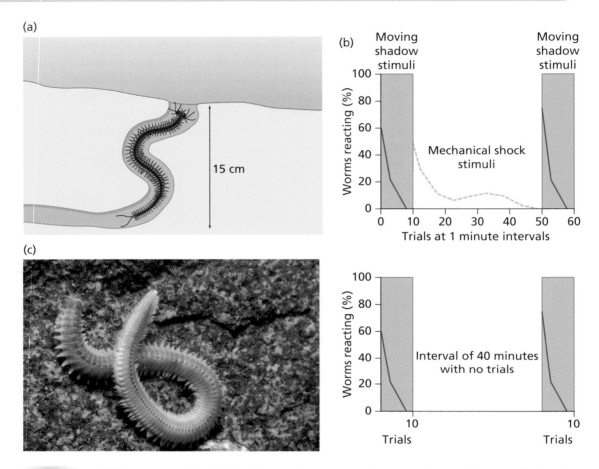

Figure 5.2 (a) The ragworm *Nereis* in its burrow. It emerges to forage on the sea floor, usually keeping its tail close to the burrow's mouth and withdrawing instantly when it senses danger. (b) The rates of habituation to two different stimuli presented to worms when they are foraging with their heads emerged. The response measured is the sudden retraction of the worm into its tube and trials are given at 1 min intervals. The shaded areas record the responses of a group of 20 worms to a moving shadow. Within 10 trials they have all ceased to respond, but switching to a mechanical stimulus (unshaded areas in upper graph) brings back the response in half the worms, and they characteristically habituate more slowly, taking more than 30 trials. The recovery of the response to moving shadow is complete after 40 min, whether the habituation trials to mechanical shock intervened (upper graph), or the worms were simply left alone (lower graph). Clearly, habituation is quite independent for these two very distinct types of stimulus.

These qualities of habituation will be clearer from an example. Again the annelid *Nereis* which leads a fairly simple and often hazardous life has provided good material for studies by Clark (1960a, b). As already mentioned, the worm burrows in mud or sand on the bed of brackish estuaries. Its head and anterior segments protrude from the tube to feed from the surface around the burrow. During its bouts of feeding a variety of sudden stimuli will cause the worm to jerk back into its tube. In the laboratory, Clark could easily get the worms to live in glass tubes in shallow basins of water. He found that jarring the basin (mechanical shock), touching the head of the worm, a sudden shadow passing over and a variety of other stimuli would all cause retraction into the tube, but the majority of worms emerged again within 1 minute. If these stimuli were repeated at 1-minute intervals the proportion of worms responding fell off until none of them were retracting; they had habituated. Clark found that habituation occurred more rapidly if stimuli were given close together. For example, with a bright flash of light it took fewer than 40 trials at half-minute intervals, but nearly 80 trials if the interval was 5 minutes. The speed of habituation also depended on the nature of the stimulus; mechanical shock, shadow, touch and light flash each produced their characteristic rates of habituation. Further, habituation was, to a large extent, stimulus-specific; Fig. 5.2b shows how the waning of retraction to repeated mechanical shock is independent from that to a moving shadow.

There are a number of other processes which may be confused with habituation because they also lead to a reduction in responsiveness. Some we discussed in Chapter 1 when describing Sherrington's work. Results such as those illustrated in Fig. 5.2b eliminate any possibility that the waning of response is due to motivational changes or to muscular fatigue, but it is often more difficult to eliminate sensory adaptation. Many sense organs eventually stop responding to repeated stimulation. We cease to be aware of our clothes within a few moments of putting them on because the tactile receptors in the skin cease to respond. In *Nereis*, Clark could also eliminate sensory adaptation as an explanation for the waning. For example, the worm soon ceased to retract when touched by a probe, but clearly still detected the stimulus because it then attempted to seize the probe with its jaws. Sensory adaptation is usually a short-lived phenomenon; a few minutes without stimulation is usually sufficient for complete recovery. We sensorily adapt to the feeling of our clothes rather than habituate to them and we should retain the term 'habituation' for a more persistent waning of responsiveness which must be a property of the central nervous system and not the sense organs.

Clark could detect some recovery from habituation within an hour or less and the worm's retraction response was completely recovered and back to full strength within 24 hours. This is not very long and, if we think in terms of ordinary memory span, not impressive, but this is to overlook the role that habituation plays in the life of animals like *Nereis*. It is, in fact, a vital process for adjusting an animal's behaviour to the minute-to-minute events in its environment. Habituation enables it to concentrate on important changes there and ignore the others. Small prey animals lead hazardous lives but they cannot spend too long skulking in shelter – they must normally be out feeding. Although, at first, it is best to retreat rapidly in response to sudden shadows, not every shadow means a bird or a fish overhead: it might just as well be a floating piece of seaweed. In general, frequent shadows are more likely to be seaweed than predators and it is certainly adaptive to cease responding when a repeated stimulus has no attendant consequences. Just as certainly, it is dangerous to ignore shadows from then on. By the next incoming tide the seaweed will have moved and a shadow could now signal a predator's approach. The same cautious testing out must begin again.

This is not to imply that all habituation is so short-lived. Newly hatched chicks begin to feed by pecking at any object which contrasts with the background. Responses to inappropriate objects rapidly habituate and do not return – rather, the growing chick begins to learn positively which objects represent food.

In nature, we must expect to find sensitization, sensory adaptation and habituation almost inextricably combined, so that the time course and persistence of diminished responsiveness depend on a variety of factors. Habituation processes are shown by all animals, but predators and prey animals, for example, are unlikely to respond in the same ways because the costs and rewards will be so differently balanced.

 ## Associative learning

Habituation is generally regarded as a simple form of learning because it involves the waning of a response that is already there. We tend to think of learning as a process whereby we acquire new responses and new capacities. For this reason, the basic characteristics of associative learning are immediately familiar to us. A previously neutral stimulus or action has sufficiently important consequences to be singled out from other such events. After some repetitions followed by the same consequences, a long-term association is built up between the event and its result and the animal's response changes accordingly. Having found sugar solution there previously, a honey-bee picks out the blue dish from an array of dishes laid out by an experimenter. Rodents exploring new territories quickly learn the shortest routes to shelter for use when a hawk or an owl swoops.

Perhaps the most familiar type of associative learning situation is the conditioned reflex, inseparably linked with the name of the great Russian physiologist, I.P. Pavlov (see Pavlov, 1941, and Fig. 5.3) whose school was active around the turn of the twentieth century. Everyone has heard of Pavlov's dogs who salivated to the sound of a bell! So far has his influence run that the type of learning he studied is often referred to as classical conditioning. Pavlov's influence on behavioural studies and neurophysiology in Russia was pervasive but, perhaps because the whole 'reflex theory' he developed around his experimental work attracted little favour here, his influence in the West has been less. Pavlov and Sherrington were working at the same time, but from completely different viewpoints. Sherrington studied the organization of reflexes in the isolated spinal cord of dogs and cats, having deliberately cut off influences from the higher centres. Pavlov worked with intact animals and considered that, just as simple reflexes are a property of

Prof. Pavlov in his laboratory at the Institute of Experimental Medicine, Dec. 1922. The temperature of
the rooms is so low that overcoats must be worn
and at times I have seen frost form on his beard.

Figure 5.3 I.P. Pavlov surrounded by his group in 1922. The handwritten caption vividly portrays 'the atmosphere'! Note the dog at the left and compare with Fig. 5.4.

the spinal cord, so conditioned reflexes are the particular property of the higher centres of the brain, especially the cerebral hemispheres. Pavlov's aim was to study 'the physiology of higher nervous activity', but most of his experiments were, in modern terms, pure experimental psychology. Indeed, Pavlov was really one of the founders of experimental psychology; he was applying objective techniques to the study of learning years before J.B. Watson, whose book *Behaviorism* (1924) had such a huge influence on American psychology. Pavlov's work remains important because it systematized for the first time many of the basic phenomena involved in the learning process.

Pavlov's classical experiments with dogs often involved the 'salivary reflex'. Dogs salivate when food is put into their mouths and Pavlov could measure the strength of their response by arranging a fistula through the cheek from the salivary duct, so that

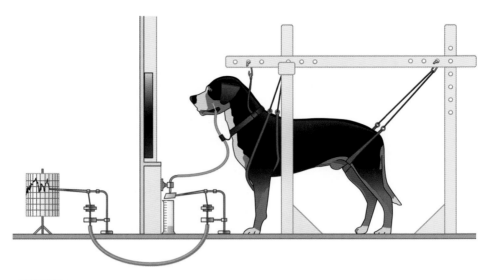

Figure 5.4   A typical experimental set-up in Pavlov's laboratory. The dog is gently restrained on a stand facing a panel and, under experimental conditions, is well insulated from external disturbances. Tactile, visual or sound stimuli can be presented in a carefully controlled fashion. Tubes run from the fistula in its cheek, which collects saliva as it is secreted. A simple arrangement of a hinged plate, onto which the saliva drips, and a measuring cylinder where it collects, enables the amount of saliva to be measured together with the intensity and duration of its secretion. Not shown is the device for blowing a controlled amount of powdered meat into the dog's mouth to reward the salivation response.

drops of saliva fell from a funnel and could be counted. A hungry dog was placed on a stand, restrained by a harness and every precaution was taken to exclude disturbances. In this position, it could be given various controlled stimuli such as lights, sounds or touch, and meat powder could be puffed into its mouth through a tube (see Fig. 5.4). A standard quantity of meat powder caused the secretion of a certain amount of saliva. Now Pavlov preceded each ration of powder by, say, the sound of a metronome ticking. At first, this stimulus caused no response, save perhaps that the dog pricked up its ears momentarily. However, after five or six pairings of metronome followed by food, saliva began to drip from the dog's fistula soon after the metronome started and before the meat powder arrived. Eventually the amount of saliva produced to the metronome alone was the same as that which was given to the meat powder.

The dog had learnt to respond to a new stimulus, previously neutral, which Pavlov called the conditioned stimulus (CS). The salivation response to the CS is the conditioned response (CR). Prior to learning, only the meat powder or unconditioned stimulus (UCS) produced salivation as an unconditioned response (UCR).

Pavlov found that almost any stimulus could act as a CS provided that it did not produce too strong a response of its own. With very hungry dogs even painful stimuli, which initially caused flinching and distress, quite soon evoked salivation if paired with food. The CR is formed by the association of a new stimulus with a reward and in the same way a CR for withdrawal can be formed by associating the CS with punishment. An electric shock to the foot causes a dog to lift its paw; if a metronome is paired with the shock, the dog soon raises its paw to the sound alone.

Pavlov carried out exhaustive tests on the precision with which a particular stimulus was learnt. He found that if a dog was conditioned to salivate when a pure tone of, say, 1000 Hz was sounded, it would also salivate when other tones were given, but to a lesser extent. It **generalized** its responses to include stimuli similar to the conditioned one, and the more similar they were the more the dog salivated. The process opposite to generalization is **discrimination**. Dogs naturally discriminate to some extent, otherwise they would salivate equally to all sounds, but their discrimination becomes refined after repeated trials when only one particular tone is followed by reward. We can accelerate discrimination if, as well as rewarding the right tone, we slightly punish the dog when it salivates to others.

This conditioned discrimination method has been of enormous value for measuring the sensory capacities of animals. After training to one particular stimulus – it may be a colour, brightness, shape, texture, sound, smell, weight, etc. – we then test to see how far the animal can discriminate this stimulus from others. We present it together with another stimulus of the same type and reward only responses to the former, perhaps giving slight punishment for incorrect responses. The two stimuli are made increasingly similar until there comes a point beyond which the animal can no longer learn to discriminate between them which marks the limit of its sensory capacities as measured by its behaviour. This was the method was used by von Frisch (1967) in his classical studies of the colour vision of bees and it has been used for a wide variety of animals and for all sensory modalities.

Conditioned reflexes of the type investigated by Pavlov have been observed in many different animals, from arthropods to chimpanzees. For example, birds learn to avoid the black and yellow caterpillars of the cinnabar moth after one or two trials which reveal their evil taste. They associate this with the colour pattern and generalize from cinnabar caterpillars to wasps and other black and yellow patterned insects. Because predators generalize, it is advantageous for different distasteful insects to resemble one another – the phenomenon of Müllerian mimicry. Poisonous or distasteful vertebrates use the same colour scheme to deter their predators – bright reds, yellows and black are found here also (Fig. 5.5).

In nature we rarely observe reflexes in as 'pure' a form as in the laboratory. When foraging naturally, bees do not just learn to associate a colour with the nectar reward: they also learn the position of the group of flowers with respect to their hive, and learn what time of day the nectar secretion is highest, directing their foraging trips accordingly. Pavlov, despite his scrupulously controlled environment, found that his dogs certainly learnt more than one particular response to one particular stimulus. A hungry dog familiar with his laboratory would run ahead of the experimenter into

(a)

(b)

(c)

(d)

Figure 5.5 Müllerian mimicry. Black and orange for danger! (a) A hornet (*Vespa*); (b) distasteful caterpillars of the cinnabar moth (*Tinia jacobaeae*); (c) the gila monster (*Heloderma suspectum*), a poisonous lizard, and (d) a 'poison arrow' frog (*Dendrobates leucomelas*). Such bright conspicuous colours serve to warn potential predators that they are noxious or dangerous. Bright reds and blues also feature in other species. In every case, animals with such warning coloration make no attempt to conceal themselves, quite the opposite. Each probably derives a certain degree of protection from bad experiences predators may have had from other similarly coloured animals. Hence, the selective value of converging on the same colours and patterns. Tests have shown that predators such as birds very rapidly learn to associate such colours with bad results.

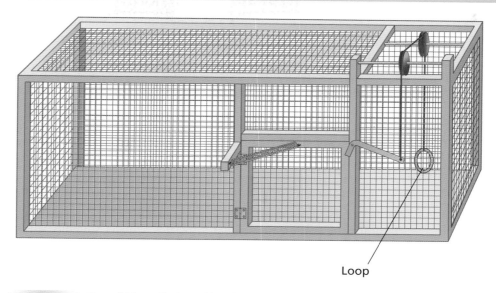

Loop

Figure 5.6 One of Thorndike's problem boxes. A cat is confined inside the cage and must learn to pull the string loop to open the door.

the test room and jump up on to the stand, ready to be hooked up to the apparatus and wagging its tail with every sign of expectancy! Animals, then, don't just hang around waiting for a stimulus to signal reinforcement: they try actively to put themselves into the situations or perform actions which lead to reward or escape from punishment. As we shall discuss later, Rescorla's (1988) account of modern work on classical conditioning shows that animals are not simply making stimulus–response (S–R) connections, but searching for regularities in the sequence of events which face them so as to arrive at reasonable predictions of the best outcome.

Pavlov's dogs running ahead into his testing room introduces us to another type of association which animals can make. This is not to link reinforcement with a novel stimulus, but with some action they themselves have just performed. Thorndike, one of the pioneers of American experimental psychology, investigated this type of learning with cats, using a series of 'problem boxes' of the type illustrated in Fig. 5.6. For example, we may have a box with a spring door which can only be opened from the inside by depressing a lever. A cat is shut in and tries hard to escape; it moves around restlessly and after a time – by chance – it steps on the lever and the door opens. The second trial may be a repetition of the first and also the third, but soon the cat concentrates more attention on the lever and eventually it moves swiftly across the box and presses the lever as soon as it is confined. Thorndike gave the descriptive name 'trial and error' to this type of learning. The cat learns to eliminate behaviour which led to no reward and increases the frequency of behaviour which is rewarded, but in the early stages there is little system to its activity – the first reward is obtained by pure chance.

The famous 'Skinner Box' (Fig. 4.8), named after B.F. Skinner, who used it extensively for the study of learning, is basically a problem box of a convenient form in which an animal learns by trial and error that pressing a bar or pecking a key yields a small reward. Because the animal's own 'spontaneously generated' behaviour is instrumental in its gaining a reward, such learning has been called instrumental conditioning (Skinner also used the term operant conditioning), but it is no different in principle from trial and error.

Skinner, whose book (1938) was grandly entitled *The Behavior of Organisms*, has been a very influential figure in learning theory and because his school has concentrated largely on instrumental conditioning whilst others have worked on classical conditioning, there has been a tendency to regard the two situations as leading to two rather distinct types of learning. At first sight they do seem clearly distinguishable. With classical conditioning the animal associates a novel stimulus with a response (the UCR) which was there from the outset. With instrumental conditioning or trial and error it is a novel response, not a novel stimulus, which is learnt.

However, close examination of how the animal behaves in the two situations shows that this distinction may be more apparent than real, at least in some cases. The clearest examples come from pigeons, familiar subjects for instrumental conditioning studies because they can readily learn to peck at a key on the wall of a Skinner box to obtain a food or water reward. If, as is commonly the case, you simply record key presses automatically, then it looks like very conventional trial and error – the pigeon learns this new response and can use it to obtain food or water, depending on circumstances. However, there is real value in bringing an ethological approach into the Skinner box. A pigeon key-pecking for food reward does not behave in exactly the same way as when pecking for water. Close observation of the head and bill reveals that when hungry and pecking for a food reward the pigeons' eyes are partly closed and its bill open (Fig. 5.7a); when pecking for water its eyes are fully open and the bill almost closed (Fig. 5.7b). These are precisely the contrasting features which distinguish a pigeon pecking to grasp food grains when feeding from one which dips its bill into water to drink (pigeons, unlike most birds, suck up water like mammals). In other words, the pigeon which pecks instrumentally is treating the key 'as if' it were food itself in one case and water in the other (Moore, 1973).

Pavlov regarded the conditioned stimulus as coming to replace the unconditioned one, i.e. the dog treats the metronome sound 'as if' it were food. It seems that pigeons in the Skinner box respond in a similar way and such a conclusion certainly forms a bridge between classical and instrumental conditioning and forces us to look at the learning situation more from the animal's point of view. This is not to say that any distinction between classical and instrumental conditioning is meaningless, but we have to recognize that animals will come to any learning situation with some built-in predispositions which will affect the way they respond.

Once it was widely believed that, given the right opportunity, animals could learn to associate any stimulus they could perceive or any response which they could perform

(a)   (b)

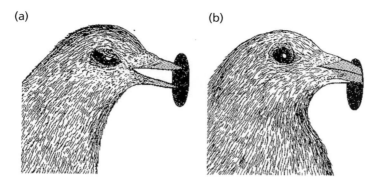

Figure 5.7   Trained pigeons in a Skinner box 'pecking' a key to obtain a reward. In (a) the reward is food and the bird's head moves sharply towards the key. As it reaches it, the bill is opened as if to seize a food item and, in a characteristic protective reflex, the eyes are almost closed. In (b) the bird is 'pecking' for water reward and the bill, almost closed, is pressed against the key more slowly whilst the eyes remain fully open.

with any reinforcement – it was the association which counted. In practice, this assumption has proved quite inadequate. In Chapter 2, we have already discussed some evidence that animals may have inherent biases to learn particular things. In conventional learning situations, we usually find that they most readily learn responses which form part of their natural behaviour to the reinforcement. Cats will quickly learn to lick or bite at a lever to get food, but it is very difficult to train them to turn a treadle wheel with their paws for the same reward. Exactly the converse happens if the reinforcement is escape from electric shock. Now the turning of a treadle is easily associated with escape, but licking a lever cannot be. We can understand these biases in terms of the normal way the cat responds to food or runs away from unpleasant situations. Essays in the volume edited by Hinde and Stevenson-Hinde (1973) review other examples of what they refer to as 'constraints on learning'.

Problems with species-specific biases are well illustrated by avoidance conditioning. One familiar form of laboratory avoidance conditioning is the 'shuttle-box'. A rat is put into a box whose floor is a grid which can be electrified to deliver a mild shock to its feet. The box is divided into two halves by a partition with a gap for easy passage between them. Each half can be electrified separately. On a signal (usually a buzzer or light) the rat has a few seconds to run through into the other half to avoid electric shock. This is not an easy task for rats because during the early trials they are being asked to take refuge in another part of the box where they have also received a shock just previously. It takes them some time to accept this as a refuge, but at least they have the right natural bias – they run in response to shock. Bolles (1970) discusses the varying fortunes of psychologists who have used other species in a shuttle-box situation. For many animals it is a nearly impossible task because their natural response to foot shock as to any alarming stimulus is to crouch or freeze. It is no use expecting a hedgehog to learn to run away from approaching danger – hence the carnage on our roads.

Sometimes the biases animals show are so strong that when placed in unnatural learning situations they will perform patterns which actually delay their getting reward. Breland and Breland (1961), pupils of Skinner, tried to put their techniques to good use by training animals to perform various eye-catching tricks for TV commercials. They knew that it should be possible to train hungry pigs to drop money into piggy banks or get chickens to ring bells, provided these actions were associated with food. It worked – up to a point – but they could not eliminate 'undesirable' side actions. The pigs would root with their snouts on the way to the bank, the chickens scratched and pecked at the ground. Spending time on this behaviour was certainly inefficient in operant conditioning terms. It was never reinforced and it delayed the arrival of food. The Brelands, recognizing the instinctive behaviour patterns which chickens and pigs use in feeding, came to accept that animals do not always simply associate response and reinforcement. They mischievously entitled their paper 'The Misbehavior of Organisms' (note Skinner's title); Seligman and Hager (1972) give many other such examples.

The fact that animals do not always learn to respond so as to minimize the delay between an action and its reinforcement was most surprising to conventional learning theorists. Contiguity – the close association in time between stimulus or response and reinforcement – was considered essential. Figure 5.8 shows the familiar picture from Pavlov's work: CS and UCS must be close or overlap if learning is to occur. If the CS ends too early, so that there is a delay before the UCS signals the reinforcement, then no association is formed. Skinner found, in general, that delays of more than about 8 s between a response like bar-pressing and its reinforcement greatly slowed learning. In practical terms the deleterious effects of delayed reward can often be overcome by introducing a **secondary reinforcement**. Suppose a rat learns that a reward is delivered when a light comes on in the Skinner box (light and reward must overlap as discussed above), then it will learn to press the bar to switch the light on. Light becomes a secondary reinforcement or a 'bridging stimulus' between the response and the primary reinforcement – food. Bridging stimuli are useful for training animals, as in a circus when it is often difficult to reward immediately after the response is made.

However, just as there were exceptions to the rule about it being possible to link any response to any reward, so we find that immediate contiguity is certainly not an invariable requirement for learning. There are certain types of associative learning which consistently occur when reinforcement is delayed for a matter of hours. Barnett (1963) describes how wild rats only nibble at small amounts of any novel foods that appear in their territory. If it proves edible, they will gradually take more on successive nights until they are eating normally. If it is poisonous, yet they survive, they avoid it completely on subsequent occasions. This type of behaviour is highly adaptive and makes poisoning rats no straightforward task. The interesting feature for our discussion is the delay that must ensue between a rat tasting poison bait (always

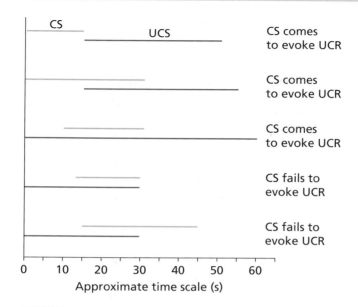

Figure 5.8 Hundreds of experiments with different CSs and UCSs using Pavlov's basic set-up revealed the general relationships of timing between the two shown here. In each case the upper orange line denotes the duration of the conditioned stimulus (CS) and the lower, blue line that of the uncontrolled stimulus (UCS). The results are given on the right; note that the CS must not end with or persist beyond the UCS if a positive conditioned reflex is to be established. If it does so, then the CS will remain neutral and may even tend to inhibit the response to the UCS.

made superficially palatable with sweet substances) and any subsequent ill effects. Few rat poisons take less than an hour to produce effects. Laboratory findings have confirmed this ability. Not only will rats learn to avoid tastes associated with sickness that sets in at least an hour later; if deprived of the vitamin thiamine they will learn to choose a diet containing it, although many hours must elapse before they can feel its benefits. Rozin and Kalat (1971) provide an interesting review in which they link the specializations of learning which are involved in such 'specific hungers' with the rat's ability to avoid poison baits (see also Rozin, 1976).

It is only when a new taste acts as the CS that reinforcement can be so delayed. In one ingenious experiment, Garcia and Koelling (1966) supplied rats with a drinking tube containing saccharin-flavoured water so arranged that when they licked the tube, bright lights flashed on. After these sessions the animals were irradiated with X-rays or, in other experiments, given an injection of lithium chloride; both treatments made them sick about an hour later. Subsequently the rats avoided saccharin taste but did not avoid flashing lights. Conversely, if they were given flashing lights plus immediate electric shock to the feet whenever they licked the tube, they subsequently avoided the light but still licked at saccharin. Rats are in some way 'prepared' to associate taste with sickness after a single trial and a long delay, but visual stimuli and sickness are not so connected. Conversely, light and electric shock are easily associated if they occur close together, but taste and shock are not. Such a result suggests that the mechanisms of association have been modified adaptively and argue against those models of learning which have sought universal principles of association between CS and UCR. However, it has been argued in defence of such models that because saccharin and nausea both involve the same sense and that the taste of saccharin may, in some way, linger this links CS and UCR in some way and general learning principles may still hold (see MacPhail and Bolhuis, 2001). However the special adaptation view is strengthened by the finding that in birds, which seek food largely by sight, visual stimuli and sickness can be associated. If delayed sickness is induced with lithium, they associate the *appearance* of novel food with this effect and avoid it later (Martin and Lett, 1985). In the wild, such an association means that they rapidly learn to avoid poisonous caterpillars which are often strikingly coloured such as those of the cinnabar moth shown in Fig. 5.5.

Hitherto, although its timing clearly can vary with circumstances, reinforcement itself seems central to the learning process. Certainly, in the situations covered by standard Pavlovian conditioning or trial and error, there can be little doubt that it is a crucial factor. If Pavlov ceased the supply of meat powder, his dogs rapidly ceased salivating to the stimuli he was presenting, a process he called experimental extinction. Extinction also follows in the Skinner box situation but, for reasons which are not clear, it usually takes much longer than with classical conditioning. Once a rat has learnt to press the bar for food, the proportion of reward to presses can be reduced to as low as 1 in 100 in some cases, and the rat will go on pressing. If rewards are stopped altogether it is a long time before the response finally extinguishes.

Pavlov realized that an extinguished CR did not just disappear and leave the animal as it was before conditioning started. In the first place, if we simply leave the animal alone for a few hours and then give it the CS again, the CR returns, i.e. it shows spontaneous recovery. This recovery is not back to the original level and the response extinguishes more rapidly, but this process of a pause followed by spontaneous recovery can be repeated several times.

A second way of reviving an extinguished response is to give a novel stimulus along with the CS. A dog which has had its conditioned salivary response to a bell extinguished salivates again if a light flashes as the bell is sounded. Similar results have been obtained with rats in Skinner boxes. Pavlov called this process disinhibition because he regarded extinction as another new learning process which inhibited the original CR. Neutral stimuli presented with the CS early in the original acquisition of the CR often 'inhibit' it temporarily and reduce its strength. Similarly, perhaps the neutral stimulus disinhibits an extinguished CR by inhibiting the new learning that takes place during extinction.

In the heyday of rat and pigeon experimental psychology, reinforcement was a central concept and was often equated with 'drive-reduction'. Animals were said to have needs for food, water, etc. and would learn tasks which reduced these needs or drives. Without drive reduction learning would not take place. This immediately raised the question of how many drives there are and how distinct they are from each other – questions which have already been touched on in Chapter 4. Reduction of 'anxiety' or some similar concept was needed to explain avoidance conditioning (see p. 253), because once a rat has learnt the shuttle-box problem it gets no more conventional reinforcement (i.e. foot shock).

The drive-reduction hypothesis also runs into trouble with what we might call exploratory learning; in Thorpe (1963) and in Munn (1950) it is called latent learning. Rats given an opportunity to explore a maze at will, without any inducement from hunger or any other drive, turn out to learn it quicker than other naïve rats when both groups are later running the maze for food. No biologist is surprised by this result – animals learn the vital features of their home ground in this way. Initially, however, it caused great controversy amongst the adherents of drive-reduction, for what drive is reduced during exploration? We could equate it with anxiety if we wished, but there seems little point to the exercise. Monkeys will learn to press a bar in order to move a shutter aside, giving them a few seconds' view into another room with a toy train set in action. Reduction of boredom seems the best hypothesis! It is better to accept that conventional reinforcement is not the only route to learning and recognize the adaptiveness of the result and the biases – sometimes highly specialized ones – which may have become built into the learning mechanisms, for learning ability is an aspect of an animal's biology which has been subject to natural selection like any other.

Specialized types of learning ability

When we look beyond the familiar rat and pigeon, prime subjects for laboratory studies of learning, we must expect to find animals whose learning systems are specially adapted to match their very different life histories. We will consider two very different examples.

Honeybees

Learning is particularly crucial to worker bees during the second half of their brief 6 weeks of life. This is the time when they act as foragers, making regular excursions outside the hive to collect nectar and pollen from flowers. von Frisch (1967) and Lindauer (1976) give good accounts of the role learning plays in the life of bees and the types of features in their environment which they learn. Menzel and Erber (1978) and Menzel *et al.* (1993, 2001) review experiments on the learning process and its underlying mechanisms. Honeybees learn rapidly and retain remarkably well the effects of even a single association between a colour and food reward. Accordingly they make excellent subjects for any teacher wanting to demonstrate animal learning to students and they can easily be trained to visit a dish of sugar solution on a table.

In one set of experiments, Menzel and Erber (1978) used a table bearing a food dish which was lit from below with light of variable wavelength, forming in effect an artificial flower whose colour could be switched at will (Fig. 5.9). Since it takes a bee a minute or two to fill its crop with sugar solution, they could easily arrange to change the colour of a 'flower' during a visit. Using naïve bees as subjects, they changed flower colour between blue and yellow at various stages and then gave the bees a choice of these two colours at their next visit to see which colour they now associated with the sugar reward. The results, shown in Fig. 5.9, were quite dramatic.

The association period is very brief, between 4 and 5 s only, and colour changes made outside this period are ignored. The timing of the period is also very precise and we can best measure it from the exact moment at which the visiting bee's proboscis touches the sugar solution, i.e. the moment when reinforcement begins. Then the colour stimulus must be present within 3 s before and 0.5 s after the start of sucking if an association is to be made. Sucking has to continue for 2 s for the bee to retain the memory, but the colour signal can be changed 0.5 s after sucking begins – it will still be associated with the reward and the bee will fly to this colour on the next visit.

So if the flower is, say, yellow as the bee approaches, switches to blue 3 s before it alights and begins to suck, switches back to yellow 0.5 s after sucking starts and remains yellow for all the time it is filling its crop – perhaps 1 minute in all – and is still yellow as it leaves, the bee subsequently associates blue with reward. This is an extraordinary result if we think in terms of conventional vertebrate learning. There we would assume that the enduring association would be with the colour most recently and most persistently associated with the reward (note, for example, the response of rats given vinegar after saccharin flavour preceding induced sickness, which is described on p. 263). However, for the specialized honeybee learning system the brief flash of blue at the beginning is the significant feature, this degree of contiguity fixes the association. By confining a worker bee to the hive after such a trial, so that it was prevented from learning any other colour associations, Menzel and his co-workers could demonstrate that a single experience of this type was retained for several days. Three such trials on the one colour and the association was retained for all the bee's remaining life (Fig. 5.10). The same kind of time scales for both learning and retention hold true for conditioning to scents as well as to colours, (Menzel et al., 1993, 2001).

In nature, honeybees use both colour and scent in guiding their responses to flowers and learn both characteristics. Colour is effective during the early approach, but once they are close, the flower's scent is a powerful stimulus for bees to alight and probe. We must remember that honeybee foragers are often directed to a flower crop by the dances of their hive mates (see Chapter 3, p. 164) and they also learn the position of each crop in relation to the hive. Most flowers do not secrete nectar uniformly throughout the day and bees are able to match their foraging activities to match the secretory cycle of flower crops and so increase their efficiency. Like many animals which use the sun or stars for

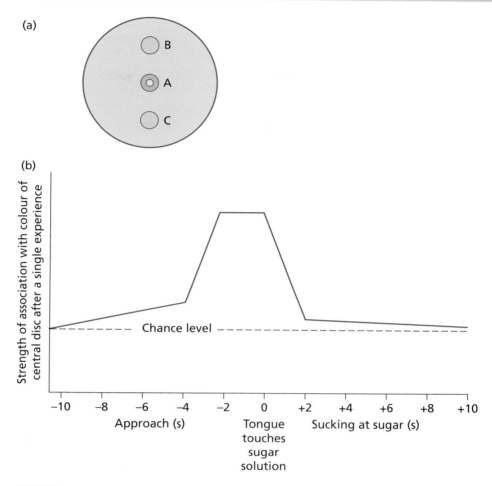

Figure 5.9 Menzel and Erber's experiments on the timing of the association of colours in honeybees. (a) A plan view of a table with artificial 'flowers' provided in Menzel and Erber's experiments on learning in honeybees. Marked bees feed at the table whose surface carries three ground glass discs which can be illuminated from beneath with either pure blue or yellow light. In the training situation, only disc A is illuminated and a bee feeds on sugar solution from a tube which opens at its centre. The colour of A, blue or yellow, can be changed instantly at will by shifting interference filters beneath the table. When the bee next returns from the hive, disc A is covered and discs B and C are now illuminated, one blue and one yellow; again, colours can be changed at will. It is the bee's choice between these two which indicates what colour association it has made whilst feeding on A. (b) A strong association between colour and reward is dependent on exact timing. As the bee approaches the disc its colour has no significance until about 4 s before the bee begins to suck sugar. Then there is a sharp peak of strong association, which has disappeared within 2 s following the onset of sucking. Apart from this brief period the colour the bee sees during the approach or for the whole time it takes to fill its crop is of no importance. Further explanation in text.

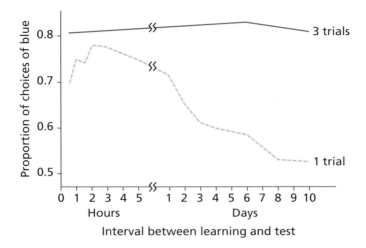

Figure 5.10 The time course of a bee's memory after training to associate colour as illustrated in Fig. 5.9. The results are expressed as choice of blue over yellow after one or three trials with disc A coloured blue. (Prior to any training blue and yellow are chosen equally.) One trial alone strongly biases choice for a day or so, three trials fixes it indefinitely.

navigation, bees have an inherent sense of time against which they regulate their behaviour through the daily cycle. Koltermann (1971; see also Bogdany, 1978) showed that it is possible to train bees to associate a particular scent at a particular time of day with food reward. More, they could learn up to eight distinct pairings of a specific scent with a specific time of day and switch their choices appropriately as the day progressed. Thus, having been trained to associate orange blossom but not lavender with food between 1000 and 1100 h, and then the converse between 1100 and 1200 h, bees change their choice at the appropriate time. Such beautifully adapted behavioural responsiveness enables bees to be extremely efficient in deploying the huge team of foragers to the best effect and thus extracting the maximum amount of food from their environment. In Chapter 3, p. 172, we have already discussed the evidence of how effective honeybee foraging behaviour is.

Food-storing birds

Among the most striking learning specializations of vertebrates is that associated with food storage. When the chance arises, a number of bird and mammal species collect more food than they can eat immediately and hide it. In this way, they can take advantage of a rich food source while it is there and reduce the degree to which they have to share it with others. Some animals make a single larder to which they return regularly, but most interesting from our point of view are those species which hide food items singly or in caches dispersed around their territory (see Sherry, 1985, for more details).

Figure 5.11   A European nutcracker caches a pine seed at the foot of a tree. It has dug a small hole to receive a few food items (2-12 usually) which it will then cover up – It carries a number of seeds in an enlarged pouch (the subgular pouch) opening into its mouth cavity.

For example, in Europe a small woodland bird, the marsh tit (*Parus palustris*), will store several hundred seeds a day in winter, secreting them in tiny crevices under tree bark or stones, etc. Careful observation in the field and experiments in aviaries have shown that the birds, though not perfectly recovering each seed, can recall with extraordinary precision dozens of dispersed hiding places for 1 or 2 days afterwards (Cowie *et al.*, 1981; Shettleworth, 1983). An American relative, the black-capped chicka-dee (*P. atricapillus*) can recall hiding places for at least 4 weeks (Hitchcock and Sherry, 1990). However, perhaps the most extreme specialists in food-hoarding and retrieval are those members of the crow family called nutcrackers, of the genus *Nucifraga*.

Clark's nutcracker (*N. columbiana*) lives in the high coniferous forests of the western North American mountains. For winter food it depends almost entirely on caches of pine seeds. During late summer and autumn there is a huge glut of food and the nutcracker hides tens of thousands of seeds in several thousand separate caches in the soil, mostly on south-facing slopes. Each cache is made by forming a small hole with its bill and depositing a few seeds which it collects in quantity and can store in a large, distensible pouch under its tongue. The cache is then covered with soil and sometimes a stone is moved over to cover it. Figure 5.11 illustrates this behaviour in the European nutcracker (*N. caryocatactes*).

Working with Clark's nutcracker, Balda and Kamil (1992) had few problems in getting some captive birds to form caches in a large room with suitable holes drilled in the floor. The exact sites of each bird's caches were recorded and, once made, they were excluded from the room for various periods. Even after a lapse of 40 weeks or more, the birds sought out the correct locations far better than by chance. It seems clear that the birds have the extraordin-ary ability to fix the position of their caches with respect to landmarks around, such as trees or rocks. They were put off if equivalent features in the laboratory room were shifted.

The abilities of the titmice and the nutcrackers require a very remarkable development of spatial learning and memory. It is the more striking because close relatives of such food-storing species – other tits such as the great tit (*P. major*) or other crows such as the North American crow (*Corvus brachyrhynchos*) or the European jackdaw (*C. monedula*) – store little or not at all and, it has been claimed, perform less well in spatial learning tasks generally. Evidence is now accumulating that the brains of food-storers differ significantly from those of their non-storing relatives. The former have relatively more neurons in the hippocampus, an area which is known to be involved with memory formation and most particularly with spatial memory. Of course, the number of neurons alone cannot be the only significant factor, one would like to know if there are differences in their interconnectivity also, but this is much more difficult to assess at the moment. However, what has added particular significance to this observation on the brain structure of hoarding birds is that similar specializations have been found in some other contexts. The hippocampus is larger in males than in females in some rodents where the males hold large territories whose spatial characteristics they learn in great detail. Conversely, the hippocampus of female cowbirds is larger than that of their males. Cowbirds are nest parasites like cuckoos and the females, but not the males, have to learn the detailed positions of several dispersed nests of their hosts. Thus it has been suggested that here we are seeing a kind of 'neuro-ecology', where brain development and function are adapted to the special needs of an animal's life history. Comparisons of this type which try to relate brain proportions to behaviour in different species are not problem-free, nor is it always easy to assess the performance of storing versus non-storing species because it is difficult to be sure that we are comparing like with like. Not everyone is convinced that we need to explain behavioural specializations by equivalent brain specializations and there are lively reviews of the arguments from both sides by Bolhuis and Macphail (2001), Macphail and Bolhuis (2001), Shettleworth (2003) and Healy *et al.* (2005).

The learning abilities of bees and nutcrackers are just two examples of a very active field of research which shows how comparative behavioural studies can contribute to an understanding of the brain and of memory mechanisms generally (see Shettleworth, 2009, for a useful review). Knowledge is very incomplete, but if we survey the animal kingdom we can expect to find more specialized learning systems. Natural selection will shape associative learning into diverse forms to match lifestyles, but it is none the less valuable to look for common features of principle. There is, at a behavioural level, a real equivalence between classical conditioning in a honeybee, an octopus, a dog and a chimpanzee – the same logical processes are required.

What do animals actually learn?

Approaching from rather different viewpoints, there have been some valuable interactions between experimental psychologists and ethologists examining what logical processes are involved when animals learn. The rigorous exploration of classical conditioning in laboratory situations is complemented by field studies on animals as diverse as the honeybees and nutcrackers we have just discussed. Rescorla (1988), Mackintosh (1994), Pearce (1996) and Bitterman (2006) review the field with many examples. As we have seen, we can reject any idea that what animals achieve when learning is a simple, almost idealized association between an event and a reinforcement which occur in some regular sequence. It is more profitable to start from the notion, put forward early in this chapter (p. 251), that what animals must do is learn which events best predict important outcomes. We must also remember that animals will come to all new situations with some inherent biases to pick on certain events as signals and to respond in particular ways.

Thus in the taste aversion situation, described on p. 255, rats learn to avoid saccharin whose flavour has been followed by sickness after about an hour. If, between tasting saccharin and being given the injection of lithium which induces sickness, the rats are given vinegar-flavoured water to drink, they subsequently avoid vinegar but not saccharin. The physical and temporal association between saccharin and sickness remains the same but the rats – sensibly enough, we may feel – choose to pick on the taste of vinegar as the key event. Because it occurred more recently, closer to the reinforcement, it is taken as the better predictor.

A rather more complicated example shows another side of the same process. Two groups of rats were given experience of electric shocks in a Skinner box where they were obtaining food by bar-pressing; not surprisingly, when the shock arrived, bar-pressing was depressed. In one group – we may call it group A – the shocks were paired with a distinctive light stimulus; in the other group, B, there was no light. Now both groups were given the light plus a sound tone stimulus together with the shock and this combination of light plus tone was certainly effective. The rats signalled that they had learnt, so to speak, because when light plus tone was given, now without any shock, bar-pressing showed a sharp decline – they anticipated a shock. But had the two groups learnt the same thing? The test came when they were given just the tone by itself. Now group B showed just as strong an effect on bar-pressing as with light plus tone. Group A did not: these rats paid little attention to the tone without the accompanying light. A's previous experience had told them that light was the best predictor of shock, so although the contiguity of tone and reinforcement were the same for both groups, they interpreted the situation differently. Experimental psychologists say that the rats' reaction to the light 'blocks' their reaction to the tone, and there have been a number of studies investigating how blocking works (see Pearce, 1996; Bitterman, 2006).

Now it may be said that such an effect is just what we would expect. We probably mean by this that the rats are behaving much as we would. But this is really the key point; the rats are not behaving like associating automata, they are behaving in a more complicated way, we might even say intelligently. This leads us directly to the following question.

Are there higher forms of learning in animals?

With this question we begin to approach those issues we referred to at the outset of this chapter, to an area which is often labelled, 'animal cognition', a loosely defined concept embracing a whole series of interrelated questions. We can list some of them here. Can animals solve new problems rapidly without prior experience? Can animals plan ahead and anticipate the outcome of their actions? Do animals recognize themselves as separate entities, understanding that other individuals may respond in different ways, having their own and distinct point of view? Do animals have thought processes along the same lines as ourselves, albeit less elaborate? All of which may incorporate a crucial speculation – are animals conscious? Most of these questions still arouse controversy and go far beyond just questions about learning abilities. However in the history of animal experimental psychology some of them first emerged when research workers began to investigate what was often called 'insight learning' and it is useful to begin there. We might regard this as an exploration of animal intellect – perhaps their 'cleverness' – as a preliminary to considering what any results may indicate about the underlying mind that controls the behaviour.

Thorpe's (1963) book includes an examination of insight learning, taking it as a separate category of learning which he defines as, 'the sudden production of a new adaptive response not arrived at by trial behaviour or the solution of a problem by the sudden adaptive reorganization of experience'. We can readily see the drawbacks of such a definition (e.g. how do we know a response wasn't arrived at by trial behaviour, etc.), which is not to say that, as humans, we do not recognize the phenomenon. Everyone can recall occasions when the solution to a problem has 'come in a flash', perhaps as the climax to several minutes of concentrated thinking. It is obviously going to be very difficult to demonstrate conclusively that there are similar processes going on in animals.

All we have to go on is what animals actually do in learning situations, but this is no reason to despair. As Dickinson (1980) discusses very well in his book, mental processes *can* be inferred from behaviour, even if it is difficult to do so. In practice, animal workers have used the term 'insight' when they observe animals solving problems very rapidly, too rapidly for normal trial and error – at least, too rapidly for the animal to carry out actual trials, but there is the possibility that it is 'thinking' about them and trying them out in its brain. This would imply that the animal can form ideas and reason, and this is why, as we have just suggested, studies on so-called insight learning are not very far from a consideration of animal reasoning and ability to plan ahead. These phenomena do not often lend themselves easily to a quantitative experimental approach but this does not diminish their importance.

Munn's 1950 *Handbook* describes a range of the early experiments with rats, most of them trying to test whether they could make detours to reach a goal or take short cuts in mazes when these were suddenly opened up for them. We may probably feel that the rat is not the most obvious animal in which to search for higher mental capacities but we must remember that it was the main and very amenable experimental subject at the time. Such 'insight' studies often have three features in common. Firstly, they include a longish period of exploration of a maze prior to any testing, so that the animals will already have learnt a good deal about the general features of the situation. Secondly, performance is judged by the speed of solution. If a rat makes mistakes or has to explore further it is assumed that a simpler, trial-and-error form of learning is involved. Thirdly, it is an interesting fact that rats and other animals which have been similarly tested are highly variable in their apparent 'insightfulness'; it is often only 20–30% that respond rapidly and we do not know what factors of prior experience or genetic constitution may contribute to such differences.

As mentioned above, diverse phenomena have been lumped together when considering insightfulness or reasoning in animals, and we must not confuse capability per se with intellectual ability. For example, a common component of human intelligence tests measures the ability to recognize complex shapes when they are rotated from their original position. However, as we mentioned in Chapter 3 (p. 139), pigeons have little difficulty with rotated shapes. Hollard and Delius (1982) showed that when tested in a Skinner box, pigeons turn out to be much quicker and more effective than we are (Fig. 5.12). Our anthropocentric attitude tends to equate human abilities with intelligence, but this will mislead us here. Pigeons, evolved for flight and intensely visual animals, must have a sensory system adapted to the recognition of the constancy of a feature whose retinal image changes. The shapes on the landscape below them will apparently rotate as they wheel above them in flight. Delius *et al.* (2001) describe a number of studies with pigeons which reveal the extraordinary facility with which they can make visual discriminations. Birds live in a very complex three-dimensional space where rapid changes are the rule. They have evolved sensory processing systems to match but, if we are to progress in a search for intellectual capacities in animals, it is probably better to get away from tests which tend to accentuate specialized sensory processing.

If we are to acknowledge real insight in animals, it must come from the identification of truly novel behavioural associations, such as we referred to in Thorpe's original definition. It is not surprising that many of the best examples come from primates. One of the first studies to become generally famous was that of Wolfgang Köhler with a captive group of chimpanzees, described in his book, *The Mentality of Apes* (1927). Fond of bananas, they would learn to overcome problems of increasing difficulty to get at them. If presented with a bunch too high to reach, they would pile up boxes to make a stand for themselves or fit two sticks together in order to pull down the bananas (Fig. 5.13). Often they arrived at this solution quite suddenly, although they benefited from previous experience of playing with boxes and sticks and showed considerable and obvious trial and error when actually building a stable pile of boxes. Köhler's chimpanzees were using knowledge obtained in one context (something of the

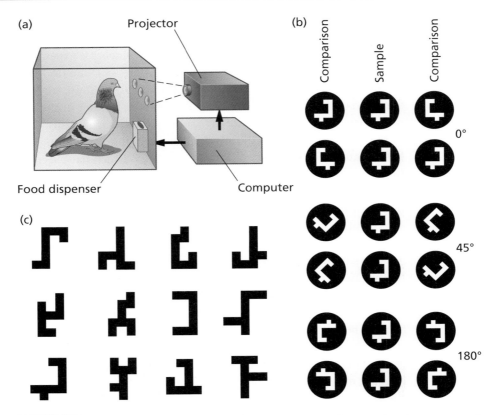

Figure 5.12 Test to measure the ability of pigeons to recognize complex shapes even when rotated from the familiar. (a) The experimental procedure, a Skinner box with three keys. Trained to respond to a shape presented on the central key, the pigeon is now given three shapes on the three keys – see the rows in (b) – with a reward for pecks to the side key whose shape is identical to the sample. When the wrong shape is a straight mirror image of the sample, as in the top two rows of (b), the task is quite easy. It becomes (for us) much more difficult when the comparison has to be made with shapes rotated from the sample orientation. (c) Some of the shapes with which pigeons were tested in this way. They are remarkably quick and efficient at such discriminations.

properties of sticks and boxes) and applying it in another. Certainly, many people would have no doubt that that apes and some other primates can show true reasoning on occasions. Many dog and cat owners will cite examples of reasoning in their pets; this is possible, but we must take care to exclude other explanations.

Before we try to extend our discussion to consider the nature and the implications of reasoning ability in animals, it is useful to reflect on why we should assume that primates will be more able to reason than dogs, and dogs more than rats. We are tacitly assuming that learning ability and 'mentality' have, like brain size, shown trends during the evolutionary history of the vertebrates.

(a)

(b)

Figure 5.13 Wolfgang Köhler's original set-up with a captive group of chimpanzees on the island of Tenerife where he was effectively marooned throughout World War I. In his classic account, *The Mentality of Apes* (1927), he describes what has been called 'insight learning'. In (a) they pile up boxes to reach bananas hanging from the roof of their cage. In (b) one slots together two rods to rake in food from outside.

The comparative study of learning

Comparative psychology has a long history (Dewsbury, 2003; Burghardt, 2005) and, at least in the early stages, researchers were working with a wide range of animals even if later the rat and the pigeon came to dominate. Psychologists have always tended to concentrate upon studies of learning and recently those with an ethological background have also come to study the diversity of learning abilities, so we have a reasonable body of data to search for evidence of evolutionary changes.

There are all kinds of inherent difficulties in making such a survey. The first and most obvious is our essential vanity concerning human intellectual powers, which often leads us to look for progression upwards amongst animals, leading towards human beings on a very detached, high pinnacle. Any biologist knows there are other pinnacles on alternative pathways. Honeybees are not to be judged by the same criteria as monkeys. Nor is it sufficient to rely on laboratory studies of learning alone; we must try to get some understanding of the role that experience plays in the natural life of different animals and also attempt to distinguish between simple and more advanced types of learning which, as we have seen, is not always easy.

One approach has been to examine the correlation between brain size and learning ability. Certainly the capacity to benefit from experience, to process information from diverse sources and to be flexible in response is greater in mammals and birds than in fish and reptiles and we can reasonably associate this difference with the evolution of a large brain – encephalization, as it is termed (Jerison, 1985, 2001). It is not, of course, just brain size that is important; whales and elephants have larger brains than ourselves but smaller, though considerable, cognitive powers. Brain size as a proportion of body size is more realistic and primates do rate highly here. It is less easy to construct any kind of series among the invertebrates, which are much more heterogeneous, but advanced insects and cephalopods have the largest brains of their respective phyla, Arthropoda and Mollusca, and also the greatest capacity for learning. However, as we mentioned in Chapter 2, the two most advanced groups of insects, the Diptera and the Hymenoptera, both have brains enlarged and much modified from the ancestral type, but they are specialized in different ways and differ markedly in learning ability.

Within vertebrates the most dramatic aspect of brain evolution is the enlargement of the cerebral hemispheres, especially the cerebral cortex, which reaches its greatest extent in the primates. However, we can reject any simple equation of cerebral cortex size with learning ability. Birds have in the past often been underestimated because, although their brains are relatively large, those parts homologous with the mammalian cerebral cortex are small. But the birds have evolved along a line separate from the mammals for some 250 million years. They have evolved a different type of brain structure, yet learning ability in some bird families, notably the crows and the parrots, is in some respects second only to that of the primates.

It would be a valuable and interesting exercise to study how different animals perform on similar learning tests and there have been many attempts, but they are fraught with difficulty. How can we devise tests which are equivalent for diverse animals who inhabit such different worlds of sensory capacities and manipulative ability? The procedures needed to measure discriminative conditioning in an octopus, a honeybee and a rat have to be very different and we can no longer be sure that problems are of equal difficulty or that the animals 'see' them in the same way. Motivation and reinforcement present further problems: the level of motivation often affects the rate of learning and may indeed determine whether the animal learns at all. How can we equate levels of hunger motivation in a rat and a fish? The latter may live for weeks without food, the former only days. A small piece of food may be an excellent reinforcement for a hungry mammal, but mean much less to a fish and less still to an earthworm. It is perhaps easier to equate punishment, since all animals 'dislike' electric shocks, though even here there are difficulties because shock or fear affects the behaviour of animals in such diverse ways. As we discussed earlier (p. 253), animals come to learning situations with a good deal of built-in bias. To equate the effects of punishment we require some knowledge of each animal's natural responses in fearful situations. Whenever possible we must try to use a reinforcement which is directly relevant to the animal being tested. Escape into a darkened area may be best for many small invertebrates. Vowles (1965) found that the best reinforcement for maze learning by ants was to get back to their nest.

In the face of all these difficulties it is not surprising that some question the validity of any comparisons of intellectual ability between different animals. MacPhail (1985, 1987), for example, probably with his tongue in his cheek, takes up a position which we might call constructive provocation. He argues that, 'there are, in fact, neither quantitative nor qualitative differences among the intellects of non-human vertebrates'. Few would take this literally, but recourse to common sense should not be our only reply: we must try to devise valid comparative tests and particularly not jump to conclusions about 'scales of intelligence' when we are not really comparing like with like (see Hodos and Campbell, 1969).

The concept of 'learning sets' developed by Harlow (1949) perhaps offers more chance of valid comparisons between animals. A learning set helps us to discover whether an animal can 'learn to learn'. It needs the ability not just to learn an individual problem, but also to grasp something of the principle behind it. There will be an increase in learning speed when given a series of similar problems. Harlow described this basic technique using primates. A monkey is presented with a pair of dissimilar objects – a matchbox and an egg-cup, for example. The matchbox, no matter where it is placed, always covers a small food reward, the egg-cup never has a reward. After a number of trials, the monkey picks up the matchbox straight away. Now the objects are changed: a child's building block is rewarded, a half tennis-ball unrewarded. The monkey takes about the same time to learn this, again the objects are changed, and so on. After some dozens of such discrimination tests the monkey learns each much more

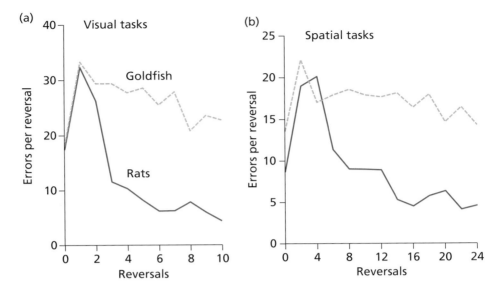

Figure 5.14 Repeated reversals for learning visual and spatial discrimination by goldfish and rats. The graph plots the number of errors to achieve the criterion of all responses to the rewarded object. The point 0 represents the original task; note how scores both for it and for the reversal (always the most difficult) are very comparable for both animals. Thereafter, with repeated reversals the rat improves much more rapidly than the goldfish.

rapidly, although viewed as an individual problem it is just as difficult as the first one. Eventually after 100 or so tests the monkey presented with any pair of objects lifts one and, if it yields a reward, chooses it for all subsequent trials. If not, it switches to the other and thereafter stays with that. It has learnt the principle of the problem or, in Harlow's terminology, it has formed a learning set.

This is one type of learning set, based on successive trials of discrimination. Perhaps a simpler version is the 'repeated reversal' problem. Here we train the animal to select object A in preference to object B. Once learnt, object B is now rewarded and A unrewarded; when this first reversal is learnt, the reward is again given with A, and so on. If the animal gets progressively quicker at learning each reversal, this again implies that it has learnt a principle. The great advantage of this system is that the details of the training and the nature of the problems can be matched to suit the animals being tested. Sometimes we can be reasonably confident that we are making a fair comparison, even if the subjects are very distant in evolutionary terms. Figure 5.14 compares rats and goldfish solving two of the easier reversal learning set situations, one a spatial task, left versus right, the other a visual brightness discrimination. Both learn the original discrimination and its first reversal – always the most difficult – with about the same number of errors, which

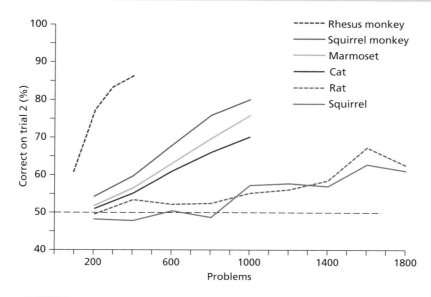

Figure 5.15 The rate at which various mammals can form the more demanding discrimination learning sets in which each problem presents them with completely novel objects. Thus, at each change the animal's choice in the first trial has to be random, but if it has learnt the principle behind the problems, trial 2 should be correct. Note how long it is before the scores of rats or squirrels on trial 2 become better than chance or 50%. Many monkeys reach almost 100% within 400 problems.

suggests that both animals find the problem roughly 'equal' in difficulty. Thereafter, however, rats improve much more rapidly than goldfish.

The ability to form learning sets was once regarded as a capability of the more advanced mammals only. A useful review by Warren (1965) summarized the comparative data to that date which suggested that all the vertebrates with the exception of fish shared this ability. In fact, as we have just described, later work (Mackintosh *et al.*, 1985) shows that goldfish can form repeated reversal sets and so can one of the most advanced invertebrates, the octopus (Mackintosh, 1965).

However, it may be significant that nobody has yet succeeded in demonstrating that fish can ever improve their rate of learning to discriminate between completely novel pairs of objects. These were used in the tests shown in Fig. 5.15, which shows that although a wide range of animals can form these discrimination learning sets, the speed with which they do so does change dramatically through the vertebrates, with subjects varying greatly in the time they take to improve. As Warren (1973) points out, there are all the usual problems in taking this result at its face value; the types of manipulation involved in testing a squirrel are very different from those for a cat or a monkey. Most people would not wish to argue too much about comparative intelligence from this sort of data. The sheer volume of testing and the time span which must be needed are likely to make comparisons difficult but despite the uncertainties, with caution we may conclude that cats and dogs do reveal more signs of intelligence than rodents and that the superiority of primates is clear. Perhaps not surprisingly, in view of their large brains, dolphins (the only cetaceans small enough for intensive study) have also proved quick to acquire learning sets (Herman and Arbeit, 1973). However, it seems safest to regard the superiority of cetaceans and primates as a quantitative one; we have no evidence that they possess any abilities which are not at least foreshadowed elsewhere among their mammalian relatives.

One cannot help feeling vaguely dissatisfied by this conclusion. Especially for the primates, the rather circumscribed artificiality of many learning experiments does not seem to do justice to the great flexibility and ingenuity of animals such as chimpanzees. Here we are undoubtedly influenced by their similarity to ourselves, particularly because they can manipulate objects much as we do and, of course, because of the rich complexities of their social organization. We may underestimate the intelligence of other animals because their structure and environment is so different from our own. The dolphins and whales are an obvious case in point; we have only just begun to take some measure of their behavioural complexities.

Social learning and culture

At the conclusion of Chapter 2 on the development of behaviour we emphasized its links with evolution and also pointed out that learning itself can be regarded as a form of development. Learning results in behavioural changes within an animal's lifetime and this can introduce a new dimension into behavioural evolution, one that – at least initially – does not involve genetic changes. We are familiar with the idea that behaviour can have a genetic basis and so change occurs by natural selection operating on inherited mutations, favouring those in which the interactions with the environment lead to greater success. This process will often be slow and constrained because many, perhaps most, mutations will have unfavourable effects. But behavioural evolution can sometimes be far less constrained.

The essential difference is that through its behavioural repertoire an animal can become more than a passive agent, having to accept – as it were – what the environment hands out during its development. By their behaviour, animals may be able to modify their habitat or vote with their feet and move to another habitat which suits them better. Thus they can sometimes actively change the nature of the selective forces which impinge upon them. Some of the ways animals and plants modify their physical surroundings throughout their lives will persist and affect succeeding generations. Soil builds up in woodlands from leaf litter decay and the actions of earthworms. This alters the environment for seed germination and for the earthworms themselves. Beavers build dams which create new bodies of deeper water, whilst animals as diverse as ants and rhinoceroses make trails which persist for years and are used by successive generations. Odling-Smee *et al.* (2003) argue that we have often failed to recognize the extent to which animals have affected their evolution in this way, a process which they call 'niche construction'.

We are considering learning here, and the most important niche for us is the social environment. Any animal which grows up in a persistent social group or simply has frequent contact with others has an environment which will profoundly affect the way its behaviour develops. Many of the other animals around – and not just its parents – will be adults. They form a constantly changing set of 'models' from which young animals may learn. Extensive observation of social groups in nature commonly reveals enduring differences in their behaviour. Individual patterns have arisen in one group or another and persist because young animals acquire them from the adults with whom they grow up. It seems entirely appropriate to refer to such persistent behavioural features as 'culture' and to the changes as cultural evolution. They are acquired and passed on in a manner similar to if simpler than those of human cultures. Laland and Galef (2009) review an active field of research.

Clearly cultural evolution is possible only among animals living in contact with others and having the ability to modify their behaviour by copying and practice. Such capacities are certainly much more widespread than once thought. The development of bird song, discussed in Chapter 2, offers some good examples and it is also clear that rodents such as rats and mice, even if not closely social, can learn through observation. Rats have been shown to acquire the tendency to dig for hidden food from watching others doing so. Having learnt this, they themselves can become models from which others learn, thus initiating a chain of cultural transmission (Laland *et al.*, 1993).

We are certainly not surprised to find other examples from primates, our closest relatives. There are now a number of well-controlled studies in which the origins and subsequent transmission of behaviour between individuals have been studied. In one (Dindo *et al.*, 2008), individual capuchin monkeys (*Cebus apella*) were taken from their group and shown how to open a food box using one of two methods involving either a sliding partition or a hinged lid (Fig. 5.16). They learnt extremely quickly, sometimes two demonstrations were enough. If now other group members were able to observe them operating the food box, the latter also acquired the skill and turn by turn the particular 'foraging method' was transmitted through the group. Capuchin groups in the wild are reported to have 'traditions' in the way they feed and these experiments suggest what is occurring there.

With the great apes, we must recognize that cultural transmission is a major factor in the development of their behaviour. McGrew (2004) brings together a whole range of the new observations on wild chimpanzees where, a priori, it is reasonable to assume that any persistent differences between groups are cultural. We have already mentioned

(a)
(b)

Figure 5.16 Social learning in capuchin monkeys. A 'tutor' and a 'pupil' using a problem box which delivers a grape. In (a) the tutor has learnt to lift a hinged lid, in (b) to slide aside a door. The pupil rapidly learns one method or the other by copying its companion.

the unique grooming behaviour of one group (Fig. 2.9, p. 45) and we can also observe striking differences in patterns of feeding. In some areas, chimpanzees 'fish' for termites by poking a twig, which they carefully fashion for the purpose (Fig. 5.17), into the termite mound and letting the insects crawl up the stick from where they can be eaten. Other populations use leaf stalks or grass stems for the same purpose and also use them to catch ants. The behaviour of the chimpanzees of the Tai Forest in the Ivory Coast differs in many ways from those at Gombe in Tanzania. In particular, the former show far more extensive hunting of monkeys for food and they also use a wider range of tools in foraging (see Fig. 5.18). It seems certain that all these differences are cultural in origin (Boesch, 1994).

Learning from others and copying can be a potent factor for change to behaviour, one which is both rapid and extensive. Such cultural evolution is not to be regarded as something quite separate from genetically based evolution via natural selection, for both processes will interact. The famous long-term studies on Japanese macaques (*Macaca fuscata*) have recorded the development of a 'food-washing' culture in some groups. It may have originated by some monkeys copying humans who were provisioning them with sweet potatoes, which they had improved by washing off the earth which clings to them (Fig. 5.19). Some coastal groups of monkeys also began to wash food in streams, but subsequently moved down to the edge of the sea to do so. Once there, some began swimming and started to eat seaweed – a remarkable extension of

Figure 5.17   A chimpanzee trims a twig which it has selected into a suitable form for termite fishing.

their normal behaviour and at once a whole new range of possibilities opens up. One can easily imagine how, given time, and consistent advantage, this behaviour might lead to various morphological and physiological adaptations to match. The context in which natural selection now operates has changed and, for instance, monkeys might

Figure 5.18   A female chimpanzee in the Tai Forest Reserve, Ivory Coast, cracks open a nut of the tree *Coula edulis* – a much-favoured food. She uses a heavy piece of wood whilst the nut rests on a stone which thus serves as an anvil. Her 1-year-old infant watches.

evolve insulating blubber and webbed feet. In this way, learning could be the foundation of changes to behaviour which might start a species along completely new evolutionary pathways – niche construction on a major scale.

 ## The nature of animal minds

We must return to questions which we raised when introducing the topic of insight learning, and addressing them forces us to attempt going beyond mere description of how animals behave in straightforward learning situations. It is surely impossible to

Figure 5.19 Some groups of Japanese macaques living by the sea have acquired the habit of washing food they have collected.

keep a pet or to observe any animal carefully without wondering whether it 'thinks' or has a 'mind' and 'consciousness'. Those words are in quotation marks because, whilst none of us doubts that we personally possess them – *Cogito, ergo sum* – they are not easy to define exactly and it is even harder to suggest how we might set about proving them to exist in other animals. Some would argue it is impossible for after all, each of us has no proof positive that they exist in any human being other than ourself; it is just reasonable to assume it because we are built the same way and behave the same way. As we mentioned at the outset, the information we have which bears on these issues for animals is not very systematic; often it is frankly anecdotal. We may consider three such anecdotes which will illustrate some of the questions which are posed.

The first comes from casual observations made in the mid-nineteenth century by the great biologist, Alfred Russel Wallace, who shared with Darwin the original formulation of a theory of evolution by natural selection. He spent years on collecting trips in the Far East and whilst staying on the island of Borneo he had opportunity to observe a captive orang-utan (*Pongo pygmaeus*). Around its cage there foraged some of the local village's domestic chickens. The orang showed great interest in them and once or twice tried to capture one, but they evaded the ape's long arms. One day Wallace saw the orang scatter grain from its food dish outside its cage, dropping some grain just outside the bars. It then sat quietly as the chickens foraging nearby found the grain and followed it right up to the cage, whereupon the orang flashed out an arm, captured a bird and killed it!

Our second example comes from an interesting article on the behavioural capabilities of working sheep-dogs.

Dogs experience particular difficulty when faced with recalcitrant ewes with lambs, one such ewe which had split off from the main flock refused to be moved and faced the dog square on, stamping its hooves. The dog returned to the main flock, cut off several sheep, and brought them over to the stubborn ewe. The ewe promptly joined this group and the dog was able to move them all back to the main flock. (Vines, 1981)

The third example comes from elephants. A personal account given to us by Howard Blight who, with his family, took pleasure in watching elephants who came to drink at a large water trough on their property in South Africa. Elephants drink by sucking water into their trunks and then, raising their heads, blow the water into their open mouths. He goes on,

One evening we noticed that a fully grown female had a short trunk. On examining her through binoculars our worst fears were confirmed. The animal had lost about 3/4 of her trunk. Perhaps this was the result of a wire snare having tightened completely restricting the blood flow. Then, to our amazement, the short-trunk elephant shuffled her way between the other elephants standing around the trough and the adult to her left lifted her trunk and offered the contents to the afflicted animal. She lifted her stump out of the way and allowed her neighbour to assist her to drink her fill. This action was repeated a number of times before the herd left. [. . .] the Warden at this time [of the game reserve] said he had observed the same elephant being fed by the others. The elephants would pull branches to the ground and the afflicted cow would then kneel and eat.

All these examples raise a host of intriguing questions. As so often, they are all one-off cases. We might see similar behaviour again, but it is the essence of the situation that the observations are not easily repeatable. This does not make them any less interesting or important. Some may argue that to have observed such behaviour at all is of profound significance for our view of animals. Others will immediately suggest caution. How are we ever to know their significance? Perhaps the orang spilt its food completely by accident – apes are not tidy feeders. Wallace's – and perhaps our – assumption of insight and imagination may be completely misplaced. So often we face difficulties both with the accuracy of such observations and with their interpretation. Similarly for the sheep-dog: what prior experience had it? Vines suggests that dogs often have trouble with individual sheep. During its training it might have seen older dogs using this strategy to deal with them. If so, would this make any difference to the way we interpreted the behaviour? Possibly we could categorize these two anecdotes as examples of that 'cleverness' to which we referred earlier, but the elephant story forces us to speculate beyond this.

Elephants are long-lived and intensely social animals and it is commonplace to observe them interacting in friendly ways, playing together or acting cooperatively for the mutual protection of young animals, but this situation is different. It is certainly of selective value for each of the individuals that the group survives. We observe the same mutual cooperation in meerkats whose social behaviour is also discussed further in Chapter 7 – individuals cannot survive outside a stable group of reasonable size. Elephants then may have powerful inherited tendencies to maintain the cohesion of their group, but here the injured animal would be unlikely to survive unaided even within the group. The others react to this novel situation in an appropriate way. Might we conclude that they understand the requirements of another individual separate from themselves – that they are showing something which in humans we might call empathy? We can keep in our minds all the questions raised by our three anecdotes as we move on first to discuss more of the historical background to such considerations.

We began this section with the early experimental psychology approaches to 'higher' learning and insight, but, of course, humans have always contemplated and speculated on the minds of other animals. When discussing the source of the adaptiveness of behaviour in Chapter 2, we noted that through history animals have sometimes been regarded as little better than automata, sometimes as fully sentient beings like ourselves; Boakes (1984) and Walker (1983) both provide interesting reviews of these changes. With the rapid rise of science in modern times, many researchers have been reluctant to speculate on the nature of animal thinking. Rather, as we have seen, attention is focused on analysing what animals actually do, in both natural and laboratory situations, and deducing the causal factors of their behaviour, its function and so on. Most serious observers of animal behaviour have always had at the back of their minds a rule proposed around the turn of the twentieth century by Lloyd Morgan, one of the founding fathers of comparative psychology, now referred to as 'Morgan's Canon':

In no case may we interpret an action as the outcome of the exercise of a higher psychical faculty, if it can be interpreted as the outcome of one which stands lower in the psychological scale. (Morgan, 1894)

It must be acknowledged that Morgan's canon has served us well. It has enabled us to approach the analysis of behaviour sensibly and to avoid the superficial anthropomorphism which led to many absurdities in the past. A familiar example came from Germany in the early 1900s. The eccentric Baron von Osten sincerely believed that many animals, and certainly horses, could perform mental arithmetic – all they needed was the right training. His best pupil 'Clever Hans', who once had pulled a milk cart, became world famous through being able to give the answer to sums presented on a blackboard by beating the ground with his hoof the appropriate number of times (see Fig. 5.20). A psychologist from a university in Berlin, Oskar Pfungst, was called in to investigate this remarkable ability. The answer came quite quickly. Hans was certainly not doing arithmetic, he was not even looking at the blackboard, rather he was responding to subtle and almost certainly quite unconscious cues from his owner – a tiny head movement seemed detectable by the human observers – which signalled him

Figure 5.20   Baron von Osten with 'Clever Hans' and minder; a photograph taken about 1904. The day's lessons for Hans are set out on the blackboard and an abacus is at hand for aid with arithmetic.

to stop the hoof beats. If the Baron himself could not see the question on the blackboard, the poor horse floundered, he just continued beating his hoof, waiting for the cue which never came. The Baron was misled by retaining very human expectations for the abilities of animals. Shortly before, another Morgan – Lewis – who studied beavers in the northern United States was similarly misled. Not recognizing the ability of natural selection to shape behavioural development, he believed that they must have an extensive understanding of hydraulics as revealed by the construction of their dams and channels. Both of these classic examples of anthropomorphism are described more fully by Sparks (1982).

There is no question that getting good, repeatable evidence on the minds of animals is extremely difficult. Some would argue that it is impossible and therefore a waste of time. This was certainly Tinbergen's view, expressed in his pioneering book, *The Study of Instinct* (1951), and the early ethologists studying their animals in the wild avoided the issue almost entirely. However some have returned to the issue more recently, spurred on by an influential book, *Animal Thinking* (1984) by Donald Griffin, an ethologist who was among the first to discover the echolocation systems of bats (Griffin, 1958). There, and in a subsequent volume (Griffin, 2001), he argues that no matter how difficult they may be to study, it is no longer sensible to deny the possibility of true thought processes and even consciousness of some sort in some mammals and birds. For ethologists concerned with the operation of natural selection in shaping a species' behavioural potential, one of the most powerful arguments for this conclusion is an evolutionary one. We might call it the 'phylogenetic hypothesis'. This does not accept that the mind and consciousness of human beings have just arisen *de novo* without any precursors in animals which were ancestral to us and probably very similar to the non-human primates which we observe now. It seems obvious that there would be a selective advantage to even some slight ability to anticipate the consequences of one's actions and predict the responses of others. This would seem particularly important for social animals and Humphrey (1976) has argued that the complexities of social life, especially in the primates, provided one of the most important selective forces behind the emergence of thinking ability. Increasing brain size, developing new populations of neurons not designated for sensory or motor processes but free for associating diverse information and for parallel processing, would be the physical basis for such abilities.

An alternative position is exemplified by Macphail (1998), who is unwilling to attribute such abilities to non-human animals. He suggests that there is a sharp distinction between us and them, and that its key is the unique capacities we possess through speech, the medium for thought, abstraction and communication, and thus consciousness. There can be no gainsaying that the current situation does reveal a major behavioural rift between us and our nearest living relatives, the chimpanzees. However this does not really constitute a satisfactory refutation of the phylogenetic hypothesis, which considers what might be the nature of intermediate stages. Roth (2001) points out that the under-lying organization of the brain in primates and ourselves is the same, the differences are quantitative not qualitative. We should remember that as recently as 55 000 years ago there were certainly four and possibly five species of the genus *Homo* living here simul-taneously. First our direct ancestors, *H. sapiens* but also *H. neanderthalensis* and the newly discovered Denisovan human who probably shared a common ancestor. Neanderthals had a brain the size of ours. Then there were also the last survivors of *H. erectus* whose ancestors had first emerged from Africa about 1.8 million years ago with a brain about three quarters of ours and lastly the remarkable *H. floresiensis*, a dwarf hominin with a brain little more than chimpanzee size. Stringer and Andrews (2005) and Stringer (2011) provide excellent modern accounts of human evolution – the story is changing rapidly with the advent of DNA evidence from even fragmentary fossils. It is clear that all of these other species had well-developed social lives and, for example, made stone tools for hunting and butchering and had control of fire. We now know that Neanderthals shared with us the version of the *FoxP2* gene (p. 51) connected with speech; we do not know the situation for *erectus* and *floresiensis*. The point to be made here is that the huge gap in brain size and intellectual complexity between ourselves and our closest living relatives, at present chimpanzees, is a very recent situation. What doubts might we be expressing about the mind of *erectus* were they available for testing?

Obviously the first place to look for precursors of human cognitive abilities will be among the primates and especially the great apes, but the phylogenetic hypothesis also requires us to look for them in other animals. One group which we have certainly underestimated in the past is the birds. As mentioned above birds and mammals have not shared a common ancestor for about 250 million years and the brain of birds has evolved differently, in particular they lack the highly developed cerebral cortex that is such a dominant feature of mammals and associated with their cognitive abilities. None the less, there must be more than one way to mediate such behaviour for we now recognize that some birds are very clever, in particular the crows and titmice in the order Passeriformes and many of the parrots (order Psittaciiformes). They are capable of great flexibility in feeding situations and some crows show remarkable use of tools. If we hesitate to consider the possibility such animals share some of our abilities we should reflect on the fact that until quite recently the employment of tools was paraded as the cut off point between animals and ourselves. Some exceptions – the Galapagos wood-pecker finch using a cactus spine to probe tree bark, gulls dropping shellfish on to rocks to break them open – were regarded as specialized instinctive behaviour and therefore not signs of cognitive abilities. Then came Jane Goodall's discovery that wild chimpan-zees use tools for feeding (see Figs 5.17 and 18) and even modify them to suit their purpose. We now know that it is a regular part of their behaviour, but if we wished to demonstrate tool use in animals then probably by far the best subject would be the New Caledonian crow (*Corvus moneduloides*) (Hunt, 1996; Hunt and Gray, 2003).

This bird not only uses tools for a large proportion of its feeding, it goes much further, fabricating the tools in several coordinated stages by shaping strips torn from the leaf edges from one of the local trees. The margins of these *Pandanus* leaves have recurved spines and when strips are shaped (see Fig. 5.21) they are beautifully adapted to probe and then extract insect larvae from their tunnels in wood. The crows also break off twigs from larger branches and use the slight hook which is left at the base (Fig. 5.21) to extract food. Captive crows have been observed to go one further, creating new tools from wire – a material they had no experience of – and bending it to fashion a hook to extract food from inside a glass tube; (Wier *et al*. 2002; Wier and Kacelnik, 2006). Our assumption regarding the remarkable extent of tool use here would be that some aspect of the New Caledonian crow's life style and environment has selected out tool-using individuals and this type of behaviour is now fixed. Of course to get some estimate of how far they are genetically biased to develop tool-making would require close obser-vation of young crows as they grow up in different environments.

However recent observations on rooks (*Corvus frugilegus*) have revealed similar cap-acities and this in a crow which is never seen to employ tools in the wild. Bird and Emery (2009a, b) found that some captive rooks would spontaneously fashion hooks from wire to get at food in tubes just as did their New Caledonian relatives, (Fig. 5.22a). They went further, responding appropriately to food which was made inaccessible in another way. Given a tube with a little water in it upon which floated a mealworm out of reach, rooks would pick up stones provided nearby and drop them into the tube until

Figure 5.21 Advanced tool-use by New Caledonian crows. At the top we see crows using hooked tools to get at food in holes or crevices. The hooks, as shown just below, are chosen when small twigs are pulled off a branch leaving a suitable hooked base. At the bottom are drawings of the remarkable three-staged manufacture of a tool, torn from the spiny edge of *Pandanus* leaves. The thick base is held in the crow's bill as shown and with the leaf's hooks facing backwards it forms an ideal probe for extracting insect larvae from their burrows.

the water level was raised sufficiently for them to reach in (Fig. 5.22b) – exactly as did the crow in Aesop's Fable! The way they did this showed signs of real goal orientation. They chose large stones over small and watched the side of the tube as the water level rose. They used no more stones than were necessary. There is little question of selection

(a)   (b)

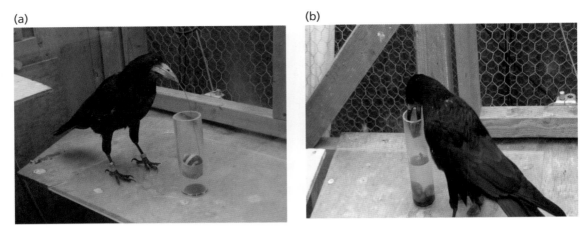

Figure 5.22   Rooks in captivity fashioning tools from scratch! On the left one bird has bent wire into a hook for lifting food from a glass tube. On the right, re-enacting the crow of Aesop's Fable! A rook is dropping a stone into the tube of water on which floats a mealworm. It will check the water level after each stone and drops in no more stones than are needed.

for tool-making ability here. Wild rooks feed on a fairly wide range of foods but none requires extra manipulative skills. We must accept that birds such as crows and parrots, which have the largest brains in relation to their body size, can be very 'clever' to use a term employed earlier.

Apart from such striking observations of tool use and tool fabrication, we also have numerous examples of complex cognitive abilities in birds. One of the most remarkable comes from Pepperberg's (Pepperberg, 2002; Pepperberg and Gordon, 2005) studies on the capacities of Alex, an African grey parrot (*Psittacus erithacus*) with whom she worked for many years (Fig. 5.23). She developed a very effective way to train this bird to associate words with objects. Of course, the most unusual feature of this particular situation is that Alex had the vocal ability to imitate human speech. He acquired a large number of words, which he associated with objects – such as wood, cork, paper, nut – many of which he liked handling with his bill and feet. He would ask for them by name in English and he responded to questions put in simple English! The fact that he seemed to 'speak' seems so extraordinary that we may be in danger of overestimating his intellectual powers, an issue we have already encountered when discussing the abilities of pigeons to recognize rotated shapes. Nevertheless, Pepperberg has proved that Alex's abilities went far beyond just naming objects. For example, he could count up to six and identify and name the quality, i.e. size, shape or colour, which a group of diverse objects – none of which he had seen before – have in common. Such abilities are probably beyond most monkeys. Pepperberg was particularly careful to avoid cueing Alex (Clever Hans-like effects) and the strength of her conclusions is supported by a very large body of tests albeit on a single individual. Even so, she has sometimes had trouble in convincing others that the bird was truly capable of what was claimed.

Figure 5.23   Irene Pepperberg poses a question to Alex.

Much of this scepticism undoubtedly relates to the fact that few people are ready to accept this level of cognitive ability from a bird; we are so mammal-centred for obvious reasons. As just described, we have to accept that some birds are capable of complex and flexible behaviour involving tools but it is extraordinary – and certainly unpredicted by most people – that Alex could handle symbolic naming, abstract qualities, such as size and number, and could respond to simple questions put to him in English.

Seed *et al*. (2009) argue that intelligent behaviour shown by crows and apes does indeed have similarities which are the result of convergent evolution from very different origins. For example, Clayton *et al*. (2007) suggest that food-hoarding scrub jays (*Aphelocoma californica*) show an extraordinary ability to judge whether other crows (which would steal their cache) have been able to see where they have hidden food, i.e. they can somehow 'put themselves' in another individual's place. We shall discuss below arguments about whether chimpanzees can do exactly this. Certainly the 'intellectual path' followed by the birds does show some hitherto unsuspected parallels with mammals but it is understandable that those pursuing the phylogenetic hypothesis will tend to concentrate on our closest relatives.

Figure 5.24   Kanzi, a male pygmy chimpanzee or bonobo, with Sue Savage-Rumbaugh using a keyboard with a large array of symbols indicating the names of objects or actions.

For a start, as MacPhail (1998) has emphasized, it is obvious that so much of human intellectual activity depends upon our symbolic language. Consequently, beginning a few decades back, there has been a proliferation of studies attempting to discover how far apes can go in acquiring language. Because they have very limited capacities to produce varied sounds they have to be trained to associate objects with non-vocal symbols, either gestural, as when chimps have been taught human sign language for the deaf, or visual, using a keyboard with symbols to which they can point. Perhaps not surprisingly, they can pick up an amazing array of names of objects and some more general terms like 'more', 'tickle', 'play', 'give me' together with numbers up to 6 or 7. It is possible to have a kind of a stilted dialogue with chimpanzees trained in this way. Some animals, especially one piygmy chimpanzee or bonobo (*Pan paniscus*) – given the name Kanzi – studied by Sue Savage-Rumbaugh and her collaborators, show an astonishing facility for such learning; his keyboard has several hundred symbols (Fig. 5.24). Kanzi added to his 'vocabulary' without direction because there can be no question that he understands a good range of vocal English. For example, he follows verbal instructions given through earphones or via a loudspeaker so that all visual cues are absent (Fig. 5.25). 'Kanzi, put the hat in the refrigerator,' is a sufficiently bizarre instruction when given for the first time, that it is hard to believe other than that the ape recognized the significance of the whole sentence. He didn't, for example, put the hat *on* the refrigerator.

Figure 5.25 Kanzi, wearing earphones through which he responds to a whole series of verbal commands when there is no possibility of him getting other clues from the experimenters.

Savage-Rumbaugh and her collaborators (1996) make claims for the language comprehension of bonobos which go well beyond any others so far. For example, they say,

Understanding of symbolic communication arises spontaneously in the bonobo. Human speech can be readily understood by apes when exposure occurs at an early age. The correct meanings are ascribed to spoken words spontaneously, without shaping or planned reinforcement by human beings. (Savage-Rumbaugh *et al.*, 1996)

Critics have pointed out that almost all the evidence for this comes from Kanzi's responses, as in the case above, where he responds to a direct instruction or when he makes a request, via the keyboard, for something he wants. In all the ape language studies, not just those involving Kanzi, the hoped-for eavesdropping on what two apes would 'say' to each other if both were equipped with symbolic means of communicating, has really come to nought! Naming things and the occasional sign for 'give me' is about the limit. No biologist should be disappointed by this; the attention to language acquisition as a means of communication reflects only a human interest. It is intriguing and definitely gives us insights into the remarkable flexibility of an ape's behaviour when it is reared in a totally unnatural and very complex environment. However, it could be very misleading and it tells us very little of how this flexibility operates in the natural environment. Although this may be physically less complex than a human environment with all kinds of gadgets and people speaking to you all the time, the social environment with many other animals all around and interacting with you is equally complex, but complex in a very different way.

There are a number of major studies on groups of chimpanzees which bring out the range of their social life. Jane Goodall's (1986) account of the early years of her research on wild chimpanzees in Tanzania and Franz De Waal's (1982) studies on a large captive

Figure 5.26 *Chimpanzee dentistry! One chimp has a loose tooth which is clearly bothering it. Another group member becomes involved and the 'patient' lies back, enabling the 'dentist' to use a small stick to dislodge the annoyance. This is surely empathy and cooperation of a fairly high order.*

group are classics of their type. These studies and many others reveal a remarkable richness of diverse behaviour. Chimps cooperating to move heavy pieces of wood which would help them to escape from an enclosure; one chimp deliberately going away from a source of food which it knows about when others, who do not know this, are watching; one chimp lying back quietly with open mouth whilst another explores a loose tooth

Figure 5.27 A juvenile chimpanzee approaches and 'hugs' a young adult who is screaming following an attack.

with a stick (Fig. 5.26); a male in a zoo enclosure who had developed the habit of throwing stones at visitors, storing up conveniently placed piles of ammunition when the zoo was closed. Then there are such obvious differences between the personalities of individuals. Most commonly they form 'coalitions' and comfort one another at times of stress (see Fig. 5.27), but not all social life is so benign. Goodall also describes the pathological behaviour of one female chimpanzee at the Gombe Stream Reserve who, with her young daughter, would seek out and kill the infants of other females.

Such complexities have such obvious human parallels that they have been taken as sufficient proof of a human-like consciousness in chimpanzees and sometimes other animals. They hark back to the issues forcibly brought to mind by the observations on elephants helping a wounded member of their group. However, we must not lightly abandon Morgan's canon and, understandably, animal behaviour workers seek out evidence which can be more rigorously based and repeatable. We cannot lightly assume that cognitive complexities can be directly equated with consciousness. Perhaps this is best exemplified by studies which concentrate on trying to discover whether any animals have a true 'theory of mind' – understanding oneself as an entity distinct from others.

Some believe that the capacity to recognize other individuals outside of yourself requires first to be able to recognize oneself! Accordingly, there have been a number of attempts to demonstrate this capacity experimentally. Practicable tests for self-recognition have to involve vision and thus can use only those animals for which vision is an important sense. Fortunately, this includes groups with large brains and complex social behaviour, the primates, cetaceans, like the dolphins, and elephants. Can an animal recognize that the image it sees in a mirror is itself? Self-recognition or being able to steer actions using a television image of one's body to reach objects which are out of direct sight, have been claimed many times for chimpanzees and orang-utans. The most systematic work has come from Gallup *et al.* (1995) and always begins by familiarizing chimps to mirrors. At first, like many other animals, they respond as to an apparent stranger, approaching and often reaching behind the mirror, but soon they get past this and settle down, though continuing to pay occasional attention and make movements which they may appear to watch in the mirror. At this stage a chimp is lightly anaesthetized and while unconscious a mark is placed centrally on its forehead.

After it is well recovered, a mirror is again presented and the commonest and almost immediate response is that the chimp begins to gaze intently at its image and then touches and explores the mark on its forehead. Every possible control test has accompanied these experiments, controls for the effect of anaesthesia on activity, controls for any skin irritation from the marking and so on. All have proved negative, the response to the mirror image is clear-cut. Even so the claim of self-recognition is contested (e.g. Kennedy, 1992; Heyes, 1995) and the argument is worth following up because it reveals the extent of resistance to any observations which may be taken to blur the line between human and non-human minds. Most will agree that it is unreasonable to doubt that the chimps are capable of self-recognition. It is much more reasonable to question whether self-recognition equates to self-awareness in a human sense.

It is surely significant that only chimpanzees and orang-utans have regularly responded to their images in this way, a few gorillas have also, but no other primates. There are of course all the familiar problems involving how very different animals 'see' the situation, but this gap may represent some kind of threshold. Certainly, the only other animals to have shown signs of self-recognition are one elephant from three tested together (Plotnik *et al.*, 2006) and several dolphins (*Tursiops* sp.) (Reiss & Marino, 2001; Marino, 2002), both representatives of those groups mentioned above with large brains and complex social lives (see Chapter 7, p. 381). These two qualities may be linked, social complexities may aid survival and they require brain and behavioural complexities to match.

So if this was regarded as a first step towards consciousness – consciousness of oneself as a distinct 'entity' – can animals go further, recognizing other individuals as distinct and understanding how they may respond? Now we can be quite certain that animals respond to others and may do so in a way that suggests they recognize what the others are doing in relation to what they themselves may do. If a hungry chicken is put in with a group which have ceased to feed, others often begin to feed again. There are many such examples of such 'social facilitation' as it is called. Here we are looking for signs of a more developed recognition of the existence of others outside oneself. There is now a large body of work addressing this question and we might summarize its aims using the title of one interesting contribution, Hare *et al.* (2001) call their paper, 'Do chimpanzees

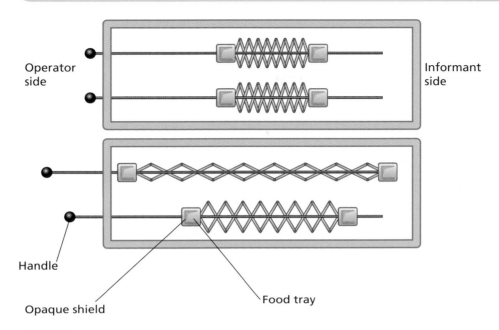

Operator side

Informant side

Handle

Opaque shield

Food tray

Figure 5.28 Apparatus developed by Povinelli and his collaborators to test whether chimpanzees could interpret the role played by another individual and adopt this role when it was their turn. Some attractive food is put into one pair of trays and is visible to a chimp on the 'information' side (upper tray). In the lower tray, the operator has pulled the handle on the side indicated by the informant and both chimps get a reward. After working in one role some chimps could immediately adopt the other when their places were exchanged. (After Povinelli *et al*.,1992.)

know what conspecifics know?' Effectively, can a chimpanzee put itself in the place of another? Many of the experiments are highly ingenious, one is illustrated in Fig. 5.29 from Povinelli *et al*. (1992) which suggests that chimpanzees are able to foresee the outcome of their own actions and the way that these will affect another individual. The apparatus requires two chimpanzees to cooperate to get a reward, with each having a distinct and different role. Povinelli *et al*. found that each role-player needed only a brief demonstration of how to operate, working with a human collaborator. After a few operations of the apparatus the experimenters swapped the two animals round so that they had to take on the other's role; three out of four chimpanzees did so without hesitation. We can scarcely explain this away as mere copying of a motor pattern from one to another because the viewpoint is so different. For instance, the operator chimp has to appreciate the significance of the informant's gestures and, after the swap, take on this gestural role. It does so spontaneously and using its own gestures, which are not necessarily the same as those it had seen used before by its partner.

This is impressive, however chimpanzees do not always succeed in situations where they need to understand another's actions. Often they do not identify human gestures well, for example pointing, where interestingly dogs may perform better. However, as

Hare and Tomasello (2004) point out in the paper cited above, dogs have long been selected for such responsiveness and their social system involves close cooperation between the members of a pack and much food sharing.

We might consider that following the gaze of another individual represents some kind of first stage of interpreting their viewpoint outside one's own. Of course, it may not imply that other animals are responding consciously as we often are. Nevertheless it is significant that Bräuer *et al.* (2005) have shown that all the great apes follow human gaze well and may move so as to allow themselves to see behind a barrier where the human's gaze is directed. Other primates also score well – e.g. the New World spider monkeys and capuchins (Amici *et al.*, 2009) – and the ability to follow the gaze of others, for instance when foraging, will have obvious advantages in social groups.

As always, we must try to interpret the results of such tests in relation to the life of the subjects. Chimpanzees, like all primates, are normally solitary foragers and try to monopolize food, further they have a very strict dominance hierarchy. As in the Povinelli task just described, Hare *et al.* have shown convincingly that chimpanzees *can* interpret the viewpoint of others but they are convinced that they do this best when competing for a food source, not collaborating to get one. In a whole series of experiments, they arrange that a dominant and a subordinate, taken from a group who know one another well, are facing each other behind glass screens on opposite sides of an arena where food can be provided. The experimenter can control who is let in to feed and when. Invariably the subordinate will defer to the dominant if there is any competition for a food item. Now screens are introduced into the arena behind which food can be placed out of sight of one or both chimpanzees. It is arranged that the subordinate can see where food is and in half the tests the dominant can also, but in the other half the dominant's window into the arena is covered at first and only the subordinate witnesses the food being placed. Then both chimps are let into the arena, with the subordinate being allowed a slight lead. In numerous permutations of such situations the subordinate shows clearly that it 'knows' which food items the dominant has seen being placed and which not. The former always goes first to food items which the dominant has not seen being placed – they are 'safe', relatively at least! Further, it is hard to argue that the subordinate is just learning, albeit very rapidly, a single association in one situation. If transparent barriers are used then the chimp realizes the dominant can see the food. Again if one dominant sees the food being hidden, but this animal is now removed and a different acknowledged dominant is put in its place, the subordinate realizes that this latter animal does not know where the food is and therefore it's safe.

In a useful discussion of their own and other's work, Hare *et al.* (2001) try to set out what we might call minimal conclusions from such tests. Rather than suggesting *either* that chimpanzees can understand fully the viewpoints of other individuals *or* that they are simply learning particular associations between – say – the position of food reward and the behaviour of a dominant animal, they argue that the mind of a chimpanzee is probably at some intermediate stage. It is able to understand what another individual has seen, so definitely capable of seeing another as a separate entity from itself, but it is not able to put itself fully into another's place. This is certainly the kind of conclusion expected on any phylogenetic hypothesis. A whole range of experiments bearing on this and other related issues concerning cooperation and conflict in primates is reviewed by Silk (2007a).

We may note that the so-called 'premotor' areas of the brains of primates have now been shown to contain a neuronal system which clearly 'recognizes' the nature of a movement in another animal (or a human for that matter). When a monkey performs a gesture, we can record that particular neurons in the prefrontal area become activated in a pattern which is specific to the movement involved. We also record from the brain of a monkey which is observing. As it watches the movement, exactly equivalent neurons in the observer's brain begin firing. The firing patterns of limb movements, facial expressions, movements of the lips and tongue – all are matched in the observer's premotor area by what are understandably called 'mirror neurons'. In some contexts, signals from the premotor areas may go to the motor area and the observer may actually copy the movement it has just seen. Mother rhesus monkeys gaze closely at their young infants and perform the friendly 'lip-smacking' gesture, which often leads to licking and face to face contact. The babies respond by lip smacking and this imitation sets up a close communication between mother and child (Ferrari et al., 2009). Interestingly, baby rhesus monkeys have been observed to put out their tongue when held up to face a human who does just that to them. Rizzolatti and Craighero (2004) and Ferrari et al. (2006) review research on mirror neurons. This system must indicate that movements, gestures, signals given by one monkey are not just perceived by others but 'understood' in some way in that the observer's brain imitates their expression internally even if going no further. Mirror neurons may be no more than the mechanism behind social facilitation: a chicken is stimulated to feed by the sight of one already feeding. However, the detailed patterning seen in the primates and the fact that it occurs across different species, suggests to some interpreters that mirror neurons are one element contributing to genuine empathy and thus the recognition of other individuals outside oneself. It must be significant that capuchin monkeys approach and interact much more closely with a human who imitates what the monkey is doing than with another who is equally active, but not imitating (Paukner et al., 2009).

There is no question of final conclusions here: indeed the co-authors of this book have had to agree to differ at times! The arguments over the interpretation of experimental results like this go back and forth in the animal behaviour literature; for example, Penn and Povinelli (2007) provide a good sceptical review. The topic never ceases to arouse strong views on both sides but it must be acknowledged that, for the most part, the general public is bemused by the controversy. Most people assume with little question that the pet animals they are familiar with have some foresight and consciousness, but it is easy to understand why the topic remains controversial for behavioural scientists. Here we have been able to touch only on selected aspects of the 'cognitive ethology' field. It is important not to confine attention to chimpanzees and other primates and Ristau's (1991) collection, inspired by Griffin's first book, is valuable in this regard, so is a more recent survey by Bekoff (2007). A valuable symposium publication introduced by Emery *et al.* (2007) includes a range of reviews of the cognitive abilities of diverse animals. For the chimpanzee work, it remains fascinating to go back to some older discussions such as Premack and Woodruff (1978) and Heyes (1998). These long reviews with the varied and extended commentary by other workers following each, offer an excellent cross section of opinions both moderate and extreme.

However admirable has been the cautious policy of behavioural scientists to avoid complex and subjective explanations for our observations, we should remind ourselves of Walker's (1983) admonition, 'Sometimes explanations can be too simple to be sensible.' We have certainly been led to underestimate some of our animal relatives in the past. It is most important that we no longer continue to do so because, as we discussed at the end of Chapter 4, everywhere there is increasing concern for the conservation and welfare of animals, both wild and domestic. It is the responsibility of behavioural scientists to provide a sensible and practical basis for our efforts (see Dawkins, 1980, 1993, 2012; Dawkins and Bonney, 2008; Fraser, 2008).

The nature of memory

We return finally to the opening sentences of this chapter reiterating that learning and memory are the two sides of the same coin. Learning is nothing without memory on some scale. We must be able to store the results of experience and recall them to our benefit later. It is surely one of the most remarkable properties of the nervous system that in some cases it can retain some representation of past events for almost a lifetime – tens of years or more.

It is now generally accepted that memory storage must be represented in a chemico/physical form and we have enough evidence to understand, in some cases, how such changes are brought about and what is the nature of the store. This is now a huge field of research and most of it goes far beyond the scope of this book. Obviously, the study of memory mechanisms will involve neurophysiological and biochemical methods, for we must investigate the fine-scale operation of neurons and the synapses between them. Yet it remains the case that much of the most crucial evidence still has to come from behavioural observations. It is what the animal actually does which provides our knowledge of what it has learnt, stored and now recalls, and from this knowledge we go on to deduce mechanisms and then check them physiologically and biochemically. Dudai (1989) provides a very clear introduction to the whole field, taking account of all these different levels from which we must approach.

Just as with learning itself, so the study of memory mechanisms has benefited from the comparison of different types of animal. Valuable evidence has come from certain molluscs, the honeybee, a few bird species and a few mammals. Human memory itself has been a rich source. Although invasive experiments are usually impossible, we have inevitably a stream of cases where the effects of brain damage arising from accidents can be studied and, of course, we can get detailed evidence from humans who, unlike animals, can tell you verbally what they can or cannot recall.

Different types of memory

Introspection tells us that not all events are stored in the same way in our memory. We look up a telephone number to make a call, but may not be able to recall it a couple of minutes later. On the other hand, we can recall the numbers of close friends after months or years. Repetition and the greater significance attached to certain events must make a difference to the way they are stored. Perhaps all events go into a **short-term** store but only events of some consequence are held in a **long-term** store.

The idea that short-term memory and long-term memory are somehow distinct is supported by some remarkable psychological and physiological evidence. People suffering from concussion or other severe shock are often unable to recall the events that led up to the accident, but their memory of events in the more distant past is unaffected. Further, if their recent memories recover – and they often do – they do so roughly in order, the most distant first, until gradually almost all memory is recovered. Commonly though, the few moments before the accident can never be recalled. This phenomenon of retrograde amnesia can also be reproduced in animals. It suggests that short-term memory and the process by which events are moved into a long-term memory store is more labile and sensitive to disturbance than the store itself once in place. More evidence to support this conclusion comes from studies of rats and chicks using a range of drugs which affect the way the nervous system functions (see Andrew, 1985). For example, administering drugs which inhibit protein synthesis just after learning a new task does not affect old memories, but the memory of the new task rapidly fades and never recovers.

Recently there have been great advances in identifying brain mechanisms involved in consolidating and then storing memories in mammals and birds. We mentioned earlier when describing food storing birds that one particular brain structure – the hippocampus – is involved in processing spatial learning, a key component of hoarding behaviour (Fig. 5.29). In mammals, this is a fairly large ring-shaped tract at the base of the forebrain – so named because its shape was taken to resemble a Roman horse-racing field! Testing different types of learning in different animals has confirmed that the hippocampus is one region which is vital for memory formation and not just spatial memories. Damage to the hippocampus often prevents any new associations being formed. There have been many tests with animals which demonstrate this, but perhaps the most compelling evidence comes from human patients whose hippocampal region has been severely damaged.

Over a century ago, the Russian neurologist Korsakov described patients who were able to recall distant events normally but had lost permanently the ability to form new memories following a head injury, a stroke or severe alcoholism, so-called **anterograde amnesia**. A common feature of such patients was damage to neural structures at the base of the forebrain and adjacent to the hypothalamus, especially the mammillary bodies and the thalamus. Since Korsakov's original descriptions, damage to other

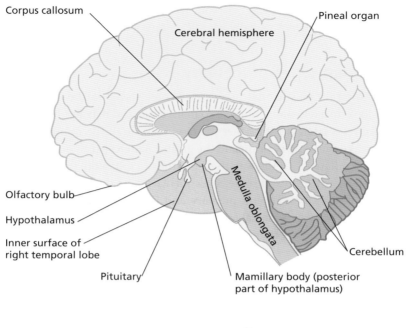

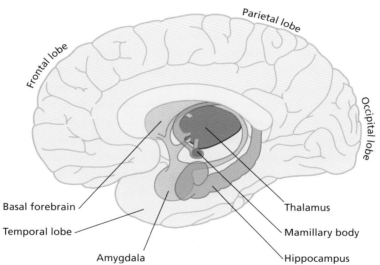

Figure 5.29 The human brain viewed (top) in straight sagittal section and (bottom) from the same aspect but with brain stem and cerebellum removed so as to reveal more of the structures involved in memory formation. Further explanation is given in the text.

closely associated brain regions, notably the hippocampus, has been shown to have the same effect. Bilateral damage to any of these structures can lead to a complete inability to store new memories and recall beyond a few seconds may be impossible. There is no better description of the behavioural consequences of such damage than the essay, 'The

Lost Mariner', in Sacks (1986). He provides a vivid account of the extraordinary and tragic existence of a person who has lost all ability to memorize current events and has only a distant past to which he can relate.

One person who became very well known to psychologists and neurobiologists was an American known, to preserve his anonymity, as 'HM'. HM, who died recently at the age of 84, suffered complete loss of short-term memory from 1953 when, at the age of 27, he suffered unforeseen side effects from a brain operation performed to stop persistent epileptic seizures. In this limited sense, the operation was a success and superficially HM might not have appeared to be handicapped. Thus he was able to hold a conversation and carry out many normal tasks. He could even learn new motor skills using tools, etc. He enjoyed newspapers, films and crossword puzzles but he had no recollection of recent events. Batteries of researchers studied HM, testing the details of his disability in a multitude of ways over many years. This might sound an intolerable invasion of his personal life, but so complete was his inability to acquire new memories that, after a few minutes, a researcher or a repeat of a test situation was greeted as totally novel. HM had no conscious recollection of them but clearly he was conscious in most senses. Like Sacks' 'Lost Mariner' he had effectively to elaborate the consciousness of his own current existence afresh from minute to minute.

People like HM have, albeit unintentionally, contributed greatly to our understanding of the mechanisms of memory. So have animals whose memory systems are specialized in particular ways – we have mentioned specializations in spatial memories already. Honeybees have likewise provided insights into the mechanisms of a tiny insect brain – about two cubic millimetres – which is specialized for remarkable types of learning during a forager's last three weeks of life (Menzel *et al.*, 2001). Imprinting in newly hatched birds, which we discussed in Chapter 2, has also proved a valuable system for identifying very precisely changes in the brain following learning. The chicks hatch with what we might call a blank slate as far as new associations are concerned. Then they can be given controlled amounts of visual experience with a 'mother object' and the degree to which their behaviour is changed compared with biochemical and even structural changes in a brain uncluttered with earlier learning experiences (see Horn, 1990).

In mammals, many new advances have followed the emergence of remarkable techniques – various types of 'tomography' – which make it possible to locate particular areas of the brain which are activated when fully conscious subjects are learning or recalling learnt events; Fig. 4.28, p. 227, revealing areas involved in tasting shows an example of how such techniques reveal brain activity. So far, our ability to locate is only on a fairly crude scale – regions involving many millions of neurons – but this will change. Rolls (2008) reviews this rapidly developing field.

In conclusion then, even though many questions remain unanswered, the new approaches do allow progress which seemed impossible only a few years back. Non-invasive techniques mean we can use human subjects and indeed it is the human memory system which leaves us with some of the most tantalizing but wonderful questions. Even allowing for the garnishings of our imagination, the richness of association in human memory is amazing – Hamlet soliloquies, a long dead grandparent's face, the creak made by a particular door in the home of our childhood, the smell of the upholstery in one's first old car. How can it all be stored? Is there any limit to the amount stored? Luria (1975) gives us a fascinating study of a Russian mnemonist who made a living on the stage as a 'memory man'. He describes a person who could memorize staggering lists of nonsense syllables, meaningless mathematical formulae, strings of numbers and, when asked without prior warning, could repeat them back again *many years later*. Luria could find no limits to this man's storage capacity and no evidence that he ever forgot anything. Such an astonishing memory was as much a burden as a blessing. Selective forgetting will always help us and our animal relatives with the management of our lives.

SUMMARY

The ability to learn from experience and store memories is one of the key ways in which animals adapt their behaviour to the environment in which they live. Such abilities are widespread through the animal kingdom including simple invertebrates. Here we begin by outlining some of the simplest forms of learning, e.g. habituation in which animals learn to cease responding to any stimuli or events which have no important outcomes. Associative learning is more specific. Animals learn that certain stimuli, hitherto neutral, may be associated with reward or punishment and modify their behaviour accordingly. Alternatively they learn that the performance of certain actions – pulling a lever, pushing a bar – yield results which are rewarding. We describe some of the systematic work of the Russian, I.P. Pavlov, and the American, B.F. Skinner, whose schools investigated the behavioural characteristics of such learning in great detail.

There certainly are some general characters which are common to all associative learning but once biologists began to study a wider range of animals than the dogs, rats and pigeons that were Pavlov's and Skinner's subjects, numerous specializations in learning ability were revealed which showed how selection could modify learning abilities to match life histories. We give an account of two such remarkable specializations, those of honeybees and of food-storing birds.

Looking at learning abilities across different groups, we can find evidence that we may have underestimated the capacities of some animals to respond appropriately to new problems. In particular, birds such as parrots and crows have been shown on occasions to behave, as we might label it, 'intelligently' at a level to match monkeys and even apes.

This leads us to consider complex or higher forms of learning and how far such studies yield evidence on what we might call the minds of animals. How far are they able to predict the outcomes of their own actions or those of others? One must be cautious not to assume that animals are responding as we would and we should never apply an elaborate explanation for their behaviour if a simpler one will suffice. There are numerous observations and experiments with a range of 'higher' animals – whales, elephants, monkeys and apes – attempting to address such issues which are both inherently fascinating and highly controversial. We attempt to cover some of the evidence which is often conflicting. It leads some people to suggest that the minds of animals will have evolved, like any other behavioural feature, to match their life histories. Others argue that there is a significant discontinuity between ourselves and even our closest relatives – chimpanzees. This gap has resulted from our acquisition of speech, absent elsewhere and which they suggest forms the basis for thought.

We conclude with a brief discussion of the nature of memory itself and evidence on how memories can be stored in the brain. Modern, non-invasive techniques are now enabling such investigations in humans as well as animals.

CHAPTER SIX

Evolution

Evolution by natural selection is the great unifying concept of biology, so much so that most biologists now feel that, without it, none of the phenomena they study really make sense. Richard Dawkins (1986) provides an excellent modern survey of the great explanatory power of what has come to be called 'neo-Darwinism'. Natural selection, operating on random, inherited variation has, over the generations, shaped animals to match the environments in which they live. Sometimes adaptation has been achieved through genes, accumulated over many generations, biasing the development of behaviour directly into appropriate responses. In other cases, animals inherit only biases to respond adaptively to their immediate environment so as to acquire, individually, appropriate responses by learning. Much behaviour develops by a mingling of such processes. All through this book we have emphasized the adaptive role of behaviour in an animal's life and so the concepts of 'evolution' and 'adaptation' have been implicit in much of what has already been said. Now we look at them in more detail.

In fact, not just one but two of Tinbergen's four questions are about the evolution of behaviour. One of them, about 'adaptiveness', is about how behaviour that we see present-day animals doing helps them to survive and reproduce. The other, about phylogeny, is about the changes in behaviour that have taken place over evolutionary time. In other words, we ask what the long ago ancestors did and how what they did gradually changed over time to give us modern behaviour, rather in the way we can ask how the limbs of the ancestors of horses or whales gradually changed over time to give modern-day hooves and flippers. Of course, these two evolutionary questions are closely linked together because what happened in the past can 'spill over' into the present. Modern animals behave as they do because of natural selection acting on their ancestors, but the two questions are also usefully distinguished. Adaptiveness is about how natural selection has acted or is still acting in the present day, eliminating some animals and favouring others, determining what future generations will be like. Phylogeny is about the past history of each species, unfolding over perhaps millions of years, the story of how animals changed over time to have the shapes and colours and behaviours that they do.

In this chapter, we begin with the adaptiveness of behaviour – what exactly this means and how we can study it. We will see that although the 'survival of the fittest' sometimes leads to conflict and aggression between individuals, it can also lead to cooperation and caring for others. Nature may sometimes be 'red in tooth and claw' but it can also, through the very same process of natural selection, lead to protection, sharing and self-sacrifice. Then we will turn to the last of Tinbergen's four questions, the one about phylogeny and look at the way in which behaviour has not only changed over evolutionary time but has also been a major influence on the evolutionary process itself.

The adaptiveness of behaviour

It is easy enough to recognize in a general sense that animals are adapted to their way of life. A camouflaged insect sitting quite immobile on a background that it matches perfectly, for instance, is clearly adapted to conceal itself from predators. However, to appreciate the full power of evolutionary processes to bring about adaptation, we need more detailed studies to show just how overwhelming the evidence for natural selection can be.

Mole crickets (*Gryllotalpa*) are large insects whose males dig burrows underground in which they sit and sing to attract females flying overhead (Fig. 6.1). As with many other

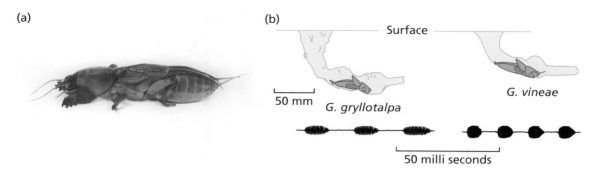

Figure 6.1 (a) A mole cricket (*Gryllotalpa vineae*). Note the massive forelegs for digging. (b) Side views of males of two species of *Gryllotalpa* sitting in their burrows, head downwards in the singing position. The shape of the burrows approximates to an exponential horn. *G. vineae*, on the right, has a smoother tunnel and a much louder song. In both cases the shape of the burrow is adapted to the song frequency so as to emit with the maximum efficiency.

species of crickets and grasshoppers, the song of each species is very distinctive and plays a role in the sexual isolation between species. The sound of the song is determined in part by the structure and size of the forewings, which are rubbed rapidly together to produce sound. A scraper or plectrum on the hind margin of the left wing rubs along the toothed undersurface of a vein on the right wing (the 'file'). In two species studied by Bennet-Clark (1970, 1987), the fundamental sound frequency within the pulses of the song (see Fig. 6.1) is about 1600 Hz in *G. gryllotalpa*, a species with small wings, shallow teeth on the file and a quiet song, but 3500 Hz in *G. vinae* which has large wings, deep teeth on its file and a much louder song.

Here, then, we can see a direct connection between behaviour (the singing) and morphology, but the mole crickets' adaptiveness goes much further than this. By comparing different, but closely related species, we can see that the male of each species excavates its burrow in a different way (Fig. 6.1b). Bennet-Clark showed that the burrow forms an exponential horn with a bulb at its base so that when the male sings with his head just at the origin of the horn, his song is amplified, as by the horn of an old-fashioned loudspeaker. However, the exact size and shape of the burrows differ between species. *G. vinae* builds a double-mouthed, horn-shaped burrow to sing in. The walls are smooth and there is a bulb that acts as a resistive load for the vibrating wings and concentrates the sound in a disc-shaped patch of sound energy above the burrow. *G. gryllotalpa* also builds a horn but it is much larger and is very effective at amplifying the much lower pitched sound that this species makes. Thus the wing structure, singing and burrow-digging behaviour of the mole crickets are all precisely co-adapted with each other so as to provide the most efficient sound production, 'aimed' at the females, which are attracted to the males of their own species.

Such attention to detail, as it were, has many more strictly behavioural counterparts. For example, a pattern of behaviour exhibited once per year and then only for 5 minutes

(b)

(a)

Figure 6.2   (a) A black-headed gull removes a piece of empty eggshell from the nest after a chick has hatched from it. As the gull flies away from the nest with the shell, it leaves the nest vulnerable to predators such as herring gulls and crows, which led Tinbergen to suggest that the behaviour itself must have some strong positive advantage. (b) To test the hypothesis that the advantage was protection from predators, Niko Tinbergen painted hens' eggs to look like gulls' eggs and then placed eggs and empty shells out on the sand dunes.

might seem to be of trivial significance (Fig. 6.2a). This is the habit of black-headed gulls of removing empty eggshells from their nests soon after their eggs hatch. Even gulls breeding for the first time carry away eggshells promptly, although they never pick up eggshells in any other situation. Niko Tinbergen (Fig. 6.2b) and his colleagues (1962), in a classic study, clearly showed that eggshell removal is vital for keeping the nest camouflaged. By placing eggs out on the sand dunes with or without eggshells near them, they showed that chicks or unhatched eggs with eggshells near them were much more vulnerable to predation by crows and herring gulls. Both aerial and ground predators discover nests and chicks more easily if empty shells, with their smell and white shell insides, are nearby. Natural selection has operated to ensure that parent gulls always perform this brief but important behaviour. Clearly, we must never write off as functionless – however trivial it may seem – any piece of behaviour which we regularly observe in a natural situation.

It is worth pausing for a moment to consider how we can know for certain that a trait is adaptive. Gould and Lewontin (1979) argued that it was all to easy to speculate about adaptation and accused biologists of making up adaptive explanations that are often nothing more than 'just-so' stories – plausible ideas that are completely untestable. So what kind of evidence can we hope to find about adaptation and, more importantly, how can we put adaptive hypotheses to the test?

The first thing we have to do is to clarify exactly what we mean when we describe behaviour (or any other trait) as adaptive. What we mean is that animals that possess that trait (such as doing a particular behaviour) survive and reproduce better than animals that do not. In the struggle for existence, the act of removing empty eggshells from the nest gives an advantage because the chicks were less likely to be taken by predators. More of them survived, carrying the favourable trait into the next generation while the chicks of those parents that did not remove the empty shells died. Adaptiveness in the animals that are still alive, therefore, may need to be defined by success relative to animals that have already died out. Since we have no direct evidence of what happened in the past, we have to look for evidence of adaptiveness in animals that are still alive today. For example, Tinbergen's demonstration that eggs with shells close to them are more at risk than eggs with empty shells at a greater distance points directly to the current adaptiveness of a behaviour that reduces this risk.

Then, if we return to the case of the mole crickets, we have two more pieces of evidence of clear adaptation to function – the shape of their burrows and the difference in shape between the burrows of the two species. A human engineer, asked to design a simple device for amplifying sound, would probably come up with an exponential horn (as in ear-trumpets or the early gramophones). The fact that the mole crickets dig burrows of precisely this shape strongly suggests that sound amplification was the feature favoured by natural selection. Crickets that built burrows of a different shape would not be able to emit such a loud sound (an engineer could tell us that) and, by implication, would be less successful at attracting females. The burrow appears to have been 'designed' to amplify the male's song.

This conclusion is confirmed by further evidence – the comparison between the two species. The sound engineers know that the exact 'best' shape for an exponential horn depends on the sound frequency it has to transmit. If a mole cricket's burrow is 'designed' to amplify with maximum efficiency, then we should expect species with different song frequencies to have burrows of different shapes with each 'tuned' to match the song frequency. This is exactly what we find. In fact, comparison of this type between species, each of which is successful but achieves success in a slightly different way, has become one of the most important methods by which ethologists can document the adaptiveness of behaviour. This method is particularly powerful when it involves large numbers of species, all convergently showing the same adaptation (Clutton-Brock and Harvey, 1984).

Figure 6.3   The Nemestrinid fly *Prosoeca peringueyi* visiting the long-tubed flowers of *Lapeirousia silenoides* (Iridaceae) in South Africa. An extreme example of mutual adaptation; yet more of the fly's tongue is inside the flower's tube as it reaches in for the nectar at the base.

Other powerful demonstrations of adaptation come when we do indeed examine two species not because they are phylogenetically related but because they have co-evolved, each adapting to the other to their mutual benefit. In South Africa, there are several families of plants which have evolved very long-tubed flowers whose nectar, secreted right at the base, can be reached only by very long-tongued pollinators. Some flies of the family Nemestrinidae have evolved just such tongues to match! Figure 6.3 shows a flower of the family Iridaceae being visited by *Prosoeca peringueyi*, just one of these amazing flies whose life is linked with that of the flowers – beautifully adapted to one another. The fly has a source of food which only it can obtain, the flower ensures that its pollen (which the fly inadvertently carries on its head after sucking nectar) has a maximum chance of being passed on to the stigma of another such flower, thus

Figure 6.4 The moth *Automeris* has cryptically coloured forewings, which provide camouflage when it is settled. However, if touched by a predator, it flashes open the forewings to reveal bright eye spots on the hind wings, draws in its legs and rocks rhythmically. It is quite startling and has induced alarm in small birds.

ensuring cross-fertilization without 'wasting' pollen (Manning and Goldblatt, 1997; Johnson and Steiner, 1997).

The wing markings of some moths and butterflies are another case where the intricacy of adaptation can be clearly seen. The so-called 'eye-spot' patterns of these insects and the ways in which they are displayed (Fig. 6.4) are a beautiful example of co-adaptation between morphology and behaviour. Birds are common predators of Lepidoptera and there is experimental evidence (e.g. Blest, 1957; Vallin *et al.*, 2005) that quite crude circular markings are better at alarming small birds than other patterns, but most Lepidoptera go much further than this. Many predatory birds, especially hawks and owls, have large, prominent eyes. They need them to detect their prey, often under low light conditions. It can be no coincidence that small birds respond with intense alarm to large eye patterns and, for example, are far more disturbed by painted eyes on a plain background than by a real owl whose eyes are covered. The development of colour patterns on the wings of Lepidoptera often takes place from foci, which means that colour tends to be laid down in whorls or circular shapes. These intrinsic forms have offered butterflies and moths an 'opportunity' to discourage small birds which try to prey on them, by presenting a sign stimulus which resembles that presented to birds by their predators – owls.

Now, of course, we are observing the wing markings with human eyes and immediately interpret them as 'eyes'. Small birds may not 'see' them in this way and it could be that they are responding simply to something which is startling. The experiments of Blest (1957) and Vallin *et al.* (2005) show that the patterns do involve display too. Eye-spots on the hind wings of a hawk moth are hidden beneath the forewings when at rest but are suddenly flashed into view if the moth is touched. Stevens (2005) reviews a range of possibilities on how eye-spots 'function' and certainly it is not always clear. The reason

Figure 6.5 The eye-spot on the underside of the hind wings of the butterfly *Caligo*. This illustrates how closely natural selection has been able to shape colour patterns. The resemblance to large vertebrate eyes, such as those of owls, is remarkable.

they are interesting to us here as examples of adaptation is because of the situation in which they evolve. Modifying the detail of a pattern involves only pigment changes to some of the minute scales which cover the wings. It seems unlikely that this is costly in metabolic terms nor will it affect the functioning of the wings in any way. For once, then, natural selection can operate with few, if any, constraints and there can be no better illustration of what can be achieved than the eye-spot pattern of the Brazilian butterfly, *Caligo* (Fig. 6.5). *Caligo*'s markings mimic with extraordinary precision large bulging vertebrate eyes set in deep sockets. There are even a few white scales arranged in an arc added, as would a human portrait painter, to suggest the highlights reflected off the glistening curved surface of an eye. For *Caligo*, there has been no barrier to selection operating over hundreds of thousands of generations to bring these extraordinarily precise details into play. It is certainly very hard to believe that the resemblance to a large vertebrate eye which we humans recognize has not played a part in its evolution, even if puzzles still remain as to how *Caligo* 'presents' them (Stevens, 2005). Presumably each minute change will have brought only a tiny advantage, but there has been time for even the tiniest advantage to pay off in terms of slightly more effective scaring of predators and hence result in the survival of a few more of those butterflies which bear it.

Adaptation or perfection?

Such examples, and they are legion, reveal how powerful natural selection can be in shaping behaviour. But 'adaptive' does not mean that every aspect of an animal's behaviour is perfect. After all, some animals do get killed by predators or die of starvation, and over 90% of all the species that have ever lived are now extinct. Natural selection is an on-going struggle for survival and there are many reasons why perfection is never achieved (Dawkins, 1982; Ridley, 1993).

One important reason for a lack of perfect adaptation is that no behavioural trait evolves in isolation, but has to be seen in the context of everything else the animal is doing. This means that selection on one behaviour or structure can rarely be without effects elsewhere, so what the animal does may end up being a compromise. For example, adaptations by males to attract females, such as elaborate displays or loud songs, are also likely to attract predators, so there will be an adaptive compromise between reproductive success and danger from predation. Most male birds reduce or even stop singing as soon as their mate begins to incubate and some lose their breeding plumage at the autumn moult. A most dramatic example of compromise is shown by the neotropical Túngara frog (*Physalaemus* (now called *Engistomops*) *pustulosus*) (Ryan *et al.*, 1982). Females are attracted to the sounds of males calling at the edges of ponds. But frog-eating bats use the males' calls to home in on the males and eat them. Unfortunately for the frog, those aspects of male vocal behaviour that are most attractive to females (calling intensely and producing calls with 'chuck' sounds) are also the ones that most increase the predation risk.

It is clear, then, that we must expect in most cases to find less than perfect adaptations because animals are subject to many different selection pressures and the overall 'best' animal may be less than perfect if the individual components of its behaviour are looked at separately. Sometimes this results in dramatic examples of failure, as with the hapless Túngara frogs. However, this is not argument against natural selection. Rather, it is the result of natural selection acting simultaneously in two different ways. Selection to have a loud and conspicuous call to attract females is balanced by selection to avoid predation by remaining quiet. No solution would satisfy both selection pressures simultaneously so that some imperfections, sometimes, are inevitable.

Another important reason why animals may be less than perfectly adapted is that the world is not static but constantly changing. Traits that are adaptive at one time may become less so over time but the animals may not be able to evolve rapidly enough to continue to survive. The changes in temperature that took place at the beginning and end of ice ages left many animals unable to keep up with changes in their physical environment and food supply. But while many adaptations concern the way animals survive in the face of adverse physical conditions, such as heat, cold, drought or salinity, by far the most important reason why animals may be less than perfectly adapted is that other organisms are evolving too. The struggle for existence is not just a private fight between an animal and its physical environment. It is a very public battle in which adaptations by one animal (for example, to avoid being eaten by becoming cryptic) set up counter-selection pressures on others (for example, their predators, to see through the camouflage). The more closely a caterpillar is selected to resemble its background and thus gain the advantage of not being eaten by birds, the more selection pressure there will be on the visual system of the birds to become better able to discriminate the caterpillars from twigs, which in turn sets up yet further selection pressure to look even more like a twig and so on. Neither the caterpillar's camouflage nor the birds' vision will be perfect because both are locked into a never-ending arms race (Ridley, 1993).

Genes and behavioural evolution

The evolution of any adaptive trait by natural selection implies that there exists – or at least there has once existed in the past – genetic variation between individuals. Evolution is about changes in gene frequency and so to call a trait 'adaptive' implies that the individuals showing this trait are in some way genetically different from those that do not. If the differences between gulls that removed eggshells and those that did not or between frogs that call loudly and those that call softly are entirely environmental, then even though these behaviours might have a major effect on survival and reproduction, they are not adaptations in the sense we have been discussing because the traits are not passed on to the next generation. Only if one genotype becomes more frequent in the population because it makes the body it is in taller, fatter, able to run faster or more successful in some other way are we really seeing evolution in action and only then are we entitled to talk about the adaptiveness of the trait.

In this section, we consider the implications of this intimate connection between adaptation and genes. To start with, studying the adaptiveness of behaviour does not in any way commit us to a crude genetic determinism in which possession of a particular gene determines, in a fixed and irrevocable way, how an animal will behave. On the contrary, as we saw in Chapter 2, the way genes affect behaviour is usually much more subtle and complex than this. We saw there that, whilst it is possible to study some examples of how genes control the way a nervous system develops and so exert their effects on behaviour directly, in other cases, the route from genes to behaviour is more elusive, operating though hormone levels or through complex interactions with the environment such as diet, temperature or interactions with parents, amongst many other factors. Then, in Chapter 5, we saw how behaviour can be radically modified through learning, and this raises an even more intriguing possibility, namely, that natural selection may affect behaviour not just directly but also by affecting an ability to learn. By favouring animals that learn rapidly, or even learn particular things, natural selection opens up a new route to 'adaptive behaviour': successful animals inherit genes that affect learning ability, but exactly what they learn and how they behave will depend on what they encounter during their own lifetimes. The point here is that, although different animals may behave differently partly because of the different things they have learnt about, what they learn may, in turn, depend upon genetic differences in speed of learning, memory or sensitivity to different environmental stimuli or rewards. The variation between individuals may thus owe a great deal to the environ- ment but still have a genetic base that allows evolution to occur.

We now have good evidence of precisely the sort of genetic variation within popula- tions that is needed for the evolution of adaptiveness. One of the clearest examples of this comes from looking at how humans have been able to change domestic animals by selectively breeding from extremes in a population which shows variation in the expression of some behaviour. It is interesting to reflect that in Darwin's (1859) *Origin of Species*, Chapter 1 is entitled 'Variation under domestication'. Darwin was a keen pigeon fancier and knew how dramatically the body form and the behaviour of the domestic pigeon, *Columba livia*, had been shaped in diverse ways by human selection (Fig. 6.6).

Genetic variability is also widely present in natural populations. Using well-standard- ized tests with that most famous of ethological animals, the three-spined stickleback, Bakker (1986) found considerable variation in levels of aggression between individuals all of which had been raised in isolation from wild caught parents. Selective breeding proved that a good proportion of this variability was genetic. Interestingly, when he selectively bred for high and low levels of the aggressiveness shown by males when defending their territory, Bakker found that selection for lower levels of aggression was more effective than for higher levels. This result suggests that natural selection already keeps aggression levels close to their feasible maximum. Probably, we should expect to find such a 'ceiling effect' when we are dealing with behavioural characters so crucial to being successful in reproduction.

But sometimes, as we discussed in Chapter 2, there is more than one way to success. In a cricket of the genus *Gryllus*, Cade (1981) used selective breeding to demonstrate genetic variation which may well be the basis for two different behavioural strategies which we can observe naturally. Males are either 'callers' that sing a great deal thereby attracting females, or 'satellites' that remain relatively silent but may intercept females

Figure 6.6   An amazing array of domestic pigeon breeds such as intrigued Darwin. Most of these varieties concern plumage but some have been selected for behaviour too. The fantail (6) and the pouter (24) have much exaggerated and prolonged displays which form part of the male's courtship behaviour. The various breeds of tumbler (1–4) look fairly normal but have an inner ear defect which means that they become unbalanced and 'tumble' in flight.

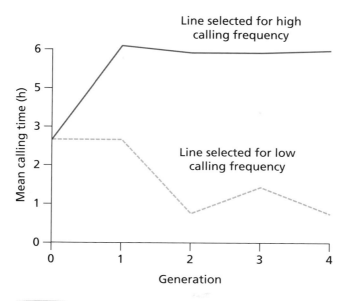

Figure 6.7 The length of time per day spent in bouts of calling by male crickets selectively bred for either high calling or low calling over four generations.

on their way to the calling males. Calling males attract more females than satellites but suffer more from parasitic flies, which are also attracted by the males' songs. Cade (1981) reared males in isolation in the laboratory and recorded how much each male sang. By measuring the mean calling rate per night from the 7th to 16th day of life, he showed that there was a bimodal distribution in the amount of singing. He then selected two to four males from each end of the distribution and mated them with non-sister females whose brothers were of the same type, and continued this process for four generations. The result was a divergence in the amount of singing shown by the two groups (Fig. 6.7), showing that there must have been a considerable genetic component to the variation in the original population. Genetic differences are not the sole determinants of calling rate, which is certainly not an all-or-nothing trait. The amount a male calls can be influenced by his environment. Callers that are surrounded by other callers tend to sing less than when they are on their own, and 'satellites' may be quite vocal if no other cricket is singing nearby. Yet Cade's results show how, if favoured by selection, genetic changes can change the bias in the way calling behaviour develops.

One of the most striking examples of genetic variation within natural populations and its significance for behavioural evolution comes from the work of Berthold and his group on the migratory behaviour of some European warblers. They have succeeded in getting blackcaps (*Sylvia atricapilla*), among others, to breed easily in captivity. Berthold (2003) reviews a number of studies on the patterns of migratory restlessness phenomenon in different populations of blackcaps. Most fly to Africa for the northern winter, but in summer breed from the Canary Islands northwards over much of Europe. This wide distribution obviously means that the length of migratory flights varies greatly between different populations and breeding experiments show that birds from different populations differ genetically in how far they are prepared to fly. Nestling blackcaps from four different breeding areas (southern Finland, southern Germany, southern France and the Canary Islands) were hand-reared and kept in aviaries away from adults. When, in their first autumn, the time came when they would naturally be undertaking their migration to Africa, the captive birds showed migratory restlessness, fluttering and jumping in the aviaries. There was a correlation between how far the birds would have had to travel had they been left in their natural populations and the degree of migratory restlessness (its intensity and how many days it persisted). Birds from the Canary Islands, which are close to the wintering areas in Africa, showed much less migratory restlessness than birds from the Finnish and German populations (Fig. 6.8). The genetic basis of these differences was clinched by cross-breeding birds from different populations. Hybrids showed intermediate levels of migratory restlessness.

In extensions to this work (see Berthold, 1993; Berthold and Helbig, 1992), artificial selection experiments have revealed just how rapidly natural selection could change migratory behaviour. In populations of blackcaps breeding in the south of France, most birds migrate south in winter but about 25% stay put. Berthold and his collaborators selectively bred from aviary-reared birds which either showed migratory restlessness or alternatively did not show it. Within three or four generations they had populations which were totally migratory or totally non-migratory (Fig. 6.9). Obviously blackcaps in nature are genetically 'poised' to respond rapidly to climate change. There is good evidence that this has already happened in Western Europe for now the German populations have diverged. One more central population still migrates south in our winter, but in western populations more and more blackcaps now migrate *northwest* and spend the winter in the British Isles. This has required a shift in direction, duration and probably the timing of their migratory behaviour. Pulido *et al.* (1996) have shown that the tendency to migrate and the duration of migratory restlessness have a common genetic basis on which selection will have been operating. Less is known about the genetic control of the direction of flight. It is possible that young birds acquire this by moving with the parental generation during their first autumn.

In summary, whilst we may not yet understand the details of how behaviour such as aggression or learning ability develop within the individual, we do have good evidence that many of the behavioural differences between individuals or species are traceable back, ultimately, to genetic differences. We can also begin to understand how natural selection brings about behavioural change as it 'searches' to produce the best fit to an animal's environment.

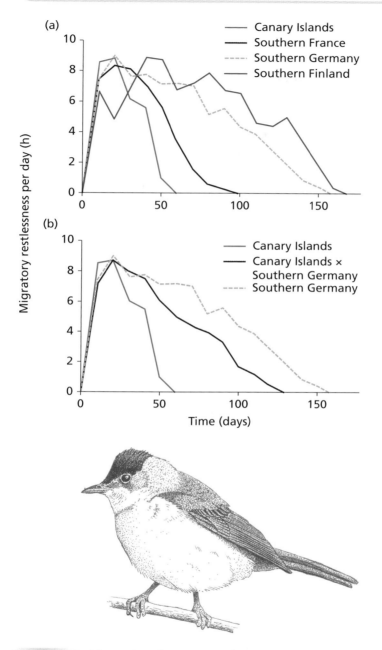

Figure 6.8 (a) Duration of migratory restlessness in blackcaps whose parents had come from different parts of Europe and the Canary Islands off the west coast of Africa. The lines represent the mean values for groups of birds ranging from 6 to 26 in number adjusted so that the first night of restlessness for each group coincides. Note that the length of time each group spends 'migrating' coincides with the distance the parent population would migrate to Africa even though the birds had been hand-reared. (b) Shows that hybrids between Canary Island blackcaps and those from Southern Germany show an intermediate degree of migratory restlessness.

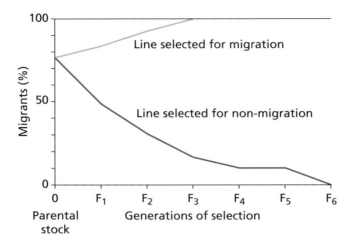

Figure 6.9   The dramatic result of a two-way selection for blackcaps which migrate (i.e. show migratory restlessness in the northern autumn and spring) and those that do not migrate. Starting with birds from a population in southern France which showed about 75% of migrants, within three generations the migrant line was fixed with 100% migrating. Within six generations the low line shows virtually no birds migrating.

Kin selection and inclusive fitness

Adaptation is about the evolution of behaviour and structures that lead to success in evolutionary terms. Up to now, we have discussed 'success' in terms of individual survival and leaving offspring that themselves reproduce. But the success of an individual is short-lived and ephemeral. In sexually reproducing species, an individual does not survive for more than one generation. As the previous section emphasized, it is genes that are passed from one generation to the next and our adult bodies could simply be regarded as the elaborate packaging that protects them. As far as a gene is concerned, the body it happens to be in at a given moment is a useful, if temporary, vehicle for getting itself passed on into the next generation, as R. Dawkins (1989) puts it. Samuel Butler's aphorism that 'A chicken is an egg's way of producing another egg' can be rewritten as 'An animal is a gene's way of producing more copies of that gene'.

Many people object to this way of looking at animals and feel uncomfortable with the idea of dispelling the sovereignty of the individual in favour of a gene-centred view of evolution. But consider something as basic and fundamental as parental care. We take it for granted that parent animals should feed and protect their young, but parental care itself is the result of strategies genes have for perpetuating themselves. Genes that help to make the bodies they are in more effective at defending their young will perpetuate themselves in the bodies of the protected young. Variation in genes affecting a tendency to defend young – perhaps mediated through variations in the level of a hormone – will result in variation in the numbers of offspring that survive to pass on the favoured genes, and so on down the generations.

But there is a twist to this tale. The direct line of parents to offspring is the only way that genes are passed on into the future, but direct parental care is not necessarily the only genetic strategy that will be successful. Helping a brother, sister, or other relative to reproduce may also enable genes to perpetuate themselves. Full brothers and sisters share, on average, half their genes (although the vagaries of Mendelian segregation mean that particular pairs of siblings may have much more or much less than this). So a genetic tendency to help a sister to reproduce could be favoured by natural selection because the sister, being so closely related, has a high chance of having the same genetic tendency. 'Genes for' helping sisters thus help copies of themselves in the sisters of the body they are in and perpetuate themselves through the children of those sisters. W.D. Hamilton (1964) showed how genes for 'care of relatives' (not necessarily direct offspring) could spread and Maynard Smith (1964) suggested the term 'kin selection' to describe selection which takes account of other relatives as well as immediate descendants.

There is a very important point which has to be made here about kin selection. Helping a given relative to reproduce will only be favoured (genes for it will only be spread) if the benefit – that is, the increase in reproductive chances of that relative as a result of the help – more than makes up for the cost – that is, the decrease in reproduction the helper incurs as the result of its action. For example, there is no point in helping a brother if the help does not enable the brother to have any more children and at the same time prevents the helper having several children of his own. Hamilton generalized the circumstances in which relative-helping of various sorts would evolve into the equation: $rb - c > 0$.

r is the coefficient of relatedness and expresses how closely two individuals are related to each other, b is the benefit and c the cost of the relative-helping genotype. The net benefit minus the cost must be positive and greater than zero for the behaviour to be favoured. In order to work out whether or not the equation does work out greater than zero, we have somehow to calculate values for the three terms, r, b and c.

'r' does not usually cause too many problems. From basic genetics we can work out that full siblings, and parents and offspring have a 50% chance of sharing a given rare gene ($r = 0.5$); nieces and nephews have 25% chance of sharing with an uncle or an aunt ($r = 0.25$), and so on.

In the past, it was very difficult to discover how closely related animals are to each other in nature. They had to be studied over a long period of time to find out which were the offspring of which adult and even then there was often considerable doubt, particularly about paternity. Now, however, DNA fingerprinting has revolutionized the determination of r. All that is needed is a small sample of blood, or even a hair or a feather and in many cases it is possible to determine the relationships among animals with considerable accuracy. This is made possible by the fact that throughout the genomes of many species there are regions of DNA in which the patterns of base pairs are repeated over and over again but the repeat patterns are slightly different in different individuals (Queller *et al.*,

1993). The results have often been quite surprising. For example, in a recent study of starlings, which have always been considered, like most birds, to be monogamous, DNA fingerprinting showed that one brood of chicks was fathered by no fewer than three different males (Pinxthen *et al.*, 1993). Behavioural observations had not revealed any extra-pair copulations and it was only the genetic evidence which showed that these must have been occurring. We now know that this situation is commonplace. Even socially monogamous birds regularly have extra-pair copulations.

It is important to remember, however, that it is not just the degree of relatedness that matters but the number of relatives that are helped. If it were possible to help a large number of distant relatives, it would be just as beneficial as helping a direct offspring – hence J.B.S. Haldane's famous after dinner remark, 'I am prepared to lay down my life on behalf of four grand-children or eight first cousins'!

Although the r component of Hamilton's equation is now readily accessible, b and c are somewhat more problematic. If we observe one animal helping another to rear its young, how do we know that the parent wouldn't have been just as successful without the help? And how do we know what the cost to the helper was in terms of the offspring it would have had if it hadn't been helping someone else? We seem to be dealing with mythical potential offspring that don't exist (the cost of helping) and extra offspring that do but are indistinguishable from the others (the benefits of helping). We can look at some examples to see how estimates of b and c are arrived at in practice.

In two original and very important papers, Hamilton (1964) addressed himself to a problem which had long puzzled zoologists. The social insects, the Hymenoptera (ants, bees and wasps) and the Isoptera (termites) show extreme altruistic or helping behaviour. There is usually just one reproductive female (the queen) and large numbers of sterile workers. The workers – both males and females in termites, but solely females in the ants, bees and wasps – perform all the tasks of the society such as foraging, rearing young, nest construction and defence, and do not reproduce at all themselves. The remarkable fact is that, in the Hymenoptera, this ultimate form of self-sacrifice – for the sterile workers do not, of course, reproduce themselves – appears to have arisen independently at least 11 times and perhaps more. Clearly, it has not been easy to evolve social life or more different types would be expected to have done so, yet somehow this one order of Hymenoptera seems predisposed to achieve it.

Hamilton drew attention to the significance for r – the coefficient of relatedness – of the Hymenopteran unique form of sex determination. Ants, bees and wasps exhibit 'haplo-diploidy' which means that the males are haploid and develop from unfertilized eggs, whereas the females are diploid and develop in the normal fashion from fertilized eggs. All the sperm from one male are therefore identical – a simple copy of the male's own haploid chromosome set. When the queen bee fertilizes eggs with this sperm, all the resulting daughters receive the same paternal chromosomes and so have half their genes in common (the half donated by their common father). In addition, they share, on average, half the genes inherited from their common mother and so their degree of relatedness, r = 0.75 (0.5 from father plus 0.5 × 0.5 = 0.25 from mother).

This means that although the workers are sterile themselves, many of the genes that they share with the young queens (which are also their sisters) will be passed on to the next generation. The sterile workers benefit, not because they help other workers which are closely related to them but are a reproductive dead-end, but because they care for the small number of their sisters which will develop into young queens. As Hamilton puts it, in gene terms, it is more advantageous for a female Hymenopteran to stay and help to rear her closely related reproductive sisters than to leave and attempt to rear less closely related daughters of her own.

While haplo-diploidy appears to predispose the Hymenoptera to the high degree of sociality they show, it is clearly not the only factor because, as we have mentioned, a similar degree of sociality is also shown by the termites (Fig. 6.10). They have an ordinary diploid mating system and sterile workers of both sexes have a degree of

Figure 6.10 The mounds built by termites can reach astonishing sizes.

relatedness to the young reproductives of only 0.5. King and queen termites are long-lived and monogamous. A queen may lay up to 36 000 eggs a day and in some species she can live for 60–70 years. The king and queen together may have literally millions of offspring in their lifetime, the vast majority of which will never reproduce at all. In order to understand why so many termite workers should be sterile, we have to remember that there are three variables in Hamilton's equation. A high r predisposes towards helping because it increases the effective benefit, but even a relatively low value of r would lead to helping if the benefits were high enough and the costs low enough.

The cost to each worker of being sterile is the loss of those offspring it would have had if it had not been helping the colony. Given that a pair of termites on their own would probably not survive, let alone reproduce even modestly, the costs of helping must also be small (no hope of reproducing means no cost to losing it). The monogamous nature of the termite breeding system means that the workers are guaranteed a long series of full brothers and sisters to take care of. Since their own parents offer no care to them whatsoever, without the workers the young termites would die. The workers, therefore, make a substantial difference to the survival chances of close relatives, even though the benefit, b, accruing to each worker is only a fraction (because it is shared with the other workers which have helped) of the output of each reproductive.

Nevertheless, it appears that the benefits of helping are significant, the costs low and r at least as high as between diploid parents and offspring. Termites often live in deserts or in very dry regions and can only do so because the mounds (Fig. 6.10), built by the collective labours of millions of workers, enable them to create their own microenvironment. A single pair of termites, removed from this specialized microenvironment, would stand no chance, and this fact effectively reduces the costs and increases the benefits of helping the royal pair in the home colony.

Until quite recently, worker sterility and a caste system of sociality were thought to be features exclusive to insects. Now we know that at least one mammal, the naked mole rat (*Heterocephalus glaber*), also has a very termite-like social system (Jarvis, 1981). These extraordinary looking rodents (Fig. 6.11) live in underground colonies in dry desert regions of southern Africa. They are almost hairless and almost poikilothermic, but their burrows provide them with a safe environment of almost constant temperature so that their inability to regulate their own body temperature is not a problem. Their burrow systems can be extremely extensive – up to 3.5 km – and shared by 70 to more than 200 individuals. There is usually just one breeding female and up to three reproductive males. The rest of the colony consists of workers, most of which never reproduce throughout their lives. Instead, they defend the colony and, working like a chain-gang, dig tunnels through the hard earth to find food for the rest of the colony. They depend primarily on the very large tubers of desert plants. These are scarce but the rats feed mostly on the inside of the tubers, which allows the plant to regenerate, and one large tuber will support a colony for months or even longer. The reproductive female does not take part in the food-gathering activities and has food brought to her by the workers. By cooperating with the colony and contributing to the reproduction of

(a)

(b)

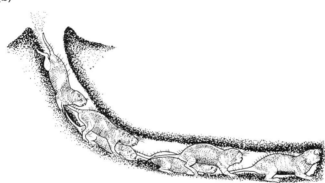

Figure 6.11   Naked mole rats live underground in colonies of closely related individuals. The workers cooperate to tunnel through the hard earth to find food.

others, at least the workers get a 'part-share' of the offspring produced. Tests show that the mole rat's monogamy plus a considerable degree of inbreeding results in an average relatedness in the colony of 0.81. The reproductive female produces huge litters of up to 27 pups, giving the sterile workers lots of closely related young to look after. This harsh environment where food is so hard to come by means that a pair of mole rats on their own would be unlikely to survive. Reproduction is hazardous and is only possible with the help that workers can give, and these factors seem, as with the termites, to have tipped the cost–benefit equation in favour of the ultimate in self-sacrifice, life-long worker sterility.

A second species, the Damaraland mole rat (*Cryptomys damarensis*), offers an interesting comparison here and indicates how the eusociality of the naked species may have evolved (Jarvis *et al.*, 1994). This species is larger, covered with hair and lives in subtropical areas of Africa where the soils are softer and the burrow environment is more variable. Although, like the naked mole rat, it is eusocial and has non-breeding workers, *Cryptomys* colonies are much smaller, usually numbering about 16 individuals. This seems to be related to the fact that the soils where they live are softer than the rock-hard desert soils faced the naked species and so fewer individuals are required to dig. The benefit that Damaraland mole rat workers gain by staying to help their parents rear young seems to be less and the workers have a much greater chance of leaving the colony and founding one of their own where at least some of them become reproductive. Their cost–benefit equation is still tipped in favour of some worker sterility, but less strongly.

Comparisons with various species of birds show how fine the balance can be. None, as far as we know, goes to the lengths of being totally sterile, but in over 200 species, the parents are helped in some way by other individuals, often their own young from previous years. Later, these helpers frequently become parents in their own right but the selective advantage to their juvenile helping bears a strong resemblance to the helping seen in termites and mole rats. Firstly, most birds not only nest in the same territory year after year but are also monogamous (although, as we have seen, not always perfectly so), which means that the coefficient of relatedness between siblings of different years is high. Secondly, helpers can contribute significantly to the viability of their siblings. They may help in nest-building, territorial defence and feeding, and they are often very important in keeping predators away. Emlen and Wrege (1989) showed that for white-fronted bee-eaters (*Merops bullockoides*), each helper enables the parents to raise, on average, half an extra chick. Woolfenden (see Woolfendon and Fitzpatrick, 1984) has shown that in Florida scrub jays (*Aphelocoma coerulescens*), which also have sibling helpers, the majority of helpers are males and are the sons of the birds they help (Fig. 6.12). They appear to increase the number of chicks the parents can raise to fledging largely through their active defence of the nest. So these helpers gain through having an increased number of young siblings and they seem to have a further benefit in that they tend to take over part or all of their parents' territory in the future, when they themselves start to breed. The cost of helping, that of postponing reproduction for a year or so, may not be all that great, since there appears to be a shortage of nest sites, making it very difficult for young birds to breed at all. Until a young male can secure a suitable breeding territory, he appears to be better off helping his parents to rear extra siblings than trying, and then failing, to reproduce himself (cf. the situation of the honeybee worker).

A similar benefit of helping, although in this case through supplementary feeding, can be seen in meerkats which are highly sociable mongooses found in the arid regions of southern Africa (Fig. 6.13). They typically live in groups of about 15 individuals, which tend to be related. A dominant male and a dominant female are responsible for most of the reproduction, with the rest of the group acting as helpers. Female helpers even start to lactate and feed young that are not their own offspring and the extra food they are able to give to the meerkat pups sets them up for life. The more the pups are fed when they are young, the heavier they are when they become independent at 3 months and the heavier they are at independence, the more likely they are to survive to reproduce themselves (Russell *et al.*, 2007).

Figure 6.12 Three adult Florida scrub jays at a nest. All were colour-ringed and could be identified as the parents of the chicks and a yearling bird from a previous brood. This young bird is now helping its parents to rear the current brood.

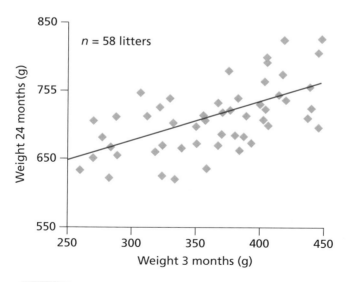

Figure 6.13 The beneficial effect of helpers in meerkats. The extra food provided by the helpers gives the young a higher body weight at independence (3 months) that is still seen in adults at 2 years of age. Higher body weight is correlated with greater lifetime reproductive success through earlier reproduction and greater chance of being at the top of the social hierarchy.

So we can see that kin selection, of which 'helping at the nest' is a clear example, is not a separate, special sort of selection, in some way different from natural selection, but a logical extension of it. Evolutionary success or fitness is not just the number of offspring reared to reproductive age by an individual, it also includes the effects that an individual might have on how many offspring its relatives have as well as how many offspring it fails to have itself as a result of its helping. This wider definition of fitness that includes effects on the reproductive success of different sorts of relatives is called 'inclusive fitness' (Hamilton, 1964; Dawkins, 1979; Grafen, 1984). Inclusive fitness includes reproductive success but allows us to see why individuals may sometimes do things that do not contribute to their own personal reproductive success but rather increase that of their relatives. What appears to be genetic suicide from the individual's point of view (such as not reproducing) can, from the gene's point of view, be a way of getting many copies of itself into the next generation.

Harming relatives and infanticide

But relatives will not always help each other, nor will parents always put themselves at risk for every offspring. Although Hamilton's equation shows us the conditions under which helping of relatives can occur, if those conditions are not met, then relatives may harm and even kill each other. This is because while an animal may be very closely related to its offspring and its siblings and other relatives, it will be even more closely related to itself. Apart from identical twins or clones, all animals have a higher coefficient of relatedness to themselves ($r = 1$) than to any other. This is important because whenever there are genetic differences between animals, there will be conflicts of interest in the sense that what is best for one animal may not be best for another. For example, the lifetime reproductive success of a parent may be greater if it withholds food from some offspring and gives it to others, even if the weaker ones die. Every time a parent feeds its young, it is expending effort and energy that could have been spent on rearing other offspring. From the point of view of an individual offspring, however, its best interests will be served by being given extra food and not dying of starvation. The genetic similarity between parents and offspring leads the parent to feed the offspring but the genetic differences between them leads to an underlying conflict about how that food should be distributed (Trivers, 1974; Godfray, 1995).

Owls and other birds of prey often lay more eggs than they can normally rear as an 'insurance policy'. All the chicks are reared if food is plentiful but the smallest one dies in times of scarcity. It is only a small step from withholding food to passive or even active killing. As Trivers (1974) pointed out, conflict between parents and offspring is most apparent at weaning, when it is in the offspring's best interest to demand a bit more food and in the parent's best interests to conserve resources for future children. For example, Hinde (1977) showed that as a baby rhesus monkey becomes older, the mother rejects her infant's advances and stops it making nipple contact with steadily increasing frequency until, at the end of the period of investment, all the offspring's advances are rejected (Fig. 2.23).

Figure 6.14   Great egret chick pushing its younger sibling out of the nest.

In some cases, siblings kill each other. Mock *et al.* (1987) showed that great egrets (*Casmerodius alba*) often practice 'siblicide' – active killing of nestlings by their older brothers and sisters, with the parents looking on and not intervening. A chick may stab a sibling with its beak, peck it and even push it out of the nest (Fig. 6.14). In the closely related great blue heron (*Ardea herodias*), however, siblicide is rare, and while the young may jostle each other for food, they do not harm each other. Mock *et al.* argued that the difference between the two species was in the size of the food items the parents bring back to the nest. Egrets bring small boluses of fish, enough for only one chick at a time, so that an aggressive chick can monopolize everything the parent brings. The result is that the aggressive chick that disposes of its siblings gets more of the food that the parent brings back to the nest than it would if it had to share it. In the heron, the parents tend to bring whole fish to the nest, far more than one chick can eat at once. Here, killing siblings would not have the same advantage because all the chicks can feed at each feeding visit.

Laughing kookaburras (*Dacelo novaeguineae*) go one stage further. The nestlings have a special hook on their upper beak that is lost in later life and appears to have been evolved solely to help them inflict maximum damage on their siblings (Legge, 2000). The female usually lays three eggs, but the last to hatch frequently dies either from direct attacks by its older and larger siblings or, if it survives this early onslaught, from starvation at a later stage – outcompeted in the battle for food. The female parent may even promote siblicide by modifying the sex of the first two offspring and synchronizing the hatching of the first two eggs. Female chicks grow faster than males, so if a female hatches soon after her older but slow-growing brother, she quickly catches up in size and can hold her own, whereas the third nestling cannot. Females in poor condition are more likely to hatch the first two eggs synchronously, more likely to have a female as the second to hatch and more likely to lay smaller third eggs, thus setting the stage for a siblicidal battle in which the third nestling to hatch is likely to become the loser.

Killing of unrelated young is even easier to understand in evolutionary terms. In a number of species, males kill infants that are not their own in order to further their own reproductive interests. For example, the permanent members of lion prides are the females and these are from time to time joined by groups of two or three males, usually following a fight that ousts the previous males. At each takeover, the new males tend to kill all the young cubs in the pride, particularly those that are still taking milk from the females (Fig. 6.15). The new males gain from this infanticidal behaviour because breaking off lactation brings the females more rapidly back into oestrus. Pusey and Packer (1987) found that females whose cubs had been killed at a takeover conceived again about 4.5 months later, but those whose cubs had survived did not conceive again until over 20 months later. As these new conceptions would be by the new males, it is quite clear that the more cubs a male can kill, the more quickly he will be able to father offspring of his own. Cub-killing by male lions is particularly associated with taking over a pride and ceases by the time the lionesses are bearing the new males' own cubs.

One dramatic and, at first sight, somewhat puzzling form of infanticide is the 'Bruce effect' that we discussed in Chapter 3. If a strange male mouse (not the father of the litter) comes into contact with a pregnant female mouse, that female will abort her litters. Why – in an adaptive sense – should the female effectively kill her own offspring? The answer seems to be that the female is 'making the best of a bad job'.

After mating with a female, a male mouse becomes very aggressive to any pups he encounters and remains so for about 3 weeks, after which time he becomes very protective and paternal towards mouse pups. Three weeks is the gestation period for mice so that the switch from male infanticide to paternal behaviour coincides with the time when his own young are likely to be born (Elwood, 1994). If a female is already pregnant and gives birth within 3 weeks of a strange male entering the territory, the pups would be almost certain to be killed by the male. If she aborts

Figure 6.15 Male lion carrying a cub he has just killed. Infanticide by males is common when they take over a pride. This means that the males are not killing their own cubs but those sired by the previous male.

the pregnancy by resorbing the embryos, by the time the new litter is born, there will have been a hormonal shift in the male and the young born then are likely to survive.

A similar explanation – sexual selection between males to sire as many infants as possible – has been put forward to account for infanticide among primates (Struhsaker and Leland, 1987; Pradhan and van Schaik, 2008). As with the lion, male langur monkeys (*Semnopithecus*) have been regularly observed to kill infants when they move into a new troop (Hrdy, 1974). Showing a rather curious role-reversal, in the wattled jacana (*Jacana jacana*) it is the females that kill the young. In these wading birds, females are larger than males, defend territories against other females and have several males incubating their eggs. Emlen *et al.* (1989) showed that when a female takes over a new territory, she kills young already there, and so hastens the time when the males will be incubating her own eggs.

Evolutionarily stable strategies

So far, we have considered the evolution of behaviour at two different levels; the level of genes and at the level of the individual and its relatives. The concept of inclusive fitness allows us to link the two levels. We cannot directly observe the genes in a gene pool. However, we can observe individual animals and measure their reproductive success and even use Hamilton's equation to estimate the costs and benefits of helping or harming relatives. The costs and benefits of different courses of action (such as helping or harming relatives) are devalued by 'r', the coefficient of relatedness, which can best be thought of as the probability that a given gene present in one animal is also present in the body of the helped (or harmed) individual. 'r' thus links genes and individuals by defining individuals in terms of the probability of their carrying a given gene. Inclusive fitness, which takes into account the helping and harming effects of behaviour on other related individuals, thus mirrors the shifting gene frequencies in the elusive gene pool (Grafen, 1984).

But even using 'inclusive fitness' instead of individual fitness does not give us a complete picture of adaptation. This is because neither genes nor individuals nor even individuals and their relatives operate in isolation. Whether or not a behaviour is adaptive very often depends on what other animals of the same species are doing.

For example, being aggressive might well be very advantageous if most of the other animals in the population were very unaggressive and simply ran away when threatened. The aggressive animal would obtain food, territory, mates, etc. with very little cost to itself. But if most of the other animals were also aggressive, and stood their ground, then the costs of aggression would rise. Aggressive animals would not have everything their own way and would keep getting injured. If the costs of injury were very severe, it might even be more adaptive to run away most of the time (Maynard Smith, 1982). Whether or not aggression is adaptive, therefore, depends on the *frequency* of other animals in the population that are pursuing the same behaviour or different behaviour. This frequency will often stabilize through negative feedback. The more frequent highly aggressive animals become in the population, the more likely they are to meet other aggressive animals and become injured in fights and the greater will be the advantage to the unaggressive timid animals that do not get injured. This will lead to selection against high levels of aggression, which will, over generations, cause the frequency of aggressive individuals to fall and that of unaggressive ones to rise. However, the more unaggressive animals there are around, the greater will become the advantage to being aggressive again. Over evolutionary time, the frequency of aggressive and unaggressive animals will either stabilize or fluctuate around a stable point (Maynard Smith and Price, 1973).

Theoretical biologists have developed a whole family of models to cover situations such as this, in which there is no one best adaptive strategy but where the population may stabilize with two or more strategies kept in place by such negative feedback. They call them Evolutionarily Stable Strategies or ESSs. There are many different sorts of ESS, some suitable for conflict situations such as aggression, others for situations where animals are more cooperative. What they all have in common is looking at adaptation in the context of what other animals in the population are doing. ESS models tell us not just whether a behaviour is adaptive but whether a behaviour is evolutionarily stable in the sense of whether it can hold its own when it meets other animals following either the same or different strategies.

One of the most important ways in which ESS models have contributed to our understanding of the adaptiveness of behaviour is by showing that cooperation sometimes pays. Apparent altruism to non-relatives can be evolutionarily stable and is therefore an understandable, if somewhat surprising, outcome of natural selection.

Cooperation and 'altruism' between non-relatives

'Nature red in tooth and claw' seems to be an apt summary of many of the examples we have discussed so far in this chapter. Apart from altruism between relatives, competition between genotypes has some fairly ruthless outcomes, and even relatives can be sacrificed if the balance between cost and benefit tips in a particular way. But this cannot be the whole story because we also find examples of help and cooperation even between animals that are not related to each other at all. Mutual grooming or preening are quite common, with one animal grooming parts of the body, such as the head, neck and back, that the animal itself cannot reach. This may occur between relatives, but is also found in animals that are unrelated but very familiar associates. It is particularly well developed in the primates where, as we shall discuss in Chapter 7, friendly contact helps to cement bonds which may have all sorts of other payoffs within a social group. Both parties gain from the interaction and both can break it off if the other does not participate.

Social interactions between animals can potentially have four different sorts of outcome (Fig. 6.16): mutual benefit, exploitation, altruism and spite. We focus in this section on the difference between 'altruism' (where an animal does something that is costly to itself but benefits another) and mutual benefit where both animals gain, although not necessarily equally. As we have already seen, one route to altruism is via kin selection, where an animal may do something that is phenotypically altruistic, but genotypically selfish. But we would not expect altruism to occur between unrelated animals because the costs incurred by altruists would make them less successful than their selfish rivals and so altruism would be selected against and the altruists would die out. So how do we explain the widespread occurrence of what looks like altruism between non-relatives in the animal kingdom – mutual grooming, the giving of alarm calls to warn conspecifics, food sharing, group defence, mobbing and many other examples? In this section, we will see that most examples of what looks at first sight like altruism (the actor losing out) usually turn out to be either mutual benefit (the actor gains as well as the recipients) or even exploitation (the actor is gaining and the recipients may even be losing out). The three situations we consider are: immediate selfish advantage, delayed selfish advantage and stakeholder advantage.

	Gains	**B**	Loses
Gains	Mutualism/ cooperation *selfish*		Exploitation/ parasitism *selfish*
A			
Loses	Altruism ???		Spite ???

If altruism occurs, this would be a major
problem for the theory of natural selection

Figure 6.16   Different types of social interaction between non-relatives (from A's viewpoint).

Figure 6.17   Horses groom each other on the neck and other areas where they cannot groom themselves. Some authors (e.g. West *et al.*, 2007) reserve the word 'cooperation' for such active helping behaviour and use the term 'by-product mutualism' if animals simply benefit from the presence of others without being actively helped.

Some cooperative acts give an immediate advantage to both parties. Horses cooperate by grooming each other's necks, a part of their body they cannot reach themselves (Fig. 6.17). They also stand head to tail flicking flies off each other's faces. A horse's tail is not long enough to reach its own face, but by standing head to tail with another individual, each horse gains relief from flies that it would not get on its own. If the other horse does not cooperate, it can simply move away and find another more rewarding companion.

(a)

(b)

Figure 6.18 (a) A meerkat sentinel that keeps a look-out while the others feed appears to be acting altruistically. However, the sentinel positions itself so that it is actually closer to cover than the foragers (b) and is often the first down the hole if danger threatens.

The sentinel behaviour of meerkats provides another clear example of how a whole group can gain from the behaviour of one so-called 'altruist' but the 'altruist' gains even more. Meerkats are vulnerable to predation while they feed and it is very common for one individual to spend long periods on 'guard duty' – watching out for predators from a local high point while the rest of the group feed. This looks altruistic as the guard is not feeding and looks as though it is making itself particularly vulnerable to predation. However, Clutton-Brock *et al.* (1999) (Fig. 6.18b) showed that the guards were in fact behaving selfishly. They are usually closer to the safety of a bolt-hole than the rest of the group (Fig. 6.18) and the first to reach cover. Furthermore meerkats tend not to 'guard' unless they are well fed and even when on their own, a meerkat also 'guards'. So 'guarding' is in fact what any meerkat does when it is well fed and has nothing better to do and it positions itself in the safest possible lookout place. The other members of the group certainly benefit as well (so we can call the sentinel behaviour cooperative), but the advantage they gain is a by-product (Dugatkin, 1997) of what the guard would selfishly do anyway.

Trivers (1974) and others have suggested that there may be an even more interesting form of cooperation where one animal helps another but then does not

Figure 6.19 A vampire bat laps blood from a tiny slit it has made in the side of a pig. The bat has razor-sharp incisor teeth and saliva that contains a powerful anticoagulant.

receive assistance back itself until some time later. Trivers called this *reciprocal altruism* but evidence for it is difficult to come by.

Wilkinson (1984) described a remarkable example of cooperative behaviour in vampire bats (*Desmodus*) that has been widely taken to be reciprocal altruism. Vampire bats feed on the blood of mammals, particularly domestic animals such as cattle, pigs and horses (Fig. 6.19). A bat will inflict a tiny (3 mm) wound on its host with its sharp teeth. Its saliva contains anticoagulants and the bat flicks its tongue rapidly in and out to take the blood. These bats are active during the darkest hours of the night and return to communal roosts in hollow trees or caves by day. Sometimes a bat will return to the roost without having fed and if it fails to find food for three successive nights, it may starve to death. But a bat that has not found food itself will often be fed by another bat regurgitating a blood meal to it. Starving bats are more likely to be fed than well-fed ones, and although bats feed their kin, they also feed unrelated bats, particularly those that have fed them in the past. Regurgitation of blood meals was seen only between bats who regularly (over 60% of sightings) associated with each other. The bats fulfil at least some of the criteria for successful reciprocal altruism. They return to the same roost day after day and associate with particular individuals over long periods of time and receive benefit. This means that they could potentially detect cheating individuals that do not reciprocate. The bats are also sensitive to whether another bat is starving or well fed and are able to help a starving bat without too great a cost to themselves. A starving bat loses weight at a high rate, but a well-fed one at a much lower rate, so that the giving of a blood meal from a well-fed to a starving bat gives the recipient more time until starvation than the donor loses. Finally, failing to find food on its own appears to be something that can happen to any bat at some time, so all bats could benefit from the 'insurance policy' of being fed by others in time of need. But those which donate food would be in a more favourable position to receive it when they themselves are in need of it. Cooperation, in the long run, can be selfishly the best policy.

A very important condition is that there must be repeated interactions between the participants so that failing to cooperate on one occasion (cheating) has a penalty in the future through not having the cooperative act reciprocated next time. If vampire bats never re-encountered the bats to which they had given food, we would not expect them to behave so altruistically. For this reason, we expect this sort of reciprocal cooperation to be characteristic of animals that stay together over long periods of time, long enough for the roles of donor and recipient to be exchanged many times. A second condition is that 'cheats' can be recognized and penalized. This does not necessarily mean that the cheat is singled out and victimized. It could just mean that if an animal fails to reciprocate, its partner simply moves away and starts to interact with other individuals.

These conditions for reciprocal altruism mean that we do not expect it to evolve in situations where there is a 'one-off' advantage with two animals being unlikely to meet again. Rather, it would be most likely to evolve in situations where the same animals associate over a long period of time and where a 'cheat' will be penalized because benefits will be withheld in the future if it does not reciprocate. It is most highly developed in animals which have the capacity to remember which other individuals are reciprocators and which are cheats, such as chimpanzees (De Waal, 1982) baboons (Packer, 1977) and tamarin monkeys (Hauser et al., 2003).

There is good evidence for this sort of memory in a wide range of animals, including fish. Many large fish rely on smaller fish, such as wrasse, to pick parasites off their skin and gills (Fig. 6.20). The cleaner fish tend to be territorial so that the large fish return to the same place again and again to be cleaned, rather in the way people return to a good hairdresser.

(a)

(b)

Figure 6.20   Longnose parrot fish *Hipposcarus harid* being cleaned by a wrasse *Labroides dimidiatus*. The wrasse (b) is picking parasites out of the gills of the much larger parrot fish.

Bshary and Schafter (2002) showed that the parrotfish (*Hipposcarus*) are more likely to return to a cleaner if they have previously had a positive interaction (cleaned) than if they had a negative interaction (ignored or damaged by the cleaner) (Fig. 6.21). In this way, the parrotfish continue to interact with cleaners that give good service and penalize 'cheats'. What is of interest is that, in all these examples, repeated, long-term cooperation leaves both parties better off than either would be alone.

There are thus many ways in which animals appear to be behaving altruistically but are in fact acting in their own interests. This can even be the case when the interaction appears to be very one-sided and only one animal appears to be cooperating. For example, in animals that live in small groups, such as mongooses, the

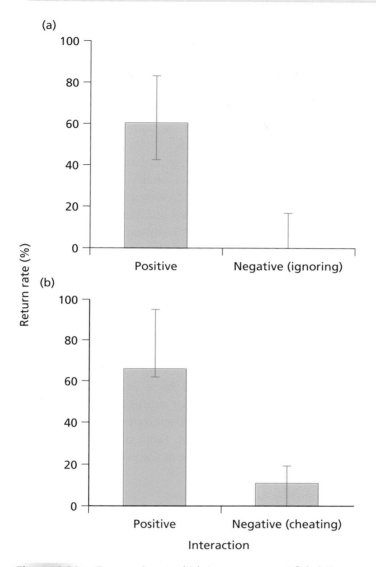

Figure 6.21 Frequencies at which Longnose parrot fish (*Hipposcarus harid*) return to the cleaning stations of the cleaner fish (*Labroides dimidiatus*) depending on whether they have been cleaned or ignored on the previous visit (upper graph) or cheated on (been bitten). The parrotfish are clearly adjusting their behaviour according to experience.

death of one member might have an immediate and negative effects on all the others. If the optimum group size for each individual is 10 and fitness declines with smaller numbers, then each individual in the group has a stake in the survival of the others. It would therefore benefit an individual to be altruistic to other group members because their survival also benefits the 'altruist' itself. Mongooses do notably attempt to rescue members of their group that have been taken by predators, often putting themselves at risk (Rasa, 1987).

Such behaviour is sometimes called pseudoreciprocity or stakeholder altruism (Roberts, 2005). Note that it is distinct from reciprocal altruism because it does not depend on the helped individual repaying the altruist and is therefore less open to cheating. If the mere presence of another individual, regardless of what it does, increases the survival of 'altruist', then the 'altruism' will benefit, although that of course takes the 'altruism' out of altruism.

Another possible example of pseudoreciprocity occurs in male lions. The core of lion society are the group of females, often relatives, that often stay together throughout their lives. Young males are evicted from the pride at adolescence and their only way of breeding is to take over another pride of females. They do this by forming coalitions with other males. A coalition is necessary because one male on his own would be unable to either take over or keep a pride all on his own. He needs the presence and cooperation of at least one other male (Legge, 1996). However, attempts to explain the interactions between males either by kin selection or reciprocal altruism have not been entirely successful. For example, males are not always closely related and although they cooperate when the pride is attacked, some males appear to 'cheat' by not participating in the attack as much as others (Grinnell et al., 1995). But instead of refusing to continue to cooperate with a cheater as would be expected with reciprocity, a male simply carries on as before. Cooperation between the males appears to depend on the mere presence of the other male and is not conditional on its behaviour, at least within a wide range. A good fighter might be a better companion than a lazy follower, but even a lazy follower is better than trying to defend a pride single-handed. Each male has a 'stake' in the continued presence of the other, as his own tenure of the pride depends on the presence of the other.

Sex and sexual selection

It will be clear by now that our understanding of the adaptiveness of behaviour is greatly helped if we look at genes as well as individual animals and that we can even understand the evolution of behaviour that may be positively damaging to the individual doing it. There is one area, however, where even a gene-centred view appeared, at least until recently, to be unable to give an adequate explanation at all. This is the realm of sexual reproduction, which is so commonplace that we tend to take it entirely for granted and fail to realize why it should be difficult to explain.

The reason that sexual reproduction is such a problem is that, in species where the male makes no investment in the young, a female could apparently pass on twice as many of her genes to the next generation if she reproduced asexually and made identical copies of herself as she could if she reproduced sexually and combined her genes with those from a male (Maynard Smith, 1978). Her sexually reproduced children are only half related to her (r = 0.5), not fully (r = 1) as they would be if she reproduced without sex. From both the gene and the individual point of view, therefore, there would seem to be a loss, a twofold disadvantage of sex. The evolution of sex remains something of a paradox so there must be some major corresponding advantage to compensate for this and explain why, despite the disadvantages, sexual reproduction is so widespread among both animals and plants. What is that advantage? The most popular explanation is that the advantage occurs when the environment is changing very rapidly: in particular, when parasites are evolving resistance to their hosts' defences. As we know to our cost, parasites and disease organisms are capable of such rapid evolution that they develop resistance to drugs that initially kill most of them. They also develop resistance to the natural defences of their hosts and when this happens, sexual reproduction may be the best counter-measure the host can take, because it leads to variation and thus to the possibility that at least some of the sexually reproduced variants might just happen to have the right combination of genes for resistance to the new strain of parasite (Hamilton *et al.*, 1990). So, on this view, the whole of sexual reproduction is driven by the selection pressure from disease organisms and parasites!

Sexual reproduction has extremely important implications for animal behaviour. Not only does it result in the evolution of two sexes so different in appearance and behaviour that they may be mistaken for two different species. It has also resulted in the evolution of elaborate and conspicuous mate-attracting ornaments and displays (Fig. 6.22). Some of these may actually endanger the life of the animal – usually a male – possessing them. For example, male guppies (*Poecilia reticulata*) have bright orange spots that are very attractive to females: males with the most and brightest spots are preferred by females. However, these are just the males that are most likely to be eaten by the two main predators of guppies, killifish (*Rivulus*) and the cichlid (*Crenicichla*) (Endler, 1980, 1983).

It is usually males that develop such ornaments because male animals produce large quantities of sperm, can mate with many females and often contribute little in the way of resources and parental care to raising their offspring. As a result, a successful male can potentially have far more offspring in his lifetime than any female, and consequently males compete for the available females. Females are expected to be more discriminating in their choice of mate than males because a mating that leads to inviable or unhealthy offspring represents a larger proportion of the female's lifetime reproductive output than it does for a male. A male can more easily mate again than a female.

(a)

(b)

(c)

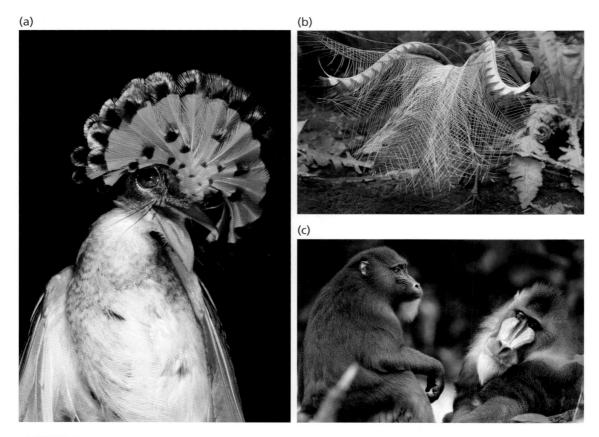

Figure 6.22   Some examples of conspicuous sexual displays. (a) Royal flycatcher (*Onychorhyncus*); (b) superb lyrebird (*Menura novaehollandiae*); (c) mandrill (*Mandrillus sphinx*).

In 1871, Darwin proposed that elaborate male ornaments such as the classic 'train' of the peacock evolved to attract females under a process he called 'sexual selection', by which he meant competition between members of one sex – usually, for the reasons given above, males – to mate with members of the opposite sex. He believed that this would lead to the evolution of traits that either helped the male to fight off other males, or made him particularly attractive to females, or both. Darwin further believed that sexual selection could be such a powerful force that such traits could evolve even though the male possessing them might not survive all that long. As long as he survived long enough to mate and leave a large number of progeny, his characteristics would be successfully passed on to the next generation. Darwin thus saw sexual selection as a special case of natural selection, with the emphasis on mating success.

For a long time Darwin's theory of sexual selection was treated with scepticism, particularly its emphasis on female choice leading to male ornamentation. Sexual selection is now, however, well established but there is still controversy about exactly how (adaptively) sexual selection operates on males and females. There are at least six overlapping theories about the evolution of male ornamentation and female choice (Andersson and Simmons, 2006).

Fisher's runaway theory (sexy sons)

In the 1930s, R.A. Fisher gave a new impetus to Darwin's theory by arguing that if there were genetic variance in male ornaments such as long tails, then female preference would, over evolutionary time, give rise to males with longer and longer tails (Fisher, 1958). This was because the more females chose long-tailed males, the more male offspring there would be with long tails like their fathers and the more female offspring there would be with a preference for long-tailed males, like their mothers. This has become known as the 'runaway' theory. Note that on the Darwin/Fisher theory of sexual selection, there is nothing else that is special about long-tailed males. It is not that they are bigger or stronger or more fertile than other males. It is simply that they possess an ornament or are able to perform a display that attracts females because there are many females with that particular preference (Kirkpatrick, 1982). We have already seen (Chapter 3) that seemingly arbitrary choice conventions may be quite widespread because of the tendency of animals to respond to sign stimuli and supernormal stimuli. Females that mate with males possessing such ornaments will benefit because, once there is a critical proportion of other females in the population with a similar prefer-ence, they will have sons with similarly attractive features and so gain grandchildren through their sons' ability to attract mates.

Sensory bias

We have already seen (Chapter 3) that the most effective signals are often those that grab the attention of a receiver by stimulating its brain and sense organs most strongly (Endler and Basolo, 1998; Ryan, 1998). The female water mite and the female jumping spider that initially respond to their males because the males attract them with food-like stimuli are responding to a pre-existing bias. It has even been suggested that the orange spots on male guppies are attractive to females because they mimic fruit, which the female likes as a food (Rodd *et al.*, 2002). In all these cases, the success of the male's signals lies exactly in the fact that it exploits the tendency of the female to respond to food anyway. And the female gains too: either she gets real food or, failing that, a willing male of the right species. Supernormal stimuli, where the animal responds even more strongly to an exaggerated version of a stimulus occurring in nature, may be the raw material required for Fisher's runaway theory.

Sensory bias has been used to explain the curious tendency of females in some species to respond more strongly to males of a different species than to males of their own. For example, female platyfish (*Xiphophorus maculatus*) prefer the long-tailed males of the closely related swordtails (*X. helleri*) to the tailless males of their own species (Basolo, 1990). Not surprisingly, female swordtails also prefer male swordtails with the longest swords (Basolo 1990; Fig. 3.14) and respond even better to males with supernormal tails (Chapter 3). In the wild this is unlikely to matter because the two species rarely meet and the sexual preference only shows up in the unnatural conditions of the laboratory.

Indicators (handicaps, 'good genes' and revealers of physical health)

A number of theories of sexual selection are based on the idea that elaborate male ornaments and displays allow females to assess a male's 'quality' (health, physical vigour, etc.), particularly genetic quality that he would be able to pass on to her offspring. The female's problem is, of course, how to choose a male possessing such 'good genes' when all she has to go on is what he looks like or how he behaves. How does she assess the honesty of the signal? Zahavi (1975) put forward the idea that elaborate male ornaments are actually a handicap and that males with such ornaments are demonstrating their physical quality by showing that they can survive despite having such a handicap. He claimed, somewhat paradoxically, that a female choosing a male with a long tail is guaranteed a high quality mate because only high quality males can afford to carry the handicap of a long tail. A low quality male would simply be unable to survive with such an encumbrance, so the signal is an honest indication of male quality. Although Zahavi's ideas were initially greeted with scepticism, they have gradually come to be more widely accepted and shown to be at least theoretically plausible (Grafen, 1990).

Hamilton and Zuk (1984) proposed that male ornaments may enable healthy males to advertise the fact that they are free of diseases and parasites. Since disease is a major source of juvenile mortality and may indeed have been the driving force behind the evolution of sex itself (p. 337), Hamilton and Zuk (1984) argued that females should preferentially choose males that have genes for resistance to parasites, which would then be passed on to their offspring. However, they did not claim that an ornament such as a long tail was a handicap in the sense that it actually lowered the male's resistance to disease. They claimed simply that it was an indicator of a male's state of health because only healthy males would have glossy, well-kept tail feathers; or only a really healthy male could keep up a display for a long time, thus 'revealing' that he was in good condition. On the other hand, growth of male sexual ornaments may be associated with increased hormone levels that depress the immune system and so do constitute a direct cost (Stoehr and Kokko, 2006; Ros *et al.*, 2009).

Resources and other direct benefits

Since the female's chances of successfully rearing offspring may be critically dependent on a reliable food supply or a safe place to rear her young, it is not surprising to find that females may choose males for the resources they have, rather than the male himself.

Figure 6.23 A male pied flycatcher singing close to a potential nesting hole. Females choose males on the basis of territory 'quality' which for a female flycatcher means a thick growth of trees and a high, safe nest site.

A good example of the importance of resources in female mate choice is that of the pied flycatcher (*Ficedula hypoleuca*), which is a summer visitor to northern Europe (Fig. 6.23). The males arrive on the breeding grounds about a week ahead of the females and set up large breeding territories. The males that arrive first tend to be the oldest males with the blackest plumage. They also have the pick of the available territories, so the fact that they are more successful at gaining a mate than late-arriving males could be either because the females like their territories or because they are attracted to some characteristics, such as the blacker plumage, of the males themselves.

Alatalo *et al.* (1986) devised an ingenious way of separating these two factors in the females' choice. They made use of the fact that pied flycatchers readily nest in artificial nest boxes and can be persuaded to nest in different areas of a wood depending on whether there is a nest box there. They forced males to defend particular territories by restricting the number of nest boxes available at any one time, and only putting up more when the first ones had been occupied. When the females arrived, Alatalo and his colleagues noted the order in which the different males attracted a mate. This time – used as a measure of success in obtaining a mate – turned out not to be correlated with how early a male had arrived or with his age or the blackness of his plumage or with any other characteristic of the male himself, but with the quality of his territory. Males that had a low density of birch trees in their territories and safe nest sites high up in trees with thick trunks were the ones that obtained mates first. Normally, older blacker males would choose such territories themselves, but when they were experimentally denied the opportunity to do so, the females could be seen responding primarily to the territory, not the owner.

This idea – that a female may sometimes be more interested in a male's territory and resources than in the male himself – has some very interesting consequences for how we can expect females to behave and how sexual selection operates. The theories listed above assume that male displays and ornaments are directed primarily at females but, as Darwin (1871) pointed out, other males may also be the targets. It is possible that many sexual displays are really male–male assessment displays and females simply choose on the basis of their ability to win fights or dominance over other males. This, of course, would be an adaptive female strategy since males that are able to compete with other males in this way are likely to be physically healthy and strong.

In some species, the female uses the male's song to assess his health and strength as well as his value as a father. The male sedge warbler produces some of the longest and most complicated of all bird songs p. 105. Males with the most variety in their songs (the largest number of syllables) find mates and start breeding earlier than males with less elaborate songs (Fig. 6.24) (Catchpole, 1980). Males with large repertoires also have larger territories, bring food to the nest

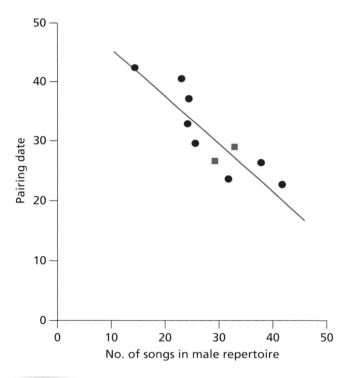

Figure 6.24 Repertoire size in sedge warblers in relation to pairing date. Males with the most songs in their repertoires attracted females earlier in the breeding season than males with smaller repertoires. (From Catchpole *et al.*, 1984; after Catchpole and Slater, 2008.)

more frequently and are less likely to be infected with blood parasites (Buchanan and Catchpole, 1997; Buchanan *et al.*, 1999). By choosing a male with an elaborate song, therefore, the female obtains a wide of variety of benefits for her offspring.

Genetic compatibility

Recently, there has been an increasing interest in the idea that one of the major genetic gains for a female from mating with a particular male is that his genes *complement* hers. This means that what constitute 'good genes' for one female may be less good for another, who might benefit from mating with a different male. Potts *et al.* (1991) showed that house mice choose mates that have different histocompatibility genes from themselves. Histocompatibility genes are concerned with the way the immune system responds to the challenge of disease and they are highly polymorphic. There are over one hundred alleles of each histocompatibility gene in mice and so by choosing individuals that have different histocompatibility genes (which they do by smell), mice ensure that their offspring are maximally disease resistant (see Chapter 2).

A very obvious kind of genetic incompatibility comes from inbreeding – that is, mating with a close relative (Tregenza and Wedell, 2002). Inbreeding is disadvantageous for two reasons. Firstly, lethal or damaging recessive genes are more likely to occur in homozygous form and therefore to have their damaging effects. Secondly, inbred offspring will be less variable than outbred offspring. They will be less different from each other and also have a smaller range of gene products within each individual. As a consequence, many animals avoid breeding with their closest relatives, either by dispersal or by active avoidance of individuals that they have become familiar with (Pusey and Wolf, 1996). Young female chimpanzees, for example, often develop strong bonds with a male in their natal group as they are growing up but will move away from the group and mate with a strange male when they reach sexual maturity (Pusey, 1990).

A more complex way of avoiding inbreeding occurs through a process known as *phenotype matching*. Here, an animal uses its experience of its siblings, parents or even itself to build up a template of what familiar relatives look, sound or smell like and then is able to recognize, and avoid mating with, an unfamiliar animal because it looks, smells or sounds similar. In other words, the animal learns something about the phenotype of its relatives and can then respond to unfamiliar individuals if they match it.

For example, bank vole females (*Clethrionomys*) recognize their brothers by their smell and in a two-choice test, are more attracted to the scent of unfamiliar unrelated males than to their brothers (Kruczek, 2007). This discrimination against the scent of brothers is shown whether or not they have been reared with their brothers and persists even if the brothers have been cross-fostered at birth and reared in a different nest. This suggests females may have two mechanisms for avoiding incestuous matings. One is to avoid mating with familiar males, which is normally enough to avoid mating with a brother since they will have interacted socially with them as they were growing up. The other is to be able to detect genetically related but unfamiliar individuals, using phenotypic matching.

Avoidance of reproduction between genetically incompatible mates can also come into play *after* copulation because, remarkably, sperm compete with one another inside the female's body. By mating with several different males, the female sets up an internal competition between the different males over which sperm are to actually fertilize her eggs. She may even be able to influence which sperm are successful and screen out genetically incompatible ones. The eggs of female crickets that have mated with only their brothers have much lower hatching success than those of females that have mated with both siblings and non-siblings (Tregenza and Wedell, 2002). By mating promiscuously, the females avoid the low egg viability associated with sibling matings.

Male–male aggression

We have so far assumed that male displays and ornaments are directed primarily at females but, as Darwin (1871) pointed out, other males may also be the targets. It is possible that many sexual displays are really male–male assessment displays and females simply choose on the basis of their ability to win fights or dominance over other males. This of course would be an adaptive female strategy since males that are able to compete with other males in this way are likely to be physically healthy and strong. In many ways, prowess in battle is a more direct indicator of male 'quality' and likelihood of possessing 'good genes' than growing a long tail. We have already seen in our discussion of honest signals in Chapter 3 that when red deer stags are in conflict over females, they signal by bellowing at each other in an escalating roaring contest where the winner is often the stag that can roar at the greatest rate (Clutton-Brock and Albon, 1979). The ability to roar at a sustained high rate thus provides both the rival male and the female with an indication of how physically strong a male is.

The huge male elephant seals, which are three or four times the size of the females (Fig. 6.25) gather harems on the breeding beaches. Only the very strongest can dominate a large area and in some breeding seasons, a mere 4% of the males are responsible for 85% of the matings (Le Boeuf, 1974). Since the males are so much larger than the females, the females apparently have little 'choice' if a male decides to mate with them. However, female elephant seals can manipulate the situation to their own advantage. When a male mounts a female, she calls loudly and the call alerts nearby males. If the male that has mounted her is a subordinate one, he is soon chased off by the bigger dominant male. The female's protest behaviour ensures that she mates with the biggest, strongest male around (Cox and Le Boeuf, 1977). Many female animals, including domestic hens (Graves *et al.*, 1985), prefer

Figure 6.25   Male and female elephant seals, showing the great difference in body size that has evolved in this extremely polygynous mammal. Mortality among males is commonplace and only a few achieve the size and strength to hold a harem of females on the beach.

dominant males, which also ensures that they mate with males that are fit and healthy enough to win and maintain their position over other males.

Conclusions on mate choice

There is still controversy about which, or which combination, of the above theories best explains mate choice. It seems certain that the balance between different factors is different in different species. Our views about sexual selection and mate choice have certainly undergone a major change over the last 20 years, mainly as the result of the information we now have from DNA fingerprinting. The traditional view of choosy females mating only once has had to be revised in the face of evidence that females in almost all species mate with many males. Females are still expected to be the choosier sex, but their 'choice' is not limited just to a choice of who to mate with but can continue after copulation.

Species isolation and species selection

As we have seen, an important element of mate choice is choosing a mate that is genetically compatible and an extreme case of genetic incompatibility would be to mate with the wrong species. Hybrids between two species, although they may show some signs of 'hybrid vigour' are usually sterile or less fertile than the offspring of within-species matings. Mules (horse × donkey) are a good example of this. Even if they are fertile, hybrids are rarely as successful as either parental type, whose genes have been selected over many generations as the best for their own environment. Dilger's lovebirds (Fig. 2.13) show how confused hybrids can be in their behaviour.

Consequently, there will be a strong selective advantage, particularly for females, in being able to recognize a mate of the same species. Behaviour, and particularly sexual signals, play a major part in this. For example, there are two closely related species of fruit fly, *Drosophila melanogaster* and *D. simulans*, that look so alike that humans find it difficult to distinguish them. However, the flies themselves clearly do discriminate, apparently by a combination of acoustic signals produced by wing vibrations and pheromones (Bennet-Clark and Ewing, 1969).

Searcy and Marler (1981) used synthetic versions of bird songs to find out exactly what it is about the songs of their respective males that the females of two similar species of bird – song sparrows (*Melospiza melodia*) and swamp sparrows (*M. georgiana*) – use to distinguish between them. The synthetic songs consisted of either song sparrow or swamp sparrow units or 'syllables', presented in either song sparrow-like or swamp sparrow-like order (Fig. 6.26). For females of both species, the song had to have both the right syllables and be in the right order, whereas the less fussy males seemed to show no particular preference for the temporal pattern of their own or the other species. Again, in red-winged blackbirds, females clearly distinguish between real red-winged blackbird song and the imitation of it produced by mockingbirds, in that they show the characteristic copulation-solicitation display much more to the real thing than to the mimic's imitation (Searcy and Brenowitz, 1988). However, males seem to be completely taken in by the deception. Playing a tape recordings of real song and mimic song to territorial male red-winged blackbirds (*Agelaius phoeniceus*) resulted in the males showing equal amounts of aggression to both (Brenowitz, 1982).

As in other examples of animal communication (Chapter 3), we often find mutual adaptation between signal and response. Male fireflies signal to females by producing flashes of light that are different in different species, and females respond only to the pattern of flashes given by males of their own species (Lloyd, 1965, 1975). The flashes also vary in colour. Lall *et al.* (1980) showed that there is a good match between the colour of light emitted by the male of a given species and the colour sensitivity of the female's eye in the same species. A comparable co-evolution, this time within one species, is described by Ryan and Wilczynski (1991) for the cricket frog. They studied two populations of this species in North America, one in New Jersey and the other 2500 km away in South Dakota. Not only did the calls of the male frogs differ in frequency between the two populations (the dominant frequencies of the calls were 3.56 and 3.77 kHz, respectively), but so did the sound frequency that most excited the auditory systems of the two sets of females. The basillar papilla – the inner ear organ that is used in the reception of calls – responded maximally to 3.52 kHz in New Jersey females and to 3.94 kHz in South Dakota females. Such differences between populations pave the way for the eventual

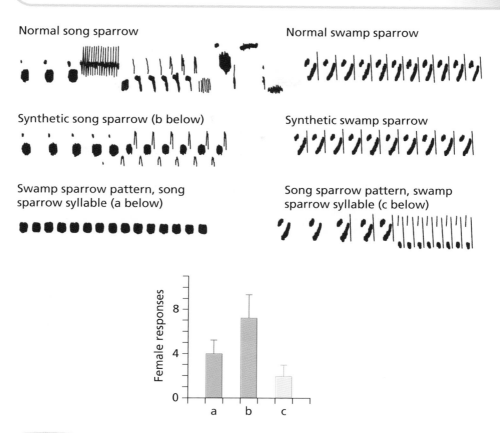

Figure 6.26 The response of female song sparrows to artificial songs played to them through a loudspeaker in their cage. The response measured was the frequency and intensity with which they adopted the copulation invitation posture with raised tail and shivering wings. At the top are sound spectrograms of natural song sparrow song and the natural swamp sparrow song. During the tests, the greatest response was to the synthesized song sparrow song, with significantly less response to either swamp sparrow pattern with song sparrow syllables or song sparrow pattern made up from swamp sparrow syllables.

evolution of new species. While geographical separation between populations may be important for them to diverge into two completely new species, this may not always be necessary. In the lakes of East Africa, almost 2000 species of cichlid fishes have evolved in the last 10 million years. Lake Malawi alone has over 500 species, all have evolved in the same lake (Kocher, 2004). The species differ in a whole range of features including jaw shape, habitat preference, colour pattern and visual sensitivity (Fig. 6.27). Different species are maximally sensitive to different wavelengths of light, which has led to the suggestion that female mate preferences might have driven the spectacular variation in male colour patterns.

Figure 6.27 Lake Malawi has over 500 hundred species of cichlid fishes, differing in size, colour, pattern and behaviour. They have all evolved in this single lake within the last 10 million years.

Tinbergen's fourth question: the phylogeny of behaviour

Over long periods of time, new species evolve and others become extinct. A courtship behaviour in the descendent species, for example, may be an exaggerated version of courtship in its ancestor, or use an ornament of a different colour or a sound of a different frequency. The last of Tinbergen's four questions is about phylogeny – that is, the evolutionary origins and antecedents of behaviour in animals we see today. We want to know how, back in the mists of evolutionary time, behaviour patterns evolved. What was the behaviour of the ancestral species like and what were the intermediates between them and our present-day species?

In attempting to reconstruct the phylogenetic history of behaviour, we are severely hampered by the fact that (with one or two notable exceptions such as fossilized dinosaur tracks or termite nests) we have no fossil record of behaviour. But we do have a very important method at our disposal. This is to look at closely related species that are alive today and to try to deduce the evolutionary changes that took place in the past from the pattern of diversity which now exists.

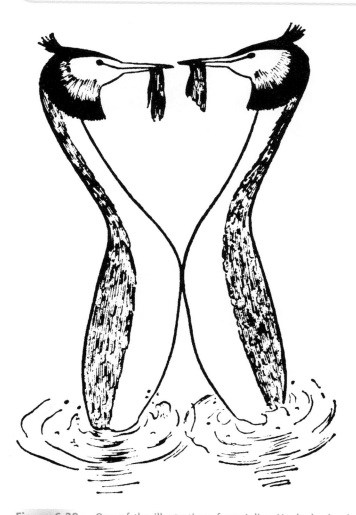

Figure 6.28 One of the illustrations from Julian Huxley's classic study of the courtship behaviour of great crested grebes. This mutual display, sometimes called the 'penguin dance' involves each bird collecting some nest material and swimming towards its mate. Both then rear up vertically, vigorously paddling with their feet to maintain their posture. They move their heads rhythmically from side to side, with crest and neck ruff raised and holding the nest material firmly in the bill.

Many of the early ethologists studied phylogeny by looking in detail at behaviour patterns to see whether the elements making them up could provide clues as to where the whole pattern came from. Julian Huxley (1914) studied the various courtship displays of the great crested grebe (*Podiceps cristatus*), one of which involves both partners rearing up out of the water and presenting nest material to each other (Fig. 6.28). Huxley argued that this display originated from elements of the grebe's nest-building behaviour which were then incorporated into the sexual display.

(a)

(b)

Figure 6.29 (a) Bank swallows (*Riparia*) excavate burrows in sandy banks. (b) Barn swallows (*Hirundo*) build cup-shaped nests of mud while (c) cliff swallows (*Petrochelidon*) build more elaborate retort-shaped nests.

Huxley coined the term 'ritualization' to describe this evolutionary change as a non-signal movement (nest-building) became used in a display context. Tinbergen (1959), too, used analyses of what animals do now to indicate evolutionary origins. By showing that the threat postures of the lesser black-backed gull had elements of both escape and attack, he argued for the 'dual motivation' origin of threat displays (Chapter 3).

Recently, however, the study of phylogeny has been completely revolutionized by molecular techniques that allow us to establish, in detail, the evolutionary relationships between species. On the assumption that the molecular composition of, say, DNA

(c)

Figure 6.29 *(cont.)*

changes at a constant rate, the differences between DNA in different species can be used as a sort of 'molecular clock'. (The relevant DNA changes are, of course, in non-functional parts of the chromosomes; genetic changes that have a phenotypic effect would not change in this clock-like manner.) The bigger the differences in the DNA of two species, the longer ago they had a common ancestor as the more time there has been for the differences to accumulate. This means that instead of speculating that the behaviour of one species looks a bit like that of another and so just might possibly have evolved from it, we can construct actual ancestral trees that show which species are closest to a common ancestor of a whole group and which are specialized late-comers. In this way, we can establish the order of evolutionary events in the past in a way that Tinbergen recognized might be possible but was without the modern tools for accomplishing.

For example, earlier in this chapter we discussed the idea of 'sensory bias' in relation to sexual selection (p. 340). One implications of sensory bias is that the 'bias' or preference for an ornament was already present before the ornament itself evolved, perhaps as tendency to respond to a supernormal stimulus. Basolo (1990) argued that the preference of playfish females for males with swords even though males of their own species do not have these ornaments was evidence for a very ancient 'bias' on the part of the females that predated the evolution of the male ornament. All female swordtails and platyfish have this preference. But how can we possibly know which came first – the ornament or the bias? Did a pre-existing bias in the females drive the evolution of the ornaments in the males or did the existence of male ornaments favour the evolution of females that had a preference for it? Did male platyfish subsequently lose their swords, leaving their females still preferring something from the past? By using a molecular family tree of the swordtails and platyfish, Basolo (1990, 1995) argued that the common ancestor of this group had no sword but that the ancestral females already had a sensory bias towards responding to long tails. This female bias for swords then subsequently favoured the evolution of swords in male swordtails, because males with longer tails attracted more females. For some reason (perhaps because there were other selection pressures, such as predation, against it) platyfish males never evolved swords, despite the fact that the females of their own species have the same sensory bias and prefer males with long tails.

Tinbergen's fourth question, for a long time neglected in favour of function, causation and development, has now been given a new impetus by modern molecular techniques. A good example is how it has elucidated the evolution of different techniques of nest-building in swallows and martins (Hirundinidae) (Fig. 6.29). This group of birds exhibits a very wide range of nesting habits. Some of them, like sand martins or bank swallows (*Riparia*), excavate burrows in sandbanks or soft earth. Others move into natural crevices in rocks and trees and yet others build mud nests that stick to vertical walls. Some of them take mud-building to the length of building domes with their own entrance corridors. A molecular phylogeny using nuclear DNA of the swallows shows that the ancestor of all Hirundines was a burrower and excavated its own nest tunnels. All the other nest-building techniques evolved subsequently and there is little evidence that any of them evolved more than once. All the mud-nesters, for example, evolved from a common ancestor and form a single clade and they now all use the same basic behaviour of bringing beakfuls of mud and adding them to the nest. They gather a ball of mud of medium consistency and then smear a bit of soft mud across the top of their beak. At the nest, they smear the soft mud across the place where the new mudball is to go thus helping it to stick to the existing structure. The swallow then vibrates its beak in the mud, which appears to weld the join (Hansell and Overhill, 2005). The basic cup nest we see in barn swallows (*Hirundo*) later evolved into the more elaborate retort-shaped nest of cliff swallow (*Petrochelidon*), which has a down-turning entrance (Winkler and Sheldon, 1993).

SUMMARY

Two of Tinbergen's four questions concern the evolution of behaviour. The first concerns the adaptive significance of behaviour and has been the object of an enormous amount of research since Tinbergen's time. As expected by the theory of natural selection, we find that, in general, animals behave in ways that optimize their survival and reproduction, but that how they do this can be quite surprising. Animals may have to compromise between different selection pressures and so behave less than optimally for any one of them. They may sacrifice their own survival for their offspring or other relatives. In the struggle to reproduce, sexual selection has resulted in many species evolving elaborate ornaments and behaviour that are risky to the survival of the individual. In other situations, animals may cooperate with non-relatives or even kill their relatives. One thing we have learnt is that genes have some amazingly intricate means of getting their selfish ways.

Tinbergen's second evolutionary question about the phylogeny of behaviour has been relatively neglected until recently. Now it is enjoying a new lease of life because we can use the DNA of different living species to give very detailed family trees of different groups. If we have an example of a living species that is ancestral to others, we can 'look back into history' and see how the behaviour of the others evolved.

CHAPTER SEVEN

Social organization

The individual in the crowd

In this final chapter, we draw together the threads of evolution, causation, development and phylogeny that have run through this book and show how they underlie one of the most striking features of animal lives – their tendency to be social. Tinbergen's four questions have stood the test of time, they are as important as they ever were and as he did in his pioneering book, *The Study of Instinct* (1951), we shall attempt a synthesis involving them all. Our knowledge of animal behaviour has grown enormously since Tinbergen's time and has spawned whole new disciplines such as behavioural ecology and neuroethology, but there is still much be gained by asking his different kinds of questions about behaviour. Asking *how* (causally) animals choose their mates, for example, is hugely illuminated by understanding *why* (in an evolutionary sense) they choose the mates they do, and vice versa. Since virtually all animals are social for at least part of their lives, social behaviour provides an obvious base for a synthesis of many aspects of behaviour across a wide range of species.

When we observe flocks of birds, swarms of insects or herds of antelope, it is easy to lose sight of the individual in the crowd. If a group stays together, it is because individual animals all benefit from staying with one another. If they respond to each other's movements so that they continue to stay together even when disturbed, this can result in spectacular group behaviour such as the aerial acrobatics of starling (*Sturnus vulgaris*) flocks (Fig. 7.1). Although the group seems to take on a life of its own its behaviour is, in fact, the result of hundreds or thousands of individual bird decisions to move in response to the behaviour of the others (Ballerini *et al.*, 2008).

What we see as social interactions between animals and describe as 'social organization' at the population or group level are the net result of selection acting on individuals and genes. 'Culture' is the net result of many individuals learning from each other. 'Mating systems' are the net result of males and females each being selected, sometimes in opposing ways, to achieve reproductive success. 'Dominance hierarchies' result when individuals each adopt different ways of dealing with the presence of other animals. What makes the study of animal groups so fascinating is that the group itself provides one of the main selective forces on the animals within it. The opportunities for social learning and the need to outwit competitors can all lead to the evolution of new types of behaviour. Furthermore, whether or not an animal benefits from behaving socially will depend critically on how the other animals in the group respond as well. A single small bird calling at an owl would have little effect and would be very likely to put itself in danger. But a group of small birds, all calling and 'mobbing' the predator, can be a very effective deterrent so that they all gain. The benefits of mobbing thus depend crucially on what other animals are doing. What would happen if some birds cheated and didn't take part in the mobbing? Are the mobbers being altruistic? Cooperative? Selfish? Is the behaviour evolutionarily stable (p. 328) or would cheaters do even better?

The term 'social organization' refers to a wide range of phenomena defined as how members of a species interact with each other; for example, whether they group together in herds, space out and defend territories, have a monogamous or a polygamous system of mating, and so on. In some instances – the various social insects, for example – social organization is fairly rigid and species-specific. In many vertebrates, on

Figure 7.1   The astonishing aerial gyrations of dense flocks of starlings before they go to a communal roost. They move almost as one. Sometimes their movements can be related to the appearance of peregrine falcons who harry the edge of the flock trying to get birds to break away and thus lose the protection such a flock provides.

the other hand, it is a much more dynamic phenomenon and may vary with changing conditions. Use of the term is not even restricted to what we might call highly social animals. Tigers, which usually live and hunt alone in large territories, avoiding contact with others except for breeding, and honeybees, which spend their entire life in a dense colony, both provide examples of social organization even though they are totally different.

As we will see in the course of this chapter, social organization among animals takes very diverse forms. Female elephants may live in the same family unit for 40 or 50 years. They clearly know and react to each other as individuals and the stability of their relationships suggests that their groups should be called 'societies'. On the other hand, the organization within many flocks of birds or schools of fish is much less complex, although individuals may stay together for months. One factor does, however, define what we mean by a social group, whatever the exact form the group takes. To be called 'social' animals must respond to each other, not just be drawn independently to a common resource such as food or water. For example, a swarm of water fleas (*Daphnia* sp.) gathered in some rich food area, or a mass of fruit flies collected on some rotten fruit, would be better described as an 'aggregation' rather than a social group or a 'society' because they are attracted to a common food source, not specifically to each other. Even they can be described as showing some social responses, however, because they react to one another's presence by spacing themselves out so that they do not touch.

Advantages of grouping

All animal groups, whether aggregations, flocks, schools or what we may wish to call true societies, must result in the individuals that are part of them being better off than they would be on their own. Silk (2007b) reviews both the benefits and costs involved. There are certainly a number of obvious disadvantages to grouping, such as having to share food and space with others, having to range further to get enough food for all and so on. Since grouping is so common there must be correspondingly big advantages although, as we discussed in Chapter 6, it is not always easy to discover exactly why (in an adaptive sense) the animals benefit. Measuring benefit implies that we can compare the survival and reproductive success of individuals in a group with those in some other situation, such as being on their own. However, because group living is so beneficial, we find that similar animals living on their own are rarely around to allow any comparison to be made.

Discovering the advantage of group living often means putting together evidence from several different sources. Specifically, (i) we can use experimental evidence (artificially creating groups or placing individuals on their own), (ii) look at naturally occurring variation (within a species some animals may be more or less likely to group than others or adopt different types of grouping), or (iii) make comparisons between related species some of which have adopted solitary lifestyles, others being social.

Allee and his colleagues (Allee, 1938) used simple experiments to show how even loose aggregations can benefit the individuals that comprise them. They showed that water fleas cannot survive in alkaline water, but that the respiratory products (CO_2) of a large group of them are sometimes sufficiently acid to bring the alkalinity down to viable levels. Thus a group can survive where a few individuals could not. Fruit fly cultures do not do well if there are too many eggs because the resulting larvae are

Figure 7.2   The defensive formation of a herd of musk oxen in the Arctic of Alaska. When a predator appears they bunch closely with the older animals at the front facing the threat.

undernourished, but they fare equally badly with too few eggs. This is because reasonably large numbers of larvae are needed to break up the food supply, encouraging the growth of yeast and making the food soft enough for all the larvae to feed easily. It is thus advantageous for a female to lay her eggs close to those of others because her own offspring will benefit.

Flocks of birds and schools of fish exemplify groups that are much more than simple aggregations because there is often a high degree of social interaction between individuals. Some, such as geese, may even migrate as families. Even here, physical factors may still count, as in the case of emperor penguins which huddle closely together as they stand incubating their eggs during the Antarctic winter. Heat is conserved and birds on the outside move more than those in the centre, leading to mixing and a reasonable distribution of shelter.

One of the most obvious advantages of a cohesive group whose members respond to each other's behaviour, is protection against predators. This could be the direct physical protection which a group can give, as in the musk oxen (*Ovibos moschatus*) shown in Fig. 7.2. As a group, even small, individually weak animals can offer a formidable defence. Colonial nesting birds – for example gulls and terns – may mob an invading predator such as a fox, even striking blows with their feet. Even though each is responding individually to defend its nest, the proximity of other birds all doing the same thing means that their combined efforts can be much more effective than that of a single bird on its own. As a result, as Göttmark and Andersson (1984) showed, the

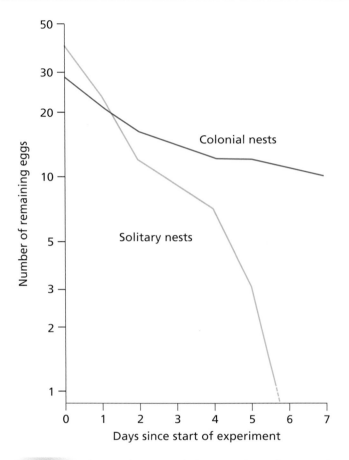

Figure 7.3 Losses due to predation over time when experimental eggs were laid out within 2 to 5 m of *either* solitary nests, *or* nests in a colony of common gulls (*Larus canus*). Predation was much higher near the solitary nests and by day 7, there were no experimental eggs at all left near the solitary nests. Eggs placed near the colonial nests were protected by the massed mobbing of the gulls in the colony.

nesting success of gulls in a large colony is considerably greater than that of gulls that nest singly or in small groups (Fig. 7.3).

Very commonly, protection comes from improved vigilance, the fact that a number of animals are on the alert means predators are detected more effectively.

There have been many studies (reviewed by Elgar, 1989) showing that birds and mammals feeding in flocks or herds spend a smaller proportion of their time being vigilant – and thus can feed more – than when on their own. It is also observed that, on average, predators are detected more quickly by groups and one alarm call or fright response will alert everyone. Even when a predator does mount an attack, the tight packing of a group makes its task more difficult and, as mentioned above, birds in flight instantly bunch when a bird of prey approaches. Hamilton (1971) showed that if each animal in a group tries to put just one other between it and a predator, then a tight bunching is the inevitable result – he calls his paper, 'Geometry for the selfish herd' which reminds us again that we must always look for individual advantage when animals group. With more elaborate societies this advantage will not come directly or at once, but crucially it may mean that the group can stay together and survive repeated challenges. Hence the now familiar image of an individual meerkat, taking its turn to keep watch and protect young whilst its other group members can feed (Fig. 7.4).

Figure 7.4 An adult meerkat keeps a vigilant eye out for predators. This member of the group remains near the nest hole 'baby sitting' whilst the others are away foraging. All members of the group and not just the parents take turns at protecting the young in this way.

Group living may also offer some other direct benefits for acquiring food. Not only can birds in flocks spend more time foraging they may also steal food from other individuals or be led to a clump of food located by others. Krebs *et al.* (1972) showed that when one member of a tit flock finds a food item, the others rapidly alter their searching strategies and concentrate their attention both on the general area and the type of niche in the trees where the food was found. Sometimes this may be a short-term disadvantage for the bird that finds the food, but by staying in the flock, it gains a longer term advantage of being able to utilize the bigger pool of resources which a flock will be able to uncover.

At other times, when the food source is so rich that there is enough for all, there is no disadvantage at all to the original finder. Brown (1986) showed that cliff swallows (*Petrochelidon pyrrhonota*) follow individuals that have located a rich source of insect food. Birds returning to the colony after a successful foraging trip were likely to be followed on their next trip out by birds that had previously been unsuccessful. But the food source – effectively a dense 'cloud' of insects – was so abundant that the finders could be at no disadvantage if they were followed. In yet other cases, it may even be of advantage to the finder to have other individuals around. Gannets (*Sula bassana*) are often found fishing in groups, and Nelson (1980) argues that a group of birds diving together so confuses the fish that they are more easily caught. Göttmark *et al.* (1986) showed a similar effect in black-headed gulls. Group fishing leads to more successful feeding for each individual than a bird could accomplish on its own because, with more gulls around, the fish are more vulnerable and more likely to be caught.

Lions (*Panthera leo*), spotted hyaenas (*C. crocuta*), Cape hunting dogs (*Lycaon pictis*) and killer whales (*Orcinus orca*) are all examples of predators that hunt together. We have to be cautious here for it is so easy to suggest cooperative hunting when we see such groups. It could be that they are each responding to prey as individuals, hunting at the same time rather than together. However, there are a number of well-observed examples where real cooperation may be involved. For example, Boesch (1994) has clear evidence that chimpanzees in the Ivory Coast regularly hunt colobus monkeys for food and cooperate effectively to do so. Lions are the only social cats and certainly appear to hunt as a group. Stander (1992) has shown that they do adopt different roles such that one individual is driving prey towards others hidden in cover. Hunting dogs often appear to take turns in running down an antelope to the point of exhaustion. Spotted hyaenas are regularly found in different sized groups depending on the size of prey they are hunting (Fig. 7.5). Kruuk (1972) showed that when hunting zebra the mean number of hyaenas in a group is 10.8, when hunting adult wildebeest the mean is 2.5 and when hunting gazelle fawns just 1.2 individuals. What is more, the differences in the size of the packs are often apparent long before the hyaenas have sighted their prey. When hyaenas set out in a large pack, Kruuk could be reasonably certain that they would end up hunting zebra even if they had to walk for miles through herds of wildebeest to find the zebra. This strongly suggests that hyaenas set

Figure 7.5 Spotted hyaenas match their forces to the prey they are hunting. Here a group of nine have managed to cut off a buffalo from its herd. One or two hyaenas may be injured or even killed, but unless it can get back to the herd, this huge animal is doomed.

out to hunt certain kinds of prey. Nor is it only humans that can predict hyaena behaviour based on pack size: zebra ignore single hyaenas and only become alarmed by large packs.

Even if we cannot be sure of true cooperation in hunting, this is not the only factor for carnivores. Packer (1986) has argued that for lions the advantage of being in a group lies in the fact that after the kill, there will be more individuals to keep scavengers and other potential thieves away from the carcass. Even if extra individuals do not contribute to hunting success they are tolerated because they help keep away vultures and hyaenas, although they eat some of the kill themselves.

It will be clear from the preceding discussion that group living confers some clear advantages but at the same time some undoubted disadvantages. Being near other individuals means increased competition for food, increased risk of disease transmission and greater conspicuousness to predators, as well as greater risks of cuckoldry, mixing and cannibalism of young, and so the balance between the advantages and individual gains as a result of being part of a group and the disadvantages it inevitably suffers from the same source may be a very fine one. Hoogland and Sherman (1976) give a long list of disadvantages suffered by bank swallows as a result of their habit of nesting communally in sand banks. These range from increased chances of picking up fleas, to having nest burrows collapse as a result of other birds nesting too close by. Nevertheless, the birds benefit from nesting in a colony because they derive safety from the massed attacks of all the other birds on potential predators. This is more than enough to outweigh the evident disadvantages of being close to other birds. The behaviour that will be favoured by natural selection – whether it is territoriality or extreme clumping – will be that which favours the reproductive interests of the individual in the long run. Nor will this necessarily mean a stable social organization for selection may shift through a year. As we mentioned when discussing mating systems in Chapter 6, many common birds, such as finches and thrushes, live in pairs and are strongly territorial during the summer breeding season but gather in flocks and forage together during the winter.

Diverse social groups

Animal groups differ widely in their complexity and the nature of the interactions individual animals have with one another. This means that it is not enough to describe a group simply by its size. Among other factors, the sex ratio, the degree of differentiation into roles, relationships with other specific individuals in the group and kinship, all will affect the structure and the costs and benefits of group living. A look at some contrasted types of groups from very different animals will illustrate what this means in practice.

Eusociality: division into castes

The most extreme form of social organization is called 'eusociality' and is characterized by reproductive division of labour, where some members of the group lose their reproductive capacity altogether and become members of a worker 'caste'. Although sterile themselves, their efforts increase the reproductive output of the colony as a whole through helping members of the reproductive caste. In Chapter 6, we discussed

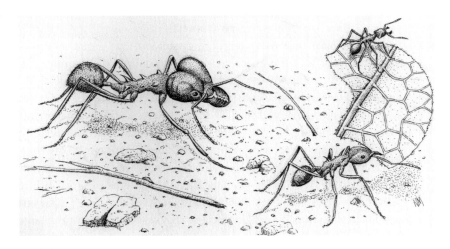

Figure 7.6 Extreme diversity of morphology and behaviour in castes of the leaf-cutter ant, *Atta cephalodes*. A 'medium' worker is carrying a piece of leaf back to the nest to add to their 'garden' from which the colony harvests a fungus. On the leaf is a tiny 'minor' worker, sometimes called a 'hitchhiker' but in fact defending its larger sister from attacks by tiny parasitic Phorid flies which attempt to lay an egg on the body of the leaf-carrying worker. (If successful the fly larvae burrow into and eventually consume the ant.) On the left a huge 'soldier' defends the colony from larger predators. Its bulging head capsule contains the muscles working the formidable jaws.

some of the extraordinary behaviour involved in the evolution of this reproductive division of labour. It is associated with a high degree of relatedness between members of all the different castes and with an overlap between generations. The most specialized insect societies are those of the termites (Isoptera) and the ants, bees and wasps (Hymenoptera) and with them the differentiation into castes may become most extreme (see Fig. 7.6). Until quite recently, eusociality was associated exclusively with the social insects, but then Jarvis (1981) described true eusociality in a mammal – the naked mole rat (Fig. 6.11). Even more recently, eusociality has been described in snapping shrimps of the genus *Synalpheus* that live in sponges on tropical reefs (Duffy *et al.*, 1999). These shrimps live in colonies of up to 300 individuals but have only one reproductive female (the 'queen'). Like termites and mole rats, the shrimps are diploid and the 'workers' are full sibs. As food seems to be abundant, the main task of the workers is to defend their sponge which becomes very crowded. On the congested reef, sponge 'homes' are sought by a wide variety of species and it takes the combined efforts of all the shrimp workers to keep competitors, including larger species, at bay.

Even though only a small number of species have taken social living to the extreme of eusociality, we can see some elements of this type of social organization in many different animals. As we saw in Chapter 6, for example, it is quite common to find young birds in their first or second year of life staying to help their parents before starting to breed themselves. Such 'helpers at the nest' have been recorded in over 200 species of birds and in a few mammals such as the black-backed jackal (*Canis mesomelas*) (Moehlman, 1979) where it is possible for offspring to help their mother to rear another litter.

In insects, too, we can sometimes see two elements required for eusociality – overlap of generations and cooperative brood care – even in species that do not show the caste divisions of fully eusocial species. The aggregations formed by insects such as cockroaches and earwigs (*Forficula* sp.) are a good example. The life span of a cockroach may be a year or more, three-quarters of which is development through a series of nymphal stages, becoming more like the adult insect. During this period, cockroaches of all ages live together in a loose aggregation near sources of food and shelter. Some species of cockroach incubate their eggs inside the female's body and bear live young, which remain in contact with the mother for some hours after birth. This contact may be important for the survival of the young, because not only are they extremely vulnerable to cannibalism when first born, but they may pick nutrients from the mother's body surface.

We know that this nutritional factor is also of major importance in the termites (which are related to the cockroaches) because, feeding largely on wood, they rely on symbiotic protozoa living in the gut to digest cellulose. These protozoa are acquired by the young termites when they feed on fresh faecal matter from the adults, and they can be transmitted only in this way. Perhaps this method of feeding was one of the factors predisposing these insects to the evolution of full eusociality.

Whatever the route, the final results are truly amazing, with the social insects showing a degree of coordination that allows some species to build huge complex nests which they defend communally. Others take 'slaves' from the nests of other species, while harvester ants cut leaves which they take back to the nest and use as a base which they 'seed' with fungal spores to grow food. The ants and termites in particular have become so successful that they now dominate some tropical ecosystems (Coleman and Hendrix, 2000). Some estimates suggest they make up well over 10% of the biomass of all terrestrial animal life! There is a huge literature on these insects, whose life histories and behaviour have attracted human attention for millennia. For an introduction to the whole literature, Wilson's masterly survey (1971) is still unmatched, Hölldobler and Wilson (1990) give a definitive account of ants, Gould and Gould (1988) deal with honeybees.

The term 'caste' is well suited to describe the division of labour in eusocial insects. It implies a rigid, limited role in society largely determined by upbringing, which seems to be the case here. One of the most important factors determining caste is what the insects are fed when young. In bees, wasps and termites, all eggs laid by the queen are potentially equal, but most larvae are fed a restricted diet and develop into workers. There is evidence that when queen ants are laying rapidly, their eggs are 'worker biased' and develop accordingly no matter how the larvae are fed. But for most of the time their eggs are also equipotential and only richly fed individuals develop into the reproductive castes.

Pheromones are also important determinants of caste. They are secreted by the insects themselves and coordinate development and social behaviour. The development of worker termites is controlled by pheromones produced by the king and queen. Similarly, the queen honeybee secretes a pheromone ('queen substance') which both suppresses the ovaries of the workers and prevents them from rearing new queens. The queen is always surrounded by attendant workers who lick her body, subsequently offering food to other workers and with it, the pheromones. The level of the pheromone must be kept up and once the source is cut off, its concentration rapidly drops, which happens if the queen becomes ill or dies. The effectiveness of the incessant food sharing in circulating queen substance is shown by the fact that some workers in the brood area of the hive exhibit changed behaviour within an hour or two of the colony losing its queen. They begin construction of 'emergency' queen cells in which some of the youngest larvae, destined in the normal course of events to become workers, are fed royal jelly throughout their larval life and become queens. On emergence from the pupa, the young queens fight to the death and only one will survive eventually to replace their mother. As colonies grow, the dilution of queen substance below a critical level is one of the factors which leads honeybees to swarm.

The honeybee is exceptional because it reproduces its colonies by swarming, but in most other social insects, new colonies are founded by a single queen (or a pair in termites). The queen begins the construction of the nest and rears the first batch of workers herself. These then take over the tasks of extending the nest and bringing food and the queen usually stays in the nest, laying eggs, from this point on. The tasks performed by the worker castes vary greatly in detail, but in most colonies they cover the main categories of foraging, rearing the young, nest construction, attending the queen and guarding the colony. Termites and ants have castes which are highly specialized both in morphology and behaviour. Among the extremes are 'soldiers' with huge jaws or other structures for colony defence and worker ants ('repletes') which act as living food stores (see Fig. 7.6 and Hölldobler and Wilson, 1990). Honeybees have only one type of worker caste but the role they play in the colony changes as they age. A worker lives for about 6 weeks as an adult and her activities are roughly synchronized with her physiology. Thus she spends the first 3 days cleaning out cells. After a day or two, she begins to feed the older larvae on a mixture of pollen and honey collected from stores within the hive. From about the 6th to 14th day of her life, the worker feeds 'royal' jelly (secreted from the pharyngeal or 'nurse' glands in her head) to the younger larvae and any queen larvae in the hive. (Royal jelly is fed to all larvae for a brief period early in their development, but those destined to become young queens are fed royal jelly throughout.) The worker then gradually changes her behaviour from feeding larvae to cell construction as her pharyngeal glands begin to regress and the wax-secreting glands on her abdomen become active. From the 18th day, she may be found guarding the hive entrance or venturing outside for a few brief orientation flights. Then from 21 days onwards, there is a major switch in her behaviour (we know something of its genetic control involving the *For* gene, p. 55) and the worker becomes primarily a forager, flying out to bring back nectar, pollen and water to the hive. Normally she will remain a forager for the rest of her life – about 2–3 weeks. Figure 7.7 illustrates this sequence of behaviour through the brief life of a worker.

This is the general sequence of events but it can be modified to suit the needs of the colony, depending on the flower crop, temperature, age of the colony or other factors. Notice in Fig. 7.7 what a high proportion of a worker's early life is spent 'patrolling' through the hive. In this way she will become aware of the colony's immediate needs. Modification of worker behaviour is augmented by the sophisticated communication systems that exist between the members of a honeybee society (see Chapter 3). We have already mentioned the importance of food-sharing as a method of communication, both as a way of circulating pheromones, but also to keep each worker directly informed of the state of the food supplies within the colony. The 'dance-language' of honeybees (p. 164) is the most sophisticated example of communication because of the detailed information it conveys about the distance and direction of a food source.

This high degree of social organization within social insect colonies and the control it gives them over their environments are based on a relatively simple series of responses to other workers in the colony and to the nest itself. There is no

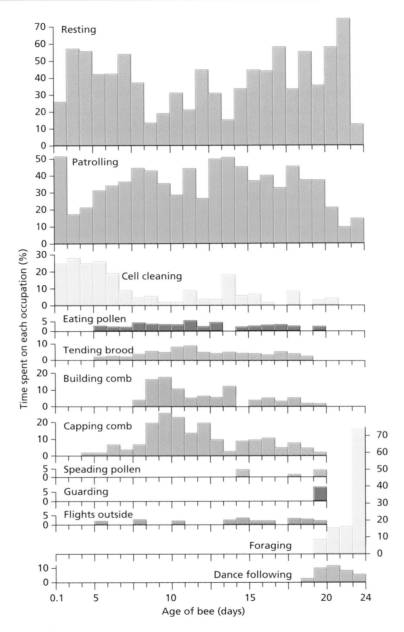

Figure 7.7   Lindauer's complete record of the tasks performed by one individual worker honeybee throughout her life. The records are classified according to the type of task. One can recognize the age-determined succession of cell cleaning, brood care, building, guarding and foraging. Note, however, the large amount of time spent in patrolling the interior of the hive and in seeming inactivity. During such periods the worker may be acquiring information on the situation in the colony and adjust her behaviour accordingly.

suggestion that they are intelligently working out what to do. To a certain extent, the social organization of the insects is flexible, as when honeybee workers change their normal sequence of tasks in response to a sudden requirement in the colony. However, the degree of flexibility is limited. Consistency of social organization within species is very characteristic of the insects and we do not find it so well marked in other groups.

Territory in the social organization of vertebrates

In contrast to the social insects, the social organization of vertebrates is not rigidly species-specific. Thus the answer to the question, 'What is the social organization of the house mouse?' is not fixed. It depends upon the nature of the food supply, the density of the population, its age, sex structure and a number of other factors. An important key to understanding the social organization of vertebrates is to remember a point we made earlier: that group living brings costs as well as benefits. Other animals are competitors for food, for mates, for nest sites and many animals have well-developed mechanisms for 'seeing off' unwelcome intruders. Not surprisingly, therefore, some animals adopt a solitary way of life or live in monogamous pairs. Many defend territories to keep other animals away from them. As mentioned above, territoriality may fluctuate through the seasons. Titmice are territorial or gregarious depending on the time of year. During the breeding season, they are strongly territorial and vigorously defend the area around their nest against others of the same species. But, during the winter, they group into flocks both with members of their own species and of others. This means that when they have hungry young to feed in the spring, they defend a private supply of insect food and in the winter, when food is scarcer, they gain the advantages of locating food and protection against predators that being part of a flock gives them.

Defence of a territory of some sort is in fact a common feature of the social organization of many vertebrates (as well as some invertebrates). Territorial behaviour is especially typical of passerine birds where often each male defends a substantial area, which will include food for himself and eventually a mate and young. He will rarely leave the territory during the breeding season. Often territories are quite closely packed and where the habitat permits, nearly all the available ground will be occupied. In birds, territorial defence is usually achieved by song and visual displays. In mammals, which may also have densely packed territories, the territorial boundaries are often defined by scent posts marked with urine, special glandular secretions or faeces.

Defending a territory against constant intrusions is clearly costly because an animal has to expend time and energy seeing off rivals and may make itself vulnerable to predators while doing so. Brown (1969) introduced the idea of 'economic defendability', pointing out that animals should only go to the time and

Figure 7.8 The golden-winged sunbird defends nectar-filled flowers against intruders if it is energetically worthwhile.

trouble of defending a territory if the resource they were defending (the food in it, for example) was worth defending and, indeed, could physically be defended. A very scattered food source might take so long to defend that the animals would lose more energy than they would gain by doing so. Gill and Wolf (1975) applied the idea of economic defendability to the territories defended by the golden-winged sunbird (*Nectarinia reichenowi*), a nectar-feeding bird found in East Africa. They calculated the energy used by a sunbird defending its territory against intruders and also the energy available in the *Leonotis* flowers they were defending (Fig. 7.8). If flowers are defended they become a better food source, because the nectar stocks rise if they are not depleted by other birds. Gill and Wolf found that the energy the birds expended in keeping other birds out of their territories was more than compensated for by the increased levels of nectar in their 'private' flowers. Defending a territory was, therefore, energetically worthwhile.

Figure 7.9   Part of a huge colony of nesting gannets on the Bass Rock, a precipitous rock stack on the Firth of Forth near Edinburgh. (The gannet was named *Sula bassana* by Linnaeus after this colony, which has been famous for many centuries.) The even distribution of the nesting territories is remarkable. The gannets obviously gain protection by nesting together on islands and from mutual defence against aerial predators, but each pair still requires a space, albeit little more than a square metre, which it defends against hostile neighbours. The result is an even spread of birds across all the suitable nesting areas.

It is clear, however, that such considerations of energy gained or lost do not apply in every case of territorial behaviour. The territories of ground-nesting sea birds, such as gulls, terns and gannets, may be as little as 1 m across and contain no food at all (Fig. 7.9). They consist simply of a small defended area around the nest which is always occupied by one member of the mated pair and are important because of the cannibalistic nature of other gulls in the colony, some of which specialize in eating eggs and chicks.

Figure 7.10 Black grouse displaying on their lek. Both black grouse and sage grouse show the extremes of sexual dimorphism in such species. The females of both species who visit the lek and 'choose' a mate are very similar and cryptically coloured (see Fig.1.4, p. 12).

Yet another kind of territory is seen in the 'lek' system of birds such as the sage grouse, which we described in Chapter 1, also in ruff (*Philomacus pugnax*) and in black grouse (*Tetrao tetrix*), (Fig. 7.10) and in mammals such as the Uganda kob (*Kobus kob*), fallow deer (*Dama dama*) and the hammerhead bat (*Hypsignathus monstrosus*). Here the males gather in a tight group or lek but within the group each male defends a small territory. There is no food in any of the territories and it is still not fully understood how these leks function, (Balmford, 1991). One possibility is that they occur when females are very spread out, perhaps because their food is widely dispersed, and the males are unable to defend a large enough feeding territory to attract them or keep them in one place (Emlen and Oring, 1977; Clutton-Brock, 2007). The lek therefore becomes a sort of 'marriage market' where the normally widely dispersed females come to choose a mate, perhaps because the sight and sound of many males all displaying together attracts females from a large area. Ryan *et al.* (1981) studied the Túngara frog, where the males aggregate into leks or choruses and showed that with bigger choruses, more females approach. Fortunately for the males, by calling in large groups they are also safer from predation by frog-eating bats.

Another possible advantage of leks is that females can make a direct comparison between different males and choose on the basis of their displays. With leks of the sage grouse, for example, Gibson and Bradbury (1985) showed that it is the males that display for the longest and in the most vigorous way that are chosen by the females. In black grouse we know – by virtue of hindsight – that females choose the males that have the highest chance of still being alive in 6 months' time. At the centre of the lek males have the smallest territories but they are tenaciously defended. This is the part the females head for and it certainly suggests that the females are using the lek to assess the health and viability of the displaying males. We cannot be sure how they choose; all we know is that the system works (Alatalo *et al.*, 1991).

If leks are one mating system adopted by males in response to the dispersal of females, another strategy may be for males to space themselves out and vigorously defend some resource in which the females are interested. This is the case in the pied flycatcher, discussed on p. 341, and also in a species of coral reef fish, the bluehead wrasse, where males defend individual protruding coral heads which attract females because they make good spawning sites (Warner, 1987; Fig. 3.36).

Mating systems and social organization

Any social organization is the result of selection acting on individuals of the two sexes and it is now abundantly clear that there will be conflict which can only be resolved as a best compromise. All sexually reproducing animals have to come together to mate but this is, as it were, only the minimal requirement. Will they disperse then or will they stay together, as a pair or in a larger group? We can understand and explain this only if we know the role of the two sexes in reproduction and how much subsequent investment each makes in the care and nurture of their offspring. This will be a major factor in determining their organization. We have just been describing some of the ways in which animals acquire resources by territorial behaviour and this may be direct – actual resources are defended – or more symbolic, as with the leks. Leks can operate because the investments of males and females in reproduction are totally unequal. Once she has mated, a female sage grouse has no further use for either the male or his territory. She leaves the lek area to rear her young alone. The male's contribution to his offspring is thus purely genetic and he will continue to try to attract other females without giving any of them assistance with rearing the young. The same is the case with most mammals where males do not assist with the care of individual offspring, although sometimes defending the group in which they live. This contrasts with the great majority of birds, a few mammals, such as marmosets and gibbons, and many species of fish where the male gives a great deal of care – feeding and protecting his offspring.

We can best begin our discussion of mating systems with monogamy, which would seem to represent a simple form. It was once assumed that individuals mate exclusively with one partner and the pair cooperate to raise young. Thus monogamy is contrasted with polygamy (mating with two, three or many more individuals). This describes the situation when one parent (usually the female) gives a great deal of parental care, while the other parent (usually the male) contributes little but sperm and mates many times. This mating system, more strictly called *polygyny*, is that shown by most mammals and a few birds such as chickens, peacocks and, as we have seen, sage grouse.

In the last few years, however, our view of monogamy has had to change. DNA fingerprinting gives us a much clearer picture of which individuals are actually the father or mother of which offspring. It has revealed a large amount of multiple mating in species that had previously been described as monogamous and in which both sexes care for the young. Starlings provide a good example. Both parents care for the young, and the pair could be described as 'socially monogamous', but it is certainly not genetically monogamous since three or four different males may be fathers in one brood! What is more, females in many species actively solicit copulations from outside males (Møller and Birkhead, 1994), so it is not simply a case of unfaithful males coercing other females into mating with them.

It is clear what advantage males gain from multiple mating because they can produce more offspring by mating with several females. It is less clear why females should gain, since reproductive success in females is limited by the number of eggs or young they can produce. One suggestion is that females gain if they can find another male available who is higher 'quality' than their own. In blue tits, it is females mated to lower quality males who are more likely to seek extra-pair copulations than those mated to high quality males (Kempenaers et al., 1992). This situation comes about because blue tits are socially monogamous and both sexes help to feed the young but they are also highly territorial. This means that when a female comes to choose a male, she may find that all the high quality males have already been mated and so has to settle for a less good male than she would ideally like. Kempenaers et al. measured 'attractiveness' of males by how many neighbouring females intruded into their territory, and 'quality' by scoring whether the males survived until the next breeding season and how many of their offspring survived. They found that males that had a nest in which there were young fathered by another male were more often abandoned by their own female when she was fertile, received fewer visits from neighbouring females, had fewer surviving young, were smaller and survived less well than males that gained through fathering extra-pair young. By mating with a high quality male, who is already mated, but having her own mate to help her care for the young, a female blue tit apparently achieves genetic quality as well as paternal care from her mate. In support of this hypothesis is the somewhat surprising finding that females of lekking species rarely copulate with more than one male (Birkhead, 1993). This could be because females on a lek have a completely free choice from the start and can choose to mate with high quality males regardless of what other females are doing.

Multiple mating by females is, of course, not in the interests of males and we often find that they take extraordinary steps to ensure their own paternity. Female dragonflies mate with a number of males and store the sperm in a special sac. In some species, the penis of the male is equipped with elaborate hooks with which the male tries to scoop stored sperm out of the female's sac before he mates with her (Waage, 1979; Fig. 7.11). The ejaculate of male mice coagulates into a hard plug, which effectively prevents the female from mating again for many hours. Among birds, female dunnocks (*Prunella modularis*) frequently form bonds and copulate with two (sometimes more) males. Davies (1992) describes how males can be seen pecking at the cloaca of females, attempting to remove the sperm of other males. Even when males do not go to quite these lengths, their sperm carry on a subtle form of male–male competition. Simply by making large amounts of sperm, males that copulate with an already mated female help

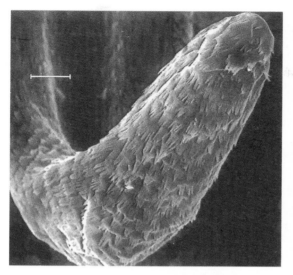

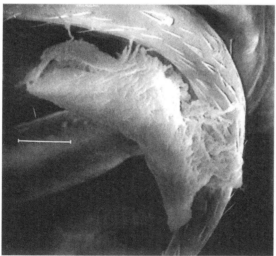

Figure 7.11 (a) Scanning electron micrograph of the penis of the dragonfly *Argia moesta* showing the cuticle densely covered in backward facing spines. (b) The same of *Ischnura ramburii* showing a clump of an earlier rival's sperm caught on the spines.

to make sure that theirs is most likely to fertilize the eggs. The remarkable aquatic warbler (*Acrocephalus paludicola*) goes one step further in sperm competition. In this species, the females can rear the young without any help from the male, and males and females lead largely separate lives, meeting only to copulate. Both sexes have multiple mates and nearly half the broods contain offspring with three or four different fathers (Birkhead, 1993). In most small birds, copulation lasts 1–2 s, but in the aquatic warbler (Fig. 7.12) it lasts about 25 minutes! The male and female lie on the ground with the

Figure 7.12   The remarkably prolonged copulation of the aquatic warbler. When a male encounters a receptive female he monopolises her and the pair lie on the ground for some 25 minutes.

male on top and the male has huge testes from which he repeatedly passes sperm to the female. The most likely explanation of this extraordinary behaviour is that, by copulating with the female for so long and making so many sperm, the male is trying to ensure that he is the winner in the battle to secure insemination.

Multiple mating by females might be a problem for males of those species where they contribute paternal care. By caring for young that are not his own, the male is 'wasting' his time and energy. Some males may be able to adjust the amount of care they give depending on their genetic relatedness to the brood and Davies (1992) found that dunnocks fed nestlings more when they were most likely to be the father. They seem to be able to get some estimate of this from the amount of contact they had with the female just before she laid eggs – the most likely time for her to be fertilized – but it is not easy for males to establish paternity. We now know that in very many cases a proportion of the offspring reared by a socially monogamous pair will be half siblings, not full siblings. Many males will have been cuckolded and so are helping to rear some of another male's offspring. One can speculate that in many insectivorous birds where a female cannot rear the brood successfully on her own, then the males have no alternative but to carry on with their parental contribution. If they do not they will lose offspring anyway. Further, although he may well be rearing another male's offspring, neighbouring males may be in an equivalent position and just because extra-pair matings are so common, social monogamy might become effectively stable again.

Huge numbers of bird species are socially monogamous, but it does not result in real uniformity because both sexes remain ultimately in competition to gain advantage. As we have seen in the case of the pied flycatcher, females may choose males because of the resources they are defending. Even where a male does not contribute to the rearing of his young by feeding them, he may nevertheless aid their survival by defending the nest against predators or intruders of the same species. This does not, however, necessarily lead to strict monogamy (one male mating with one female) since, as we have seen, a male's ability to defend resources or groups of females will also affect the nature of the mating system. One male may have several females nesting in his territory and provide resources and even some paternal care for all their broods. The 'polygyny threshold' model we discussed in the last chapter, as a way of describing the conditions under which it was advantageous for a female to mate polygamously with such a male, has turned out to be useful but something of an over-simplification. Conflict between the sexes means that the model does not always fit the data because males will always be tempted to cheat if conditions permit.

Catchpole *et al.* (1985) show how such conflict may operate in practice. The European great reed warbler is, like the pied flycatcher, a migratory species in which the males arrive on the breeding grounds before the females and defend large territories. Female great reed warblers are attracted by the quality of the males' territories and sometimes several females settle in the particularly good territory of one male. The great reed warbler provides an extreme example of sexual selection acting on song, as described in Chapter 2. Here the male's song is the auditory equivalent of a peacock's tail or a stag's antlers!

If a female is to throw in her lot with other females in a particularly good territory, then her breeding success should be equal to or greater than the breeding success of monogamously mated females in poorer territories, but in this case it is not: the second- or third-arriving females show reduced breeding success. Catchpole *et al.* argue that the females are deceived into choosing already mated males because the large territories in the reed beds prevent them from seeing the females that the male already has there. Some great reed warbler males can succeed by cheating! Alatalo *et al.* (1986) argue that pied flycatchers go one stage further and defend several different territories at once, thereby deceiving several females into apparently 'monogamous' pairings. The male benefits but the females would seem to be worse off than if they had chosen an unmated male.

Overall, it is not difficult to understand the distribution of social monogamy because it will only occur where more offspring can be raised by a pair cooperating than by one parent alone. Male birds can not only feed their young, they sometimes contribute at an earlier stage and help to incubate the eggs. (Interestingly this is taken further in many egg-laying fish. Once she has laid the eggs, the female may abandon them to the male's care.) Male mammals, on the other hand, have very limited opportunities to be active parents. The gestation of the young within the female's body and then lactation from birth means that males can contribute little. Marmosets stay as a close family and the male can at least carry the twin young around during their long period of dependence. Then there are a few carnivores such as the golden jackal where pairs stay together and the male can bring back high value food to a den which he regurgitates to feed the lactating female or weaned young, but for the most part mammals are polygynous and the males try to obtain multiple mates.

Dominance in social systems

Throughout this book, we have emphasized that natural selection often leads animals to compete with one another. Sometimes different individuals will have common interests and will show a high degree of cooperation, but even when they do, competition at the individual or gene level does not disappear. It merely takes on a new form in which individual interests are – at least for a time – best served by aiding others (Chapter 6). At other times, conflicts of interests become more overt, and animals can be highly aggressive toward one another. Male red deer and elephant seals spend much of the breeding season fighting each other in dramatic and often bloody conflict. Huge male elephant seals (*Mirounga magnirostris*) dominate stretches of beach where the much smaller females have come ashore to have their young (Fig. 6.27). The reproductive prizes are thus very great but so are the costs. The fights between males can be very damaging, many are killed before they become fully mature and few males manage to stay successful for more than one or two seasons. The same is true of red deer, where the reproductive life of a female may be up to 20 years, but that of a stag very much less. In lions, too, a coalition of male lions holds a pride of females for little more than 2–3 years before they are displaced by younger and stronger rivals.

Using signals as substitutes for fighting is one way animals have of avoiding the highest costs of aggression (i.e. death or injury) and we have seen that use of pre-fight 'assessment' signals can be an evolutionarily stable strategy (p. 328), provided, of course, that the signals used are honest and reliable. Otherwise, animals would avoid the costs of fighting but would also miss out on the benefits of the resources they might have had by winning the fight. As we discussed in Chapter 6, many authors, such as Zahavi (1991), have argued that such reliability can only come through the use of signals that are themselves costly or at least difficult to produce for any but the most vigorous and physically strong animals. Very costly signals such as the roaring of stags can be sustained if they are only in use for a brief breeding season. The stakes are very high then and it will be worth a stag's while to keep testing his 'status' against rivals.

Some stability in a social system can be advantageous because it means that the same animals interact frequently enough to learn each other's true fighting ability. Then from past experience they can avoid fights with animals that are likely to beat them. Animals using this method of avoiding overt fighting are often described as having a dominance

	Y	B	V	R	G	YY	BB	VV	RR	GG	YB	BR
Y												
B	22											
V	8	29										
R	18	11	6									
G	11	21	11	12								
YY	30	7	6	21	8							
BB	10	12	3	8	15	30						
VV	12	17	27	6	3	19	8					
RR	17	26	12	11	10	17	3	13				
GG	6	16	7	26	8	6	12	26	6			
YB	11	7	2	17	12	13	11	18	8	21		
BR	21	6	16	3	15	8	12	20	12	6	27	

Figure 7.13   A perfect linear hierarchy established within a group of 12 domestic hens. Each bird is marked on its legs by coloured rings, whose initials identify it. The number of times each bird pecked another flock member is given in the vertical columns (e.g. Y pecked B 22 times and pecked V 8 times) whilst the number of pecks received from another hen is given in the horizontal column (e.g. VV received 19 pecks from YY and 8 pecks from BB). Note that no bird was ever seen to peck an individual above it in rank. This is an artificial group and hierarchies as perfect as this are probably rare in nature.

hierarchy or 'peck order'. This phrase, which has entered the language now, we owe to the Norwegian zoologist, Schjelderup-Ebbe (1935) who observed flocks of domestic fowl. He found that a definite peck order developed amongst a group of hens, one gradually emerging as the dominant, in the sense that she could displace all others. Below her was a second-ranking bird who could dominate all except the top bird and so on down the group until at the bottom was a bird displaced by every other in the flock. Figure 7.13 shows a classic example of a peck order of this type – a linear hierarchy, as it is sometimes called. The hierarchy develops as the birds dispute and often involves a good deal of fighting in the early stages as the birds test each other out. But once it is established, subordinates usually defer without question to the approach of a more dominant bird. Chickens are somewhat unusual in often having clear linear hierarchies, but perhaps it is not surprising that in other species relationships are not so clear-cut. Particularly among primates, the situation may be more complex such that there are triangular relationships, A dominates B and B dominates C but A does not necessarily dominate C. In addition, there may be more complex alliances, in which two males cooperate to maintain a joint top position in a hierarchy. De Waal (2000) describes such an alliance that developed between two male chimpanzees at Arnhem Zoo. Neither could have achieved dominance on his own but together they could successfully ward off all challengers.

Not only chickens, but animals of many different species show a marked drop in aggression as they become familiar with each other (Chase, 1985). The animals learn that certain other individuals are reliably capable of beating them in a fight and so avoid the risks of a fight they are unlikely to win anyway, by immediately deferring to their superior. They do not even need a costly assessment signal in order to do so and the dominance hierarchy thus enables them to avoid not only the cost of fights they would be likely to lose but also the cost of giving signals. Of course, the stability of such a system (and the advantages of accepting a subordinate position) relies on the fighting ability of the different individuals remaining much the same over long periods of time, and dominance hierarchies can be expected to become unstable when the relative fighting abilities of the different members shift. This is exactly what happens. A dominant hen that becomes sick may slip down the peck order and the other individuals then fight to reposition themselves. Similarly, a young male baboon or chimpanzee that grows bigger and stronger than the dominant male begins to challenge the existing hierarchy and other relationships will change too.

Sometimes a group may be too large for animals to learn about all others individually, but we see another method of avoiding unnecessary fights while keeping signal costs to a minimum. This is to have a dominance hierarchy based not on individual recognition, but on 'badges of status', which we have already described in Chapter 3 (p. 146) on signalling. They can be instantly recognized even on unfamiliar individuals. The larger black bibs of male house sparrows or the black head feathers of Harris' sparrows distinguish dominant from subordinate animals. What is striking about them is that they do not appear to be costly and consequently would seem to be vulnerable to cheating. While we can understand why an animal might move away from a piece of food without a fight when a known dominant animal appears (it has always been beaten by this animal in the past), it is less easy to understand why it should defer to an unfamiliar animal simply because it has a dominant's status badge. What is to stop a weak animal growing black feathers and so mimicking a dominant? The answer seems to be that there are costs to status badges after all. It may be relatively easy for a weak animal to produce a status badge, but if it is constantly challenged by other dominant individuals (Møller, 1987), then it will pay a cost for its deception. The social consequences of cheating may be sufficient to keep cheats under control or at least at a low numbers. In practice, status badges thus have a higher degree of reliability than appears at first sight, although the constant possibility of cheating, albeit at a low level, may be the reason why animals do not always rely on them. They are found particularly in species such as sparrows and great tits, where animals constantly encounter such large numbers of other individuals that hierarchies based on individual recognition are unlikely.

It will be apparent by now that what we refer to as 'social organization' in animals cannot be rigidly categorized as 'territorial' or 'hierarchical', and that mating systems and other social interactions take many different forms and are affected by a range of factors. As we mentioned at the outset, for many vertebrates, the changing seasons involve a change to their social organization. In chaffinches, males tend to dominate females in winter flocks and displace them at feeding sites. But this situation is reversed in spring, when females tend to displace males as the flocks disperse to set up territories. Chickens studied in a truly feral state on an island off Queensland, Australia by McBride *et al.* (1959), alternate between a territorial system in the breeding season and a more hierarchical flock structure during the winter. The dominant male sets up a territory in the spring with a number of females in his flock. During the winter after the young birds of the year had returned to the flocks, McBride *et al.* found that the alpha male and his harem moved about over a home range with numbers of subordinate males staying at the periphery of the group, often moving between the home ranges of different alpha males. The alpha male led his flock in every sense. He it was that initiated all movements of the group, particularly across open ground, and his posture was normally more alert than those of his females. He was the first to give alarm calls and even approached predators – feral cats – while the others took cover.

Diverse mammalian social behaviour

Seasonal changes in behaviour are just as marked in some mammals. We have just been discussing signalling among red deer. The classic study, *A Herd of Red Deer* by Fraser Darling (1935, now reissued, 2008) and the more recent studies on the Hebridean island of Rum by Clutton-Brock and his colleagues (1982) give us a very full picture of the changing cycle of social behaviour in this species. Outside the breeding season, males and females live apart and range over different areas of woodland and hillside. In winter, the males live in loose bachelor herds in which there are consistent linear dominance hierarchies related to body size, usually correlated with the size of their antlers which increase with each year's new growth. Disputes may involve direct clashes (Fig. 7.14a) but high ranking stags may be able to displace others from good food patches simply by lowering their heads in a display of antlers. The females, on the other hand, live in herds, which also include young animals of both sexes. Threats and displacing other animals from food are much less common among the hinds, even in winter. In early April, the stags shed their antlers and immediately male aggression takes on a new form. The males rear on their hind legs and 'box' with their hooves (Fig. 7.14b). New antlers begin growing at once but remain soft and tender 'in velvet', which is not shed until August. The stags then become increasingly aggressive and soon go off singly to favoured display areas and begin roaring and gathering hinds, who are coming into oestrus, and the 'rut' begins. Oestrus females are attracted to displaying stags, who defend their group of females rather than any particular area. The rutting season lasts only a few weeks and soon both sexes return to their respective herds for the winter.

(a)

(b)

Figure 7.14   Stags fighting through the red deer's breeding cycle. (a) Between August and early April stags have antlers. During the rutting season such clashes can be prolonged and may result in injury or even death. (b) Antlers are shed in April and stags immediately switch to fight by 'boxing', rising onto their hind legs and kicking at each other with their forelimbs.

These are dramatic changes to social organization through the year and are linked to the deer's breeding cycle. The reproductive cycle of seals also involves major shifts in social behaviour. There is no close social organization for most of the year. Seals haul out to rest on favourite rocks or beaches from time to time, but spend much of their life at sea feeding. All changes when the females emerge on to the breeding beaches where they all give birth over a short period. The males join them, and in some species there is intense fighting and guarding of harems such as with the elephant seals described on p. 344. The true seals (Phocidae) have only a short, intense breeding season. Females suckle their young briefly, mate and return to the seas for the rest of the year. They will have conceived but after brief development, the embryo does not implant in the female's uterus for several months. Implantation and the completion of gestation is timed so that birth occurs just after the females haul out onto the breeding beaches. In the related sea lions (Otariidae) both sexes spend most of the year on the breeding sites, often isolated small rocky islands, which act as a base from which they forage. Suckling is much more prolonged with females leaving their pups in the group as they go off to feed. The males are always present and develop a dominance hierarchy which determines access to females as they come into oestrus (Riedmann, 1990; Trillmich, 1996).

In these sea mammals, there are few social interactions outside the breeding season, but others have much more permanent relationships. The selective forces which keep groups together vary but, as discussed earlier in this chapter, mutual defence against predators or cooperative hunting of large prey are often involved. Young animals are particularly vulnerable and in mammals as diverse as meerkats (see Fig. 7.4) and sperm whales (*Physeter macrocephalus*) (Mann *et al.*, 2000), some adults in a group remain vigilant whilst a mother leaves to feed. This is crucial for the whales because the adults have to make prolonged and deep dives for food but young calves have to remain near to the surface for breathing.

Cooperative hunting must have been a powerful selective force for some members of the dog family (Fig. 7.15). Three species, the dhole of southeast Asia (*Cuon alpinus*) (Venkataraman, 1998), the Cape hunting dog of southern Africa (Creel and Creel, 2002) and the familiar wolf (*Canis lupus*) (Mech and Boitani, 2003) all live in long-term and closely knit groups, commonly between 5 and 25 in size. They collaborate closely for breeding, all bringing food back to the den in their stomachs, which is regurgitated for the pups. The striking feature of these dog societies is that normally only one pair actually reproduces, the dominant male and the dominant female. The other members of the group comprise their recent offspring, but also some others. A similar group structure is shown by meerkats, already mentioned in other chapters, but these social mongooses do not hunt cooperatively. Here communal defence must be the main advantage of grouping but, as with the dogs, all members of the group feed the young of the dominant pair directly. Their prey is usually insects, small reptiles, etc. which they catch and pass over a proportion to the begging young animals. Very extensive research is in progress on the meerkats of the Kalahari region of South Africa and there

(a)

(b)

Figure 7.15   Three of the social canids, (a) Cape hunting dogs; (b) wolves; (c) the less familiar dholes of South-east Asia. All are intensely social with elaborate social greetings when the group comes together after dispersion. A single dominant pair breeds supported by the others, many of whom are offspring, especially males. The group hunt together and all bring back food to the den which is regurgitated for the juveniles.

is now a large literature. Clutton-Brock *et al.* (2004) can serve as an introduction and there is also an excellent popular account, Clutton-Brock (2007a).

In none of these mammals are groups totally closed, there is some interchange. For example, young female hunting dogs usually leave their natal pack; the males remain.

(c)

Figure 7.15 (*cont.*)

Nonetheless, group stability is the dominant feature of their social organization with all members being subordinate to the dominant pair. Subordinate animals do sometimes breed using strategies not unlike those of the 'satellites' described in Chapter 2), but usually they have very low success and often the dominant female will kill such young. Why then do the subordinates stay? This is a question with several answers, which will vary with species and with circumstances. Firstly, some of the small carnivore species, like the meerkats, have no chance alone and once they are established in a group, emigrating is very risky. Apart from predators, approaching another group is dangerous and they are quite likely to be killed! Indeed it is in their best interest to keep their own group's numbers up because small groups are vulnerable to hostile neighbours. The balance of interests is different with the social dogs. Cooperative hunting is their only way of killing large prey and whatever the obvious disadvantages – for they will have to share – the gains will outweigh the chances of surviving well on small prey items they could catch on their own. Secondly, there is often an element of kin selection (Chapter 6) involved; if your chance to set up on your own is very low, it's best to help others who share some of your genes. The male hunting dogs that stay in their natal pack are usually related to the dominant female and so are helping her to raise siblings. Thirdly, if you stay around in the group for long enough there may be a chance of inheriting 'dead men's shoes'. Long-term observations on all these species have revealed such replacements, although they are not common.

All the social groups we have just been describing do show a richness of interaction between the members. There is often much greeting, calling and posturing as all come back together after a time when they've been dispersed. Young and adults often play together and there is much excitement when adults bring prey back to the den. Yet always there are obvious tensions as the dominant pair exert their authority. Quite often as they displace

(a)

(b)

Figure 7.16   The social organization of the giants! (a) A group of African elephants, females with their young calves, in this case being visited by an adult bull (second from left) who will be solitary most of the year but visit female groups to check for oestrus cows when the males themselves are in 'musth'. Note the enormous size difference. Male elephants go on growing for decades and are rarely able to compete successfully for matings until they are 30 years old or more. (b) Part of a pod of sperm whales, a mother and infant in the foreground. There is close cooperation between females which stay together and guard young. Adult males lead more solitary lives and visit female pods only during the mating season. This whale's social organization resembles that of elephants in many ways.

subordinates there is much elaborate show of submission from the latter and, as mentioned above, infanticide usually follows birth of any but the dominant female's offspring.

The animal societies which exhibit the highest levels of richness and complexity are found in three mammalian orders, elephants, whales and primates. These animals are all large brained and potentially long lived. Some of the larger primates and especially the apes may live 30 or 40 years in the wild, elephants for 60 or 70, which is probably exceeded by many of the larger whales. Certainly one, the bowhead (*Balaena mysticetus*), is known to reach 150 and possibly over 200 years; the longest measured life of any animal! Obviously many different types of social organization are to be found among such a diverse assemblage, but in all of them, in contrast to the dogs and meerkats, the majority of adults within the group are breeding. Certainly all the females breed and most males will do so at some stage in their lives although there may be strong competition between them when they are maturing.

African elephant society is based around the matriarchal group – it is reasonable to call it a family – who move and forage together (Fig. 2.7). An old female is accompanied by her daughters and their offspring of various ages including their sons up to about 14 years. Even before this age young males begin to spend time away. Eventually they move separately and from the beginning of their maturity, around 30 years of age, they are mostly solitary, visiting female groups for mating (Fig. 7.16a). Male elephants are sexually active for only limited periods each year, coming into a state known as 'musth'. This word comes from India because musth was first identified in Asian elephants some of which are semi-domesticated. Males in musth can be extremely dangerous for their human companions.

Even brief observation of an elephant group will reveal the wonderful complexity of their social life. They are constantly alert to the behaviour of others, young animals are guarded and helped through difficult terrain, injured animals may be given special attention (see the anecdote related on p. 278) and there is much greeting, friendly contact and play. Luckily, there are a number of splendid films and well-illustrated popular accounts of elephant behaviour (e.g. Moss and Colbeck, 1992; Poole, 1997) which give some idea of the richness of elephant societies.

Elephants use a range of auditory communication within the family and each individual has a distinctive voice known to all the others. The contact calls are very powerful in what is to us the sub-sonic range (10–35 Hz) but there are higher harmonics that we can hear. It seems likely that elephants can pick up calls from several kilometres away (Poole *et al.*, 1988). During their wanderings, each family will come into contact with many other families and they too are recognized from their calls. Playback experiments have shown that families with older matriarchs, around 55 years, remember a greater range of calls and thus can discriminate better between familiar and unfamiliar calls than those led by a younger female who is around 35. In response to unfamiliar calls, families tend to bunch protectively because strangers do sometimes harass young animals. The family seems to take its lead from the old matriarchs who are 'repositories of social knowledge', as McComb *et al.* (2001) entitle their paper, and here we can recognize that age and experience, even beyond the age of reproduction in some cases, will offer advantages to all the group. Social knowledge extends over time and place and, for example, during a period of severe drought, an old matriarch may remember which water holes remained open the longest.

There is an intriguing parallel between elephant society and that of another huge, long-lived animal with a large brain, the sperm whale. Their groups, called 'pods', are made up of several females with their offspring including perhaps three generations who move together (Fig. 7.16b). Young males leave when about 5 or 6 and join with other males who move in loose pods, migrating northwards to Arctic waters. As they age they become more solitary, returning to the lower latitude waters to breed when about 25 or 30. Even in such a totally different environment, it seems likely that the benefits of social life for these marine mammals are remarkably similar to those experienced by elephants. Weilgart *et al.* (1996) refer to 'a colossal convergence' when comparing them together. Although it is more difficult to follow cetaceans in the wild, there are now a number of impressive long-term studies which reveal all the same richness of social interaction as with elephants, protection of the young, helping injured animals, play and friendly contacts (Mann *et al.*, 2000).

Dolphins are among the easiest cetaceans for study because they do not migrate over such huge distances and tend to remain in particular tracts of ocean. It has been possible to study some of the smaller dolphins, like the bottlenose, in some detail, recording the behaviour of captive animals as well as wild pods. They have a much more dynamic social structure than do the larger whales, with close groups forming and splitting but usually staying within the same area. Pods may be of one sex or both – the most consistent associations are, of course, mothers and offspring. Again there is a great deal of friendly association, but also aggressive interactions when males may compete for matings. One of the most striking characteristics of dolphins is the richness of their vocal communication. All the whales share this feature because they live in an environment where sound is crucial for maintaining contact and vision is of no use except at close quarters. Probably we are only just beginning to understand the complexity of dolphin vocalizations, but we do know that individuals have personal 'identity whistles' and can both imitate each other's calls and, in captivity, be trained to 'label' novel objects by unique calls (Mann *et al.*, 2000; Janik *et al.*, 2006).

Primate social organization

We turn last to the primates, who have inevitably cropped up before in this discussion of social behaviour. They have, of course, particular significance for us and attract attention from a number of disciplines – zoologists, psychologists, anthropologists and sociologists – all have all converged on this group, both for their intrinsic interest and also in the search for information which is relevant to exploring the origins of human societies. Recently there has been great emphasis on studies of natural primate

Figure 7.17 Japanese macaques (*Macaca fuscata*) living in the north of Japan experience extreme cold every winter and grow dense fur accordingly. In volcanic areas, they are also seen to warm themselves by sitting in hot pools.

communities in the field. Partly this is a reaction to a sense of urgency because their habitats are so threatened by human activities and many species are now listed as 'critically endangered' by the IUCN Red Data Book in its 2008 edition (see also Mitter-meier and Cheney, 1987, in this regard). In addition, although certain primates, notably the rhesus monkey, have been familiar laboratory animals for some time, it is generally accepted that studies on captive communities are by themselves quite inadequate for revealing the richness of primate social behaviour.

There is now a huge literature on the behaviour, ecology and life histories of primates. Thus there are several journals devoted to them exclusively, which include numerous behavioural studies, and a large number of books and symposium volumes.

Among the latter we may list: Smuts *et al.* (1987), Campbell *et al.* (2006), Strier (2006) and Mitani *et al.* (in press). The wide scope of these books is particularly useful because they review studies on the whole range of primate types. Individual essays cover both general topics, relevant to many species, for example, aggression or social grooming, while others give detailed accounts of the social behaviour of individual species – many of them not well-known – not just the familiar rhesus macaque and chimpanzee. Here we shall focus on a few general principles concerning primate social organization, taking a few particular examples.

Primates live in a wide variety of habitats. We tend to think of them as tropical animals, but they were more widely distributed in the recent past and two of the macaques, the Barbary 'ape' (*Macaca sylvanus*) of the Atlas Mountains and the Japanese macaque (*M. fuscata*) live in areas where snow and frost occur every winter (Fig. 7.17). The majority of primates are arboreal: some of them, like the spider monkeys (*Ateles*) of South America and the colobus (*Colobus*) monkeys of Africa, almost exclusively so. However, a number have returned to ground living, such as

Figure 7.18   Some diverse representatives of the primates together with the tree shrew *Tupaia* (a) which is not a primate but which does in its superficial appearance and life as an arboreal, nocturnal insectivore, resemble the stock of early mammals from which the order descended. (b) The tarsier (*Tarsius*) of the Philippines. (c) The ring-tailed lemur (*Lemur catta*) of Madagascar. (d) The capuchin monkey (*Cebus*) of South America exhibits the prehensile tail characteristic of a number of highly arboreal New World primates.

(e)

(f)

(g)

Figure 7.18 (*cont.*) (e) Proboscis monkey (*Nasalis laruatus*). (f) The gibbon (*Hylobates*) and (g) the chimpanzee (*Pan troglodytes*) represent the living apes, closest relatives to humans – the latter sharing about 98% of its genetic material with us.

the baboons (*Papio*) and the patas monkey (*Erythrocebus patas*) of Africa, whilst chimpanzees and gorillas also spend a lot of time on the ground. In general, the more arboreal primates are fruit and leaf eaters; the ground dwellers tend to be more omnivorous, including insects and perhaps small vertebrates in their diet, although a very high proportion of their food remains grass, seeds, bulbs, and so on. Although both mountain and lowland gorillas are almost exclusively vegetarian, chimpanzees and baboons will eat meat if they get the opportunity; certain communities of the former regularly hunt and kill monkeys for food.

The primates emerged from insectivore-like ancestors very early in the history of the mammals. They are rather unusual in that we have still living examples which represent something like the stages we believe they went through during their evolution (see Fig. 7.18). Of course, any living primates are in no sense to be regarded as intermediate stages on the way to the advanced monkeys and apes; each is equally an end point of evolution and adaptation to its environment – part of the snapshot of evolution recording the situation today. Nevertheless, they provide an opportunity to speculate about when in their history typically primate social characteristics may have appeared.

There are a wide range of morphological types, from the Madagascan lemurs which retain a long muzzle, a moist nose and claws on one of their hind toes – little modified from the ancestral primate types – to the monkeys, the great apes and humans. Throughout this series we can observe certain trends: the enlargement of the brain, the development of the grasping hand and, in contrast to many other mammals, the great reliance on colour vision as a dominant sense for exploration and communication.

It seems probable that from an early stage in their history, certainly by the lemur level of evolution, the majority of primates were social animals moving around in groups whose organization was stable. Jolly (1966) makes this point in her pioneering study of lemurs. Here we already find small mixed troops (12–20 individuals), which include several adult males and several breeding females – a very typical primate grouping. There is a dominance hierarchy within the troop (and in some lemurs females rank more highly than males), but it remains as a permanent, cohesive unit. Lemurs have group territories within their mixed woodland habitat whose boundaries are often remarkably stable. They are marked by scent in some species, and are defended by calling, which is usually sufficient to cause a neighbouring troop to retreat without further threat or fighting.

Within the troop, there are frequent minor disputes but serious fights are rare outside the breeding season. This is very brief in most lemurs – at most 2 weeks – and it is at this time that subordinate males seriously challenge the older dominant ones. Apart from this short period of strife, much of lemur social life is characterized by non-aggressive inter-actions between individuals – indeed, it is pedantic to avoid using the term 'friendly'. There is always close contact between a mother and her infant, who clings continuously to her at first and is carried around everywhere. As it grows older other adults approach and play with the infant, as they also play with each other. Lemurs have thick, dense fur and groom frequently. Mothers groom their infants and adults frequently groom each other – this being one of the commonest types of friendly contacts between individuals. Again, we find that such social grooming is common in all primates.

Thus in the behaviour of lemurs we can detect most of the elements which occur to a greater or lesser extent in all primate societies, although there are many variations on the theme. One of the most obvious types of variation concerns group size, from the single family groups of gibbons to the commoner multi-male, multi-female groups typical of howler monkeys (*Alouatta* spp.), vervets, macaques and most baboons up to the veritable herds of gelada baboons and mandrills (*Mandrillus sphinx*) which may number several hundred. These herds, although they may move together, have social sub-groupings within them. Other obvious variations in primate social life concern the social structure within the groups, the type of mating system, the extent of territoriality, and the nature of interchange between groups as animals become sexually mature. A good deal of modern primatology has been truly sociobiological in its approach and tries to develop a general theory concerning the different roles of the sexes and access to resources in order to explain the wide variation in primate social structures which we observe.

We might begin by citing one primate, and a great ape at that, which uniquely goes against the trend and leads a very solitary life. Orang-utans in Borneo and Sumatra are entirely arboreal and both sexes tend to stay in a home range over which they travel throughout the year following the fruiting of trees (van Schaik and van Hoof, 1996; Singleton and van Schaik, 2002). Females have overlapping home ranges but tend to avoid each other, travelling only with dependent offspring (Fig. 7.19). Males live alone too and there is clearly fierce competition between neighbouring males for access to females and not all survive. Moving on, although monogamy is rare amongst primates, it is well exemplified by the gibbons, who have life-long pair bonds between a male and a female. Recent work suggests that there are some social interactions between neighbouring families but as their territories are maintained by elaborate 'singing', especially at dawn, gibbons show some remarkable parallels to song birds. The South American marmosets and tamarins also have very small groups with often one, or at the most, three adults of each sex in the group together with their young. Almost all other primates give birth to single offspring, but marmosets always have twins and males help with the care of infants, carrying them for much of the time (Goldizen, 1987).

The commonest society involves groups with a number of males and females living together. Within this general structure we find considerable diversity of social organization even among closely related species such as baboons of the genus *Papio*. *P. hamadryas* is a dry country species and in their social organization several females are more or less permanently bonded to a single male, forming a so-called 'harem group'. A number of such groups band together, moving and foraging as a unit, perhaps 40 or 50 strong. The other baboons also have units of comparable size but here there are no persistent male/female bonds. Adult males form temporary consortships with females as they come into oestrus, but otherwise move generally within the group.

Chimpanzees have even more variable types of grouping. Quite often males associate and travel together more than do the females and within a 'community' – a term increasingly used to describe the population of chimps in one area – groups form and disperse for foraging on an hourly basis. Sometimes they may come together to spend the night, but groups may be widely separated and move apart for days on end. Again, there are striking differences both within chimp communities and between them and the closely related pygmy chimpanzee or bonobo most of which will be cultural in origin (see McGrew, 2004; Whiten, in press). The home range of communities may be actively defended and, although uncommon, there are sometimes lethally hostile encounters between adjacent chimpanzee communities. It

(a)

(b)

Figure 7.19 The solitary living orang utans (*Pongo pygmaeus*). (a) A female with infant and (b) a male who shows the typical facial 'disc' and dewlap which develops with maturity.

is the males which are involved in such aggression and they move as a group to cooperate in attacking and even killing neighbours (Wilson *et al.*, 2001).[1] Young females tend to leave their natal group and there is some exchange between communities as they become mature.

Faced with such diversity of social organization within a group which also displays great morphological diversity, we have to look widely for factors in the environment and life history of the different species to account for their social behaviour. Why is social behaviour such a dominating feature of the life of primates? The usual approach to answer such a general question has been to gather data from as many species as possible and to compare them for a wide range of ecological, morphological and behavioural factors. For example, the nature of the food supply, the degree of pressure from predators, body size, the extent of sexual dimorphism, the number of mature males in the group and group size – all of these and more – may be analysed to look for regular associations.

We may return to earlier arguments about the advantages versus the disadvantages of grouping. Because primates are almost exclusively vegetarian, though often including some small animals in their diet, they find food for themselves and it is not easy for them to share food extensively. Watching a group of squirrel monkeys in a tree canopy in Brazil or baboons on the ground in Africa, it is noticeable how much they are foraging as individuals, though never far from other group members. They will gain some overall advantages, as do many birds in winter, from the fact that rich patches of food will be more likely to be encountered if a spread out group is moving across the habitat. A group may be able to defend a territory containing good resources so even though individuals will have to share, there is likely to be an overall advantage.

If group foraging may be one factor, it is absolutely clear that another, and probably a dominant one is group defence. If solitary, even the largest primates would be easy prey for predatory dogs and cats with which they usually share their habitat. To some extent, life in the tree tops will protect arboreal species, but these tend to be the smaller monkeys and a favoured prey of large eagles. The extra vigilance provided by living in a group is a first line of defence, for example the vervet alarm calling discussed on p. 173, but further with monkeys like baboons the presence of several large males will protect even from big cats. A leopard will not attack baboons in a group, but one caught alone in the open is lost!

---

1 It is salutary to remember here that when Jane Goodall's pioneering results with the community of chimpanzees at the Gombe Stream Reserve in Tanzania first became well-known, the general public – and some biologists – tended to see chimp society as relatively stable and peaceful. There was competition but it was regarded as minor, disputes were settled quickly and indeed, we humans might have some lessons to learn from our closest living relatives! When news began to emerge of ferocious battles between neighbouring chimp communities with killings not uncommon, it was received with a sense of disappointment approaching rejection. This history is used to brilliant effect by William Boyd in his 1990 novel *Brazzaville Beach*.

There is no single explanation for primate sociality and probably different combinations of ecological and behavioural factors will have operated to produce the diversity we find. As with most mammals, we have to bear in mind that, in general, the two sexes have different limitations. As so often, the reproductive success of females tends to be limited by access to resources, that of males by access to females.

The arguments derived from the comparisons of the ecological and behavioural factors outlined above are quite complex and we cannot go into them here. There are challenging reviews by van Schaik (1983), Ridley (1986) and Wrangham (1987) and in the volumes cited at the opening of this section, but we shall find no complete agreement as to why primate organization is so diverse. Here we are more interested in the remarkable range of social behaviour exhibited by the primates and move on to discuss some key examples.

Watching any monkey group for a few hours, one cannot fail to be impressed by the richness and complexity of their social life and with the high level of communication that goes on between them. Each individual is constantly responsive to the posture, movements, gestures and calls of others. We have already referred to some remarkable examples of such responses in Chapters 3 and 5. This constant high level of attention to others is dramatic, and Chance (1976) suggested that the social structure of a monkey group is best regarded as a structure of attention itself, particularly as it relates to social rank, a feature which we shall discuss below.

It is necessary to be somewhat cautious in our emphasis on the complexity of primate groups. We must not automatically assume that primate societies are more organized than those of, say, ungulates or carnivores. Because we ourselves are primates, we find it much easier to identify elements in their communication system, in particular the mobility of their faces and the way they watch one another's faces for information on mood and intentions. Nevertheless there are, on the most objective criteria, good grounds for our assumption. Firstly, primates have an extended period of infancy, to which is coupled considerable longevity. The larger primates commonly live for 20 or 30 years and this means that a young primate grows up to take its place in a group where – literally – everybody knows everybody else from long experience in their company.

Secondly, there can be no doubt that primates are amongst the most intelligent of animals with a high learning ability and a correspondingly high flexibility in their behavioural response to a changing social situation. Sustaining this capacity is a large and highly developed brain. We have already discussed some of these topics in Chapter 5. We may be dealing here with a close coupling between social behaviour and evolution. Jolly (1966) and Humphrey (1976) have suggested that the complex demands of social life itself formed one of the main selective factors for the growth of brain size in primates. This evolving brain, in turn, offered still more flexibility and complexity and so a mutual evolutionary relationship was established. There is now abundant evidence that social learning, by which we mean one animal acquiring a novel piece of behaviour from observing and copying others, is widespread in primate groups (Hoppitt and

Laland, 2008; Whiten, in press). It may have been yet another factor predisposing primates to become social and develop complex social interactions in which such learning could flourish. The relationship reaches its most extreme manifestation in humans, with a brain which has evolved the capacity to mediate speech with all the richness of social interaction this allows.

What one might label the two 'highest common factors' of primate societies are in some ways opposites in their social 'content' – dominance hierarchies and grooming relationships. We have already discussed the concept of social rank in connection with other species such as chickens. In primates, dominance and subordination are always important features of their relationships with others. Ultimately, dominance always involves the threat of physical displacement or attack, even though it may be rarely seen once rank is established. However, it is most important to recognize that relationships within primate groups are as much characterized by positive interactions as negative ones. Grooming relationships are one type of the many friendly contacts between animals, as when they move and rest together. At such times, they invite grooming or offer to groom another. Mutual grooming, which we first mentioned for lemurs, is very important as a friendly or placatory gesture in all primates (see Fig. 7.20). Often a dominant animal will 'allow' itself to be groomed by a subordinate following a brief threat to which the subordinate has deferred.

The pattern of grooming relationships in a group is often a good measure of its detailed structure. It will show us which animals associate together and is a good index of the cohesive forces which maintain the group as a real social unit. For example, Seyfarth and Cheyney (1984) were able to show that frequent grooming between certain members of a vervet monkey group indicated that they were in a relationship such that one monkey was more likely to call upon another to aid it in a conflict situation, and was more likely to be successful if it did so. It is also very common to find that a mother and her offspring, particularly her daughters, form the most enduring of mutual support groups within the overall social structure. In most monkeys, it is the females with their daughters who remain and the young males who move out as they become mature.

However, notwithstanding such clear patterns of affiliation and cooperation, observations on almost all primate groups reveal the existence of a dominance hierarchy ultimately related to aggression. Certainly the role of aggression in primate societies has attracted a great deal of attention for, as may be imagined, the non-human primates have been used as evidence by both sides in the dispute about the nature of human aggression. Primates vary greatly in the degree of fighting which is observed both within and between groups. Even within a single species there may be considerable variations; langur monkey populations in northern India are apparently far more peaceful than those in the south. Population densities are much greater in the areas where the aggressive groups live and this is probably one factor involved. Persistent fighting may occur in crowded zoo colonies, which rarely give sufficient space for subordinate animals to keep out of the way of more dominant ones. Density and the

(a)

(b)

(c)

Figure 7.20   Social grooming is ubiquitous in primates. (a) Crowned lemurs (*Eulemur coronatus*); (b) Barbary macaques; (c) gorilla.

accompanying stress are obviously crucial factors, but it is unlikely that all primate species will respond to them in the same way; some may be inherently more aggressive than others.

However it is achieved, the stable situation in almost all primate groups can be interpreted in terms of a ranked hierarchy which determines to a greater or lesser extent how the individuals behave. Perhaps the best and most neutral definition of rank is that the behaviour of a high ranking or dominant animal is not limited by other individuals, whilst that of a subordinate is so limited. Fig. 3.28 shows this well. High rank in a group might determine access to food, preferred resting places and females. The latter want to mate with the top males and, as we mentioned earlier in this chapter, when they come into oestrus, female baboons move close to a dominant male and form a temporary 'consort relationship' with him. When resources are in short supply, the harsh reality of a dominance hierarchy may become extreme. One of the most striking examples comes from the long-term study of toque monkeys (*Macaca sinica*), a species of macaque endemic to Sri Lanka. Dittus (1977) found that only dominant animals could get access to the best parts of fruiting trees and during prolonged drought food was – literally – taken from the mouths of lower ranking animals by larger dominants. A dominant animal would approach a subordinate, force its mouth open and pull out the contents of the hapless monkey's cheek pouches!

Nevertheless, we must not overemphasize the disadvantages of low ranking, for the most conspicuous feature monkey groups is their stability. Certainly, it has often been observed that while rank may be determined initially by threat and fighting, once established it is maintained as much by the deference of subordinate animals as by any display of threat by the dominants. It is easy to understand how such a regular system of rank might arise from all individuals behaving to their own best advantage. The cohesion of the troop is vital for everyone's survival and some predictability about social interactions will be advantageous not only to the top animals but to low ranking ones too. Fighting is reduced, and probably stress also, since subordinate animals can keep clear of others to whom they will predictably lose in competitive encounters. Normally, there is very little disputing between animals widely separated in the hierarchy – the subordinate avoids or immediately defers to its superior – but animals of similar rank are much more disputatious. Johnson (1989) found that close ranking juvenile baboons would pick a fight over food items that would never normally lead to a dispute. It may be worth their while because a win will enhance their future prospects when more important resources are at stake. It will always be worth a low ranking animal biding its time because no hierarchy remains stable indefinitely. Subordinates are often observed to take over higher ranking positions as previously dominant animals become weaker with age.

Some of the most interesting complexities of primate social behaviour are revealed by studies of how ranks are acquired and changed. Linear hierarchies are by no means universal and, particularly at the top end where the adult males of a group are usually competing, we may find shifting triangular relationships. One male may be dominant over either of the next two individually, but these latter may frequently gang up to form an alliance and displace him. Subsequently he, in turn, may try to break up this pair by making conciliatory approaches to one of them, and so on.

Such relationships are nowhere better described than in de Waal's fascinating account of his long-term studies on chimpanzees in the Arnhem Zoo, which he called, appropriately, *Chimpanzee Politics* (1982). Observations of wild communities of chimpanzees in Africa confirm that alliances which form, break up and re-form are a normal part of their social organization. They produce a shifting, dynamic pattern of social life which greatly modifies the nature and the effects of any straightforward rank order.

The acquisition of rank is also a complex process, differing between the sexes, which relates to their very different life histories. As mentioned above, in many species of monkey, males never breed in the group into which they were born. The stability of primate groups might lead to a potential problem with inbreeding, but this is avoided with the transfer of newly mature males between groups. By contrast, in these same species, the females invariably remain in their natal troop and in close association with their mothers and sisters.

The rank of young females is determined by that of their mother. In most cases, with remarkable regularity, daughters slot in below their mothers in reverse order of age, i.e. the youngest daughter, once beyond infancy, ranks just below her mother and above her sisters. This is often because the mother gives her the most support which makes good evolutionary sense because, if she survives the dangers of infancy, a young daughter has a longer potential reproductive life ahead of her than an older one. It will pay mothers, on average, to apportion their support accordingly (see Gadgil, 1982, for a full discussion of this idea).

The rank of young males develops along a different route. As sons grow up, they become independent of their mother and their rank amongst other members of the group is largely determined by their size. In baboons, Johnson (1987) found that juvenile males defeated juvenile females in almost all disputes, no matter what the rank of their mothers. There is a rise in disputing between young males as they become sexually mature; they may even challenge the adult males and because they are rarely successful, this is the commonest age for them to leave the group. They wander briefly, but soon approach other local groups where they are usually accepted quite rapidly.

Many of the examples we have been discussing have been drawn from those primates such as baboons or vervet monkeys whose social organization involves several adult males and adult females living together with no permanent male/female bonds. This is perhaps the commonest type of primate society, but we have already mentioned others. The hamadryas baboons of north eastern Africa have societies in which male/female bonds are more permanent, based upon 'harem groups' consisting of a single adult male, several females and their offspring. A number of highly arboreal species, such as colobus monkeys, also have groups with a single breeding male, but in these cases each group moves around within its own defended territory. Since males are driven out from

Figure 7.21 A male and a female hamadryas baboon.

the group as they mature, such an organization is bound to leave a surplus of adult males. These usually form all-male bands, members of which occasionally challenge those with harems.

However, the most conspicuous examples of such harem group organization are to be found in the hamadryas baboons, a primate with very marked sexual dimorphism (Fig. 7.21) and the gelada, another baboon-like monkey of the Ethiopian uplands. In these species, a number of harem groups associate together and in the gelada one can observe hundreds of animals moving around together in a loose herd structure. Hamadryas harem groups also move together during the day, although they may spread out and split up to comb an area for food. They always come together as a troop at night to sleep on cliffs, which provide protection from predators.

For a concluding example to illustrate the subtlety and flexibility of primate social behaviour, we cannot do better than to stay with these hamadryas baboons and describe some remarkable field experiments by Kummer and his collaborators (Kummer *et al.*, 1974; Bachmann and Kummer, 1980). Within the one-male units, the male closely herds his, perhaps 3–6, females, who move with him at all times and are not allowed to stray (Fig. 7.22). The herding of females into a close unit is accomplished by threat. If, as she forages, a female strays too far from the male he first threatens her and, if she does not return to his side, chases and bites her. Customarily females return to the male at once when he 'stares' at them – a low intensity threat gesture.

Another closely related species of baboon – the yellow baboon (*Papio cynocephalus*) – lives in the same area, but with a different social organization. Like most other species, females do not associate with particular males save at their brief oestrous period when they form a temporary consort relationship with a dominant male. For most of the time they move freely and associate with many different members of the troop. Kummer

Figure 7.22 Three long-caped male hamadryas baboons with their closely-herded harems of females.

(1968) could utilize this key difference to explore something of the development of the hamadryas system.

He transplanted a few yellow baboon females into a troop of hamadryas. Of course they appeared unattached and very quickly males began to herd them into their own units. At first the females wandered away and, if they responded at all to a male's threat, it was to flee. This, of course, produced the opposite effect to that which they intended. The male instantly pursued them and drove them back to his harem. Kummer found that within a few hours the yellow baboon females had learnt their harsh lesson and stayed close to their male!

Such results suggest vividly that some of the variation in social organization which we observe between primate species is of cultural and not genetic origin. (There are also some genetic factors because hybrids between hamadryas and yellow baboons show roughly intermediate behavioural tendencies.) The young hamadryas baboon grows up in a harem group, the young yellow baboon in the less restricted social climate of the larger group. They may have similar potential but they develop quite differently and can change later in life if needs must.

Finally, hamadryas baboons can offer us a remarkable illustration of the sheer power of cultural mores in primates. As we have noted any unattached young females are rapidly acquired by males who are always on the look out to add to their harem. It is all the more striking, therefore, that within a troop there is scarcely ever any attempt to seduce females away from other harem units. This remains the case even if one male is completely dominant over another in terms of access to food or resting places – a subordinate's females remain his own.

Kummer and his team caught some baboons of both sexes and confined them in cages on site. If two males were kept together and an unattached female introduced, they would threaten and sometimes fight each other and eventually the female became attached to one male – usually, of course, the winner of the conflict. However, it was possible to bias the situation by first caging an otherwise subordinate male alone along with the female, although very close to and in full sight of the dominant one. This gave the former two a chance to interact and, depending on his attractiveness, the female would bond, more or less strongly, with the subordinate male. Now the three animals would be put all together. If she was not strongly paired to the subordinate, the dominant would attempt to attract the female, but if she was so paired he made no attempt whatsoever. Further, it is extraordinary to observe that such a male introduced to an established pair appears 'ill at ease', sitting as far away from the pair as possible and with his head turned away from them. There is complete social inhibition of poaching between males of a troop; dominance is not an absolute character, even where it could easily be expressed in order to gain a resource. It is entirely constrained by the social context – the strict rules of hamadryas society.

Because of their close phylogenetic relationship to ourselves, the primates will always hold a particular fascination. Apart from this intrinsic interest, their variety and complexity offer boundless scope for speculation on the selective forces operating on our hominid ancestors which eventually led to the emergence of *Homo sapiens* from Africa around 200 000 years ago. It seems likely that we have only just begun to comprehend the extent of their diverse behavioural potential. Alas – we may never succeed, for it is a tragic irony that the overwhelming increase of one species of ape threatens the survival of almost all other primates in the wild.

SUMMARY

Social life is a conspicuous feature of many animals. It must be studied for its own sake, but it also offers an opportunity to recapitulate on the main themes of this book, based upon Tinbergen's four questions. They will all arise when reviewing the rich diversity of social organizations animals display. The first essential is never to lose sight of the individual in the crowd. Natural selection is no respecter of sociality itself. Animals will not group unless they gain individual advantage by doing so. Since there are some obvious disadvantages to grouping – you have to share resources and space with others – there must be considerable advantages which swing the balance so firmly and we review these: protection, greater vigilance, improved searching of the habitat for food, and so forth.

We begin a review of types of social grouping with the eusocial insects: termites, ants, bees and wasps. These are social organisms of a unique type in which a rigid social system operates involving a single or very few reproductives, supported by large numbers of sterile offspring, sometimes very specialized into 'castes' whose morphology and behaviour fit them to fulfil a single role within the colony. In certain species the colony is effectively a single family and kin selection will have been involved in their evolution. We describe briefly two other examples of eusocial groups – naked mole rats and a coral shrimp.

The social systems of vertebrates are much less rigidly organized for the most part. There will be a basic plan but there may often be variations across the seasons and there may be alternative end points. We take a few general themes in social organization for discussion – territory, mating systems and the role of dominance hierarchies.

We continue with some more behavioural descriptions of organization in the social lives of meerkats, dogs, elephants and whales, which show both contrasts and some common features. Finally, we briefly introduce the range of social life within the primates where we find some of the most complex societies. Many are long-lived animals and their interactions can illustrate the way cultural differences determine the way individuals behave. Understandably the diversity of primate societies offers evidence – even though it is often conflicting – on the manner in which human societies evolved.

REFERENCES

Akessen, S. (1994). Comparative orientation experiments with different species of passerine long-distance migrants: effect of magnetic field manipulation. *Animal Behaviour* **48**: 1379–93.

Alatalo, R.V., Högland, J. & Lundberg, A. (1991). Lekking in the black grouse: a test of male viability. *Nature* **352**: 155–6.

Alatalo, R.V., Lundberg, A. & Glynn, C. (1986). Female pied flycatchers choose territory quality not male characteristics. *Nature* **323**: 152–3.

Allee, W.C. (1938). *The Social Life of Animals*. New York: Norton.

Altmann, J. (1980). *Baboon Mothers and Infants*. Cambridge, MA: Harvard University Press.

Altmann, S.A. (1962). A field study of the sociobiology of rhesus monkeys, *Macaca mulatta*. *Annals of the New York Academy of Sciences* **102**: 338–435.

Amici, F., Aureli, F., Visalberghi, E. & Call, J. (2009). Spider monkeys (*Ateles geoffroyi*) and capuchin monkeys (*Cebus apella*) follow gaze around barriers: Evidence for perspective taking? *Journal of Comparative Psychology* **123**(4): 368–74.

Andersson, M. (1982). Female choice selects for extreme tail length in a widowbird. *Nature* **299**: 818–20.

Andersson, M. & Simmons, L.W. (2006). Sexual selection and mate choice. *Trends in Ecology and Evolution* **21**(6): 296–302.

Andrew, R.J. (1985). The temporal structure of memory formation. *Perspectives in Ethology* **6**: 219–59. New York: Plenum.

Arey, D. (1992). Straw and food as reinforcers for prepartural sows. *Applied Animal Behaviour Science* **33**: 217–26.

ASAB/ABS (2006) Guidelines for the treatment of animals in behavioural teaching and research. *Animal Behaviour* **71**: 245–53.

Aschoff, J. (1981). *Biological Rhythms. Handbook of Behavioural Neurobiology 4*. New York: Plenum.

Aschoff, J. (1989). Temporal orientation: circadian clocks in animals and humans. *Animal Behaviour* **37**: 881–96.

Axelrod, R.D. & Hamilton, W.D. (1981). The evolution of cooperation. *Science* **211**: 1390–6.

Bachmann, C. & Kummer, H. (1980). Male assessment of female choice in hamadryas baboons. *Behavioral Ecology and Sociobiology* **6**: 315–21.

Baerends, G.P. (1976). The functional organization of behaviour. *Animal Behaviour* **24**: 726–38.

Bakken, G.S. & Krochmal, A.R. (2007). The imaging properties and sensitivity of the facila pits of pitvipers as determined by optical and heat-transfer analysis. *Journal of Experimental Biology* **210**: 2801–10.

Bakker, T.C.M. (1986). Aggressiveness in sticklebacks (*Gasterosteus aculeatus* L.): a behaviour-genetic study. *Behaviour* **98**: 1–144.

Balda, R.P. & Kamil, A.C. (1992). Long-term spatial memory in Clark's nutcracker, *Nucifraga columbiana*. *Animal Behaviour* **44**: 761–9.

Ballerini, M., Cabbibo, N., Candelier, R. *et al.* (2008). Empirical investigation of starling flocks: a benchmark study in collective animal behaviour. *Animal Behaviour* **62**: 201–15.

Balmford, A. (1991). Mate choice on leks. *Trends in Ecology and Evolution* **6**: 87–92.

Baptista, L.F. & Morton, M.L. (1988). Song learning in montane white-crowned sparrows: from whom and when. *Animal Behaviour* **36**: 1753–64.

Barfield, R.J. (1971). Gonadotrophic hormone secretion in the female ring dove in response to visual and auditory stimulation in the mate. *Journal of Endrocrinology* **49**: 305–10.

Barinaga, M. (1996). Guiding neurons to the cortex. *Science* **274**: 1100–1.

Barlow, H.B. (1972). Single units and sensation: a neuron doctrine for perceptual psychology? *Perception* **1**: 371–94.

Barnett, S.A. (1963). *A Study in Behaviour*. London: Methuen.

Barnett, S.A. (1964). *The Concept of Stress. Viewpoints in Biology 3*. London: Butterworth, pp. 170–218.

Basolo, A. (1990). Female preference predates the evolution of the sword in swordtail fish. *Science* **250**: 808–10.

Basolo, A. (1995). A further examination of pre-existing bias favouring a sword in the genus *Xiphophorus*. *Animal Behaviour* **50**: 365–75.

Bastock, M. & Manning, A. (1955). The courtship of *Drosophila melanogaster*. *Behaviour* **8**: 85–111.

Bateson, M., Nettle, D. & Roberts, G. (2006). Cues of being watched enhance cooperation in a real-world setting. *Biology Letters* **2**: 412–14.

Bateson, P.P.G. (1983). Genes, environment and the development of behaviour. In *Animal Behaviour, Vol. 3. Genes, Development and Learning*, ed. T.R. Halliday and P.J.B. Slater. Oxford: Blackwell, pp. 52–81.

Bateson, P. (1990). Is imprinting such a special case? *Philosophical Transactions of the Royal Society of London* **329**: 125–31.

Bateson, P. & Gluckman, P. (2011). *Plasticity, Robustness, Development and Evolution*. Cambridge: Cambridge University Press.

Bateson, P., Barker, D., Clutton-Brock, T.H., Deb, D. *et al.* (2004). Developmental plasticity and human health. *Nature* **430**: 419–42.

Bateson P., Mendl, M. & Feaver, J. (1990). Play in the domestic cat is enhanced by rationing of the mother during lactation. *Animal Behaviour* **40**: 514–25.

Baylis, G.C., Rolls, E.T. & Leonard, C.M. (1985). Selectivity between faces in the reponses of a population of neurons in the cortex in the superior temporal sulcus of the monkey. *Brain Research* **342**: 91–102.

Beach, F.A. (1942). Analysis of the stimuli adequate to elicit mating behavior in the sexually inexperienced male rat. *Journal of Comparative Psychology* **33**: 163–207.

Beekman, M., Fathke, R. & Seeley, T. (2006). How does an informed minority of scouts guide a honeybee swarm as it flies to a new home? *Animal Behaviour* **71**: 161–171.

Bekoff, A. & Kauer, J.A. (1984). Neural control of hatching: role of neck position in turning on hatching leg movements in post-hatching chicks. *Journal of Comparative Psychology* **145**: 497–504.

Bekoff, M. (1995). Play signals as punctuation: the structure of social play in canids. *Behaviour* **132**: 419–29.

Bekoff, M. (2007). *The Emotional Lives of Animals*. Novato, CA: New World Library, Green Press.

Bekoff, M. & Byers, J.A. (1998). *Animal Play: Evolutionary, Ecological and Comparative Perspectives*. Cambridge: Cambridge University Press.

Bennet-Clark, H.C. (1970). The mechanism and efficiency of sound production in mole crickets. *Journal of Experimental Biology* **52**: 619–52.

Bennet-Clark, H.C. (1987). The tuned singing burrow of mole crickets. *Journal of Experimental Biology* **128**: 383–409.

Bennet-Clark, H.C. & Ewing, A. (1987). Pulse interval as a critical parameter in the courtship song of *Drosophila melanogaster*. *Animal Behaviour* **17**: 755–9.

Bentley, D.R. & Hoy, R.R. (1970). Postembryonic development of adult motor patterns in crickets: a neural analysis. *Science* **170**: 1409–11.

Bentley, D. & Keshishian, H. (1982). Pathfinding by peripheral pioneer neurons in grasshoppers. *Science* **218**: 1082–8.

Berridge, K.C. (2004). Motivation concepts in behavioural neuroscience. *Physiology & Behaviour* **81**: 179–209.

Berthold, P. (1993). *Bird Migration*. Oxford: Oxford University Press.

Berthold, P. (2003). Genetic basis and evolutionary aspects of bird migration. *Advances in the Study of Behavior* **33**: 175–229.

Berthold, P. & Helbig, A.J. (1992). The genetics of bird migration: stimulus, timing, and direction. *Ibis* **134** (Suppl. 1): 35–40.

Berthold, P. & Querner, U. (1981). Genetic basis of migratory behaviour in European warblers. *Science* **212**: 77–9.

Bird, C.D. & Emery, N.J. (2009a). Insightful problem solving and creative tool modification by captive nontool-using rooks. *Proceedings of the National Academy of Science* **106**: 10370–5.

Bird, C.D. & Emery, N.J. (2009b). Rooks use stones to raise the water level to reach a floating worm. *Current Biology* **19**: 1410–4.

Birkhead, T. (1993). Avian mating systems: the aquatic warbler is unique. *Trends in Ecology and Evolution* **8**: 390–1.

Bitterman, M.E. (2006). Classical conditioning since Pavlov. *Reviews of General Psychology* **10**: 365–76.

Blakemore, C. & Cooper, G.F. (1970). Development of the brain depends on the visual environment. *Nature* **228**: 477–8.

Blakemore, R.P. (1975). Magnetotactic bacteria. *Science* **190**: 377–9.

Blanchard, R.J. & Blanchard, D.C. (1981). The organisation and modelling of animal aggression. In *The Biology of Aggression*, ed. P.F. Brain and D. Benton. Aphen aan den Rhijn: Sijthoff & Noordhoff, pp. 529–61.

Blass, E.M. & Epstein, A.N. (1971). A lateral preoptic osmosensitive zone for thirst in the rat. *Journal of Comparative Physiology and Psychology* **76**: 378–94.

Blest, A.D. (1957). The function of eyespot patterns in the Lepidoptera. *Behaviour* **11**: 209–56.

Blüm, V. & Fiedler, K. (1965). Hormonal control of reproductive behavior in some cichlid fish. *General and Comparative Endocrinology* **5**: 186–96.

Boakes, X. (1984). *From Darwin to Behaviourism*. Cambridge: Cambridge University Press.

Boesch, C. (1994). Cooperative hunting in wild chimpanzees. *Animal Behaviour* **48**: 653–67.

Bogdany, F.J. (1978). Linking of learning signals in honeybee orientation. *Behavioral Ecology and Sociobiology* **3**: 323–36.

Bolhuis, J.J. (1991). Mechanisms of avian imprinting: a review. *Biological Reviews* **66**: 303–45.

Bolhuis, J.J. (1995). Development of perceptual mechanisms in birds: predispositions and imprinting. In *Neuroethological Studies of Cognitive and Perceptual Processes*, ed. C.F. Moss and S.J. Shettleworth. Boulder, Colorado: Westview Press, pp. 150–76.

Bolhuis, J.J. & MacPhail, E.M. (2001). A critique of the neuroecology of learning and memory. *Trends in Cognitive Science* **4**, 426–33.

Bolles, R.C. (1970). Species-specific defense reactions and avoidance learning. *Psychology Review* **77**: 32–48.

Borbely, A.A. & Achermann, P. (1999) Sleep modulation and models of sleep regulation. *Journal of Biological Rhythms* **14**: 557–68.

Bossema, I. & Burger, R.R. (1980). Communication during monocular and binocular looking in European jays *Garrulus g. glandarius*. *Behaviour* **74**: 274–83.

Bowlby, J. (1969). *Attachment and Loss. Vol. I Attachment*. London: Hogarth.

Bowlby, J. (1973). *Attachment and Loss. Vol. II Loss*. London: Hogarth.

Bracke, M.B.M. & Hopster, H. (2006). Assessing the importance of natural behaviour for animal welfare. *Journal of Agricultural and Environmental Ethics* **19**: 77–89.

Bräuer, J., Call, J. & Tomasello, M. (2005). All great ape species follow gaze to distant locations and around barriers. *Journal of Comparative Psychology* **119**(2): 145–54.

Breland, K. & Breland, M. (1961). The misbehavior of organisms. *American Psychologist* **16**: 681–4.

Brenowitz, E.A. (1982). Aggressive response of red-winged blackbirds to mockingbird song imitation. *Auk* **99**: 584–6.

Brines, M.L. & Gould, J.L. (1982). Skylight polarisation patterns and animal orientation. *Journal of Experimental Biology* **96**: 64–81.

Brockmann, H.J. (1980). The control of nest depth in a digger wasp (*Sphex ichneumoneus* L.). *Animal Behaviour* **28**: 426–45.

Bronson, F.H. (1979). The reproductive biology of the house mouse. *Quarterly Review of Biology* **54**: 265–99.

Broom, D.M. (1991). Animal welfare: concepts and measurement. *Journal of Animal Science* **69**: 4167–75.

Broom, D.M. & Johnson, K.G. (1993). *Stress and Animal Welfare*. London: Chapman & Hall.

Brown, C.R. (1986). Cliff swallow colonies as information centres. *Science* **234**: 83–5.

Brown, J.L. (1969). The buffer effect and productivity in tit populations. *American Naturalist* **103**: 347–54.

Brown, J.L. & Eklund, A. (1994). Kin recognition and the major histocompatibility complex: an integrative review. *American Naturalist* **143**: 435–61.

Brown, R.E. & Macdonald, D.W. (eds.) (1985). *Social Odours in Mammals*, Vols. 1 and 2. Oxford: Clarendon Press.

Bruner, J.S., Jolly, A. & Sylva, K. (eds.) (1976). *Play: Its Role in Development and Evolution*. London: Penguin.

Bshary, R. & Schafter, D. (2002). Choosy reef fish select cleaner fish that provide high quality service. *Animal Behaviour* **63**: 557–64.

Buchanon, K.L. & Catchpole, C.K. (1997). Female choice in the sedge warbler, *Acrocephalus schoenobaaenus*: multiple cues from song and territory quality. *Proceedings of the Royal Society B* **264**: 521–6.

Buchanan, K.L., Catchpole, C.K., Lewis, J.W. & Lodge, A. (1999). Song as an indicator of parasitism in the sedge warbler. *Animal Behaviour* **57**: 307–314.

Burghardt, G.M. (1970). Defining 'communication'. In *Advances in Chemoreception*. Vol. I, ed. J.W. Johnston, D.G. Moulton and A. Turk. New York: Appleton Century Crofts, pp. 5–18.

Burghardt, G.M. (2005). *The Genesis of Animal Play: Testing the Limits*. Cambridge, MA: MIT Press.

Burkhardt, D. (1989). UV vision: a bird's view of feathers. *Journal of Comparative Physiology A* **164**: 787–96.

Burley, N., Krantsberg, G. & Rudman, P. (1982). Influence of colour-banding on the conspecific preference of zebra finches. *Animal Behaviour* **30**: 444–55.

Burt de Perera, T. (2004). Fish can encode order in their spatial map. *Proceedings of the Royal Society B* **271**: 2131–4.

Byers, J.A. & Walker, C. (1995). Refining the motor training hypothesis for the evolution of play. *American Naturalist* **146**: 25–40.

Byrne, R. & Whiten, A. (eds.) (1988). *Machiavellian Intelligence*. Oxford: Clarendon Press.

Cade, W.H. (1981). Alternative mate strategies: genetic differences in crickets. *Science* **212**: 563–4.

Campbell, C.J., Fuentes, A., MacKinnon, K.C., Panger, M. & Bearder, S.K. (eds.) (2006). *Primates in Perspective*. New York: Oxford University Press.

Camhi, J.M. (1984). *Neuroethology: Nerve Cells and the Natural Behavior of Animals*. Sunderland, Massachusetts. Sinauer Associates.

Camhi, J.M., Tom, W. & Volman, S. (1978). The escape behavior of the cockroach *Periplaneta americana*. II. Detection of natural predators by air displacement. *Journal of Comparative Physiology* **128**: 203–12.

Cannon, W.B. (1974). *The Wisdom of the Body*. London: Kegan Paul.

Capellini, I., Nunn, C.L., McNamara, P., Preston, B.T. & Barton, R.A. (2008). Energetic constraints, not predation influence the evolution of sleep patterning in mammals. *Functional Ecology* **22**: 847–53.

Capranica, R.R. & Moffat, A.J.M. (1975). Neurobehavioral correlates of sound communication in anurans. In *Advances in Vertebrate Neuroethology*, ed. J.P. Ewert, R.R. Capranica and D. Ingle. New York: Plenum, pp. 701–30.

Caro, T.M. (1994). *Cheetahs of the Serengeti Plains: Group Living in an Asocial Species*. Chicago: University of Chicago Press.

Caro, T.M. (1995). Short-term costs and correlates of play in cheetahs. *Animal Behaviour* **49**: 333–45.

Carroll, S.B., Grenier, J.K. & Weatherhouse, S.D. (2005). *From DNA to Diversity*. 2nd edn. Oxford: Blackwell.

Catania, K.C. & Remple, F.E. (2004). Tactile foveae in the star-nosed mole. *Brain, Behaviour and Evolution* **63**: 1–12.

Catchpole, C.K. (1980). Sexual selection and the evolution of complex songs among warblers of the genus *Acrocephalus*. *Behaviour* **74**: 149–166.

Catchpole, C.K. & Slater, P.J.B. (2008). *Bird Song. Biological Themes and Variations*. 2nd edn. Cambridge: Cambridge University Press.

Catchpole, C., Leisler, B. & Winkler, H. (1985). The evolution of polygyny in the Great Reed Warbler, *Acrocephalus arundinaceus*: a possible case of deception. *Behavioral Ecology and Sociobiology* **16**: 285–91.

Chaloupkova, H., Illmann, G., Bartos, L. & Spinka, M. (2007). The effect of pre-weaning housing on the play and agonistic behaviour of domestic pigs. *Applied Animal Behaviour Science* **103**: 25–34.

Chance, M.R.A. (1976). The organisation of attention in groups. In *Methods of Inference from Animal to Human Behavior*, ed. M. Von Cranach. The Hague: Mouton Aldine, pp. 213–35.

Chase, I.D. (1985). The sequential analysis of aggressive acts during hierarchy formation: an application of the 'jigsaw puzzle' approach. *Animal Behaviour* **33**: 86–100.

Cheney, D.L. & Seyfarth, R.M. (1980). Vocal recognition in free-ranging vervet monkeys. *Animal Behaviour* **28**: 362–7.

Cheney, D.L. & Seyfarth, R.M. (1982). How vervet monkeys perceive their grunts: field playback experiments. *Animal Behaviour* **30**: 739–51.

Cheney, D.L. & Seyfarth, R.M. (1986). The recognition of social alliances among vervet monkeys. *Animal Behaviour* **34**: 1722–31.

Cheney, D.L. & Seyfarth, R.M. (1990). *How Monkeys See the World*. Chicago: University of Chicago Press.

Clark, D.L. & Uetz, G.W. (1992). Morph-independent mate selection in a dimorphic jumping spider: demonstration of movement bias in female choice using video-controlled courtship behaviour. *Animal Behaviour* **43**: 247–54.

Clark, R.B. (1960a). Habituation of the polychaete *Nereis* to sudden stimuli. I. General properties of the habituation process. *Animal Behaviour* **8**: 82–91.

Clark, R.B. (1960b). Habituation of the polychaete *Nereis* to sudden stimuli. II. Biological significance of habituation. *Animal Behaviour* **8**: 92–103.

Clarke, A.S., Wittwer, D.J., Abbott, D.H. & Schneider, M.L. (1994). Long-term effects of prenatal stress on HPA axis activity in juvenile rhesus monkeys. *Developmental Psychobiology* **27**: 257–69.

Clayton, N.S., Dally, J.M. & Emery, N.J. (2007). Social cognition by food-caching corvids: the western scrub jay as a natural psychologist. *Philosophical Transactions of the Royal Society B* **362**: 507–22.

Clubb, R., Rowcliffe, M., Lee, P. *et al.* (2008). Compromised survivorship in zoo elephants. *Science* **322**: 1649.

Clutton-Brock, T. (2007a). *Meerkat Manor: Flower of the Kalahari*. London: Weidenfeld & Nicholson.

Clutton-Brock, T.H. (2007b). Sexual selection in males and females. *Science* **318**: 1882–5.

Clutton-Brock, T.H & Albon, S.D. (1979). The roaring of red deer and the evolution of honest advertisement. *Behaviour* **69**: 145–70.

Clutton-Brock, T.H., Guinness, F.E. & Albon, S.D. (1982). *Red Deer: Behaviour and Ecology of Two Sexes*. Edinburgh: Edinburgh University Press.

Clutton-Brock, T.H. & Harvey, P.H. (1984). Comparative approaches to investigating adaptation. In *Behavioural Ecology*, 2nd edn., ed. J.R. Krebs and N.B. Davies. Oxford: Blackwell Scientific Publications, pp. 7–29.

Clutton-Brock, T.H., Russell, A.F. & Sharpe, L.L. (2004). Behavioural tactics of breeders in cooperative meerkats. *Animal Behaviour* **68**: 1029–40.

Clutton-Brock, T.H., O'Riain, M.J., Brotherton, P.N.M. *et al.* (1999). Selfish sentinels in cooperative mammals. *Science* **284**: 1640–4.

Cohen-Salmon, C., Carlier, M., Robertoux, P. *et al.* (1985). Differences in patterns of pup care in mice. V. Pup ultrasonic emissions and pup care behavior. *Physiology and Behavior* **35**: 167–74.

Coleman, D.C & Hendrix, P.F. (2000). *Invertebrates as Webmasters in Ecosystems*. Wallingford, UK: CABI Publishing.

Cowie, R.J. (1977). Optimal foraging in great tits, *Parus major*. *Nature* **268**: 137–9.

Cowie, R.J., Krebs, J.R. & Sherry, D.F. (1981). Food storage by marsh tits. *Animal Behaviour* **29**: 1252–9.

Cox, C.R. & Le Boeuf, B.J. (1977). Female incitation of mate competition: a mechanism of mate selection. *American Naturalist* **111**: 317–35.

Creel, S. & Creel, N.M. (2002). *The African Wild Dog: Behavior, Ecology and Conservation*. Princeton, NJ: Princeton University Press.

Cresswell, W. (1994). Song as a pursuit-deterrence signal and its occurrence relative to other anti-predator behaviors of the skylark on attack by merlins. *Behavioral Ecology and Sociobiology* **34**: 217–23.

Crews, D. (2003). Sex determination: where environment and genetics meet. *Evolution and Development* **5**: 50–5.

Crews, D. & Groothuis, T. (2009). Tinbergen's fourth question, ontogeny: sexual and individual differentiation. In *Tinbergen's Legacy*, ed. J.J. Bolhuis and S. Verhulst. Cambridge: Cambridge University Press, pp. 54–81.

Dagan, D. & Volman, S. (1982). Sensory basis for directional wind detection in first instar cockroaches, *Periplaneta americana*. *Journal of Comparative Physiology* **147**: 471–8.

Dall, S.R.X., Giraldeau, L.A., Olsson, O., McNamara, J.M. & Stephens, D.W. (2005). Information and its use by animals in evolutionary ecology. *Trends in Ecololgy and Evolution* **20**: 187–93.

Dane, B., Walcott, C. & Drury, W.H. (1959). The form and duration of display actions of the 'goldeneye' *Bucephala clangula*. *Behaviour* **14**: 265–81.

Darwin, C. (1859). *The Origin of Species*. London: John Murray.

Darwin, C. (1871). *The Descent of Man and Selection in Relation to Sex*. London: John Murray.

Darwin, C. (1872). *The Expression of the Emotions in Man and Animals*. London: University of Chicago Press (1965 edn.).

Davies, N.B. (1992). *Dunnock Behaviour and Social Evolution*. Oxford: Oxford University Press.

Davies, N.B. & Brooke, M. de L. (1989a). An experimental study of co-evolution between the cuckoo *Cuculus canorus* and its hosts. I. Host egg discrimination. *Journal of Animal Ecology* **58**: 207–24.

Davies, N.B. & Brooke, M. de L. (1989b). An experimental study of co-evolution between the cuckoo *Cuculus canorus* and its hosts. II. Host egg markings, chick discrimination and general discussion. *Journal of Animal Ecology* **58**: 225–36.

Davis, W.J., Mpitsos, G.J., Pinneo, J.M. & Ran, J.L. (1977). Modification of the behavioural hierarchy of *Pleurobranchaea*. I. Satiation and feeding mechanisms. *Journal of Comparative Physiology* **117**: 99–125.

Dawkins, M.S. (1980). *Animal Suffering: The Science of Animal Welfare*. London: Chapman & Hall.

Dawkins, M.S. (1990). From an animal's point of view: motivation, fitness and animal welfare. *Behavioral and Brain Sciences* **13**: 1–61.

Dawkins, M.S. (1993). *Through Our Eyes Only? The Search for Animal Consciousness*. Oxford: W.H. Freeman.

Dawkins, M.S. (2008). The science of animal suffering. *Ethology* **114**: 937–45.

Dawkins, M.S. & Bonney, R. (eds.) (2008). *The Future of Animal Farming. Renewing the Ancient Contract*. Oxford: Blackwell Publishing.

Dawkins, R. (1979). Twelve misunderstandings of kin selection. *Zeitschrift für Tierpsychologie* **51**: 184–200.

Dawkins, R. (1982). *The Extended Phenotype*. London: Longman.

Dawkins, R. (1986). *The Blind Watchmaker*. Harlow, Essex: Longman.

Dawkins, R. (1989). *The Selfish Gene*, 2nd edn. Oxford: Oxford University Press.

Dawkins, R. & Krebs, J.R. (1978). Animal signals: information or manipulation? In *Behavioural Ecology: An Evolutionary Approach*, 1st edn, J.R. Krebs and N.B. Davies. Oxford: Blackwell Scientific Publications, pp. 282–309.

Deag, J. & Crook, J.H. (1971). Social behaviour and agonistic buffering in the wild barbary macaque *Macaca sylvana* L. *Folia Primatologia* **15**: 183–200.

D'Eath, R.B. (1998). Can video images imitate real stimuli in animal behaviour experiments? *Biological Reviews* **73**: 267–92.

De Coursey, P.J. (1960). Phase control of activity in a rodent. *Cold Spring Harbor Symposia in Quantitative Biology* **25**: 49–55.

Delius, J.D., Siemann, J.E. & Xia, L. (2001). Cognitions of birds as products of evolved brains. In *Brain Evolution and Cognition*, ed. G. Roth and M.F. Wulliman. New York: Wiley, pp. 451–90.

den Hartog, P.M., de Kort, S.R. & ten Cate, C. (2007). Hybrid vocalizations are effective within, but not outside, an avian hybrid zone. *Behavioural Ecology* **18**: 608–14.

Dethier, V.G. (1953). Summation and inhibition following contra-lateral stimulation of the tarsal chemoreceptors of the blowfly. *Biological Bulletin* **105**: 257–68.

Dethier, V.G. (1957). Communication by insects: physiology of dancing. *Science* **125**: 331–6.

De Jonge, F.H., Tilly, S.L., Baars, A.M. & Spruijt, B.M. (2008). On the rewarding nature of appetitive feeding behaviour in pigs (*Sus scrofa*): Do domesticated pigs contrafreeload? *Applied Animal Behaviour Science* **114**: 359–72.

De Waal, F. (1982). *Chimpanzee Politics*. Baltimore, MD: Johns Hopkins University Press.

De Waal, F.B.M. (2000). Primates: a natural heritage of conflict resolution. *Science* **289**: 586–90.

Dewsbury, D.A. (2003). Comparative psychology. In *Handbook of Psychology. Vol.1. History of Psychology*, ed. D.K. Freedman. New York: Wiley, pp. 67–84.

Diamond, J.N., Karasov, W.H., Phan, D. & Carpenter, F.L. (1986). Hummingbird digestive physiology: a determinant of foraging bout frequency. *Nature* **320**: 62–3.

Dickinson, A. (1980). *Contemporary Animal Learning Theory*. Cambridge: Cambridge University Press.

Dijkgraaf, S. (1962). The functioning and significance of the lateral-line organs. *Biological Reviews* **38**: 51–105.

Dilger, W.C. (1962). The behavior of lovebirds. *Scientific American* **206**(1): 88–98.

Dindo, M., Thierry, B. & Whiten, A. (2008). Social diffusion of novel foraging methods in brown capuchin monkeys (*Cebus apella*). *Proceedings of the Royal Society B* **275**: 187–93.

Dingemanse, N.J., Both, C., van Noordwijk A.J., Rutten, A.L. & Drent, P.J. (2003). Natal dispersal and personalities in great tits (*Parus major*). *Proceedings of the Royal Society B*: **270**: 741–7.

Dittus, W.P.J. (1977). The social regulation of population density and age-sex distribution in the toque monkey. *Behaviour* **63**: 281–322.

Doty, R.W. and Bosma, J.F. (1956). An electromyographic analysis of reflex deglutition. *Journal of Neurophysiology* **19**: 44–60.

Douglas-Hamilton, I. & Douglas-Hamilton, O. (1975). *Among the Elephants*. London: Collins and Harvill.

Dowsett-Lemaire, F. (1979). The imitative range of the song of the marsh warbler *Acrocephalus palustris*, with special reference to imitations of African birds. *Ibis* **121**: 453–68.

D'Souza, F. & Martin, R.D. (1974). Maternal behaviour and the effects of stress in tree-shrews. *Nature* **251**: 309–11.

Dudai, Y. (1989). *The Neurobiology of Memory*. Oxford: Oxford University Press.

Dudink, S., Simonse, H., Marks, I., de Jonge, F.H. & Spruijt, B.M. (2006). Announcing the arrival of enrichment increases play behaviour and reduces weaning-stress-induced behaviours in piglets directly after weaning. *Applied Animal Behaviour Science* **101**: 86–101.

Duffy, J.E., Morrison, C.L. & Ruben, R. (1999). Multiple origins of eusociality among sponge-dwelling shrimps (*Synalpheus*). *Evolution* **54**: 503–16.

Dugatkin, L.A. (1997). *Cooperation Among Animals. An Evolutionary Perspective*. Oxford: Oxford University Press.

Dunbar, R.I.M. & Dunbar, P. (1988). Maternal time budgets of gelada baboons. *Animal Behaviour* **36**: 970–80.

Duncan, I.J.H. & Kite, V.G. (1987). Some investigations into motivation in the domestic fowl. *Applied Animal Behaviour Science* **18**: 387–8.

Duncan, I.J.H. & Wood-Gush, D.G.M. (1972). Thwarting of feeding behaviour in the domestic fowl. *Animal Behaviour* **20**: 444–51.

Dusenberry, D.B. (1992). *Sensory Ecology*. New York: W.H. Freeman.

Dyer, F.C. (2002). The biology of the dance language. *Annual Review of Entomology* **47**: 917–49.

Ehrhardt, A.A. & Baker, S.W. (1974). Fetal androgens, human central nervous system differentiation, and behavior sex differences. In *Sex Differences in Behavior*, ed. R.C. Friedman, R.M. Richart and R.L. Van de Wiele. New York: John Wiley, pp. 33–57.

Elgar, M.A. (1989). Predator vigilance and group size among mammals: a critical review of the evidence. *Biological Reviews* **64**: 1–34.

Elsner, N. (1981). Developmental aspects of insect neuroethology. In *Behavioural Development*, ed. K. Immelmann, G.W. Barlow, L. Petrinovitch and M. Main. Cambridge: Cambridge University Press, pp. 474–90.

Elwood, R.W. (1983). *Parental Behaviour of Rodents*. Chichester, UK: John Wiley.

Elwood, R.W. (1994). Temporal-based kinship recognition: a switch in time saves mine. *Behavioral Processes* **33**: 15–24.

Emery, N.J., Clayton, N.S. & Frith, C.D. (2007). Introduction. Social intelligence: from brain to culture. *Philosophical Transactions of the Royal Society B* **362**: 485–8.

Emlen, S. (1975). The stellar orientation system of a migratory bird. *Scientific American* **233**(2): 102–11.

Emlen, S. & Oring, L.W. (1977). Ecology, sexual selection and the evolution of mating systems. *Science* **197**: 215–23.

Emlen, S.T. & Wrege, P.H. (1989). A test of alternate hypotheses for helping behaviour in white-fronted bee-eaters of Kenya. *Behavioral Ecology and Sociobiology* **25**: 303–20.

Emlen, S., Demong, N.J. & Emlen, D.J. (1989). Experimental induction of infanticide in female wattled jacanas. *The Auk* **106**: 1–7.

Endler, J. (1980). Natural selection in color patterns in *Poecilia reticulata*. *Evolution* **34**: 76–91.

Endler, J. (1983). Natural selection on color patterns in poeciliid fishes. *Environmental Biology of Fishes* **9**: 173–90.

Endler, J. (1993). The color of light in forests and its implications. *Ecological Monographs* **63**: 1–27.

Endler, J. & Basolo, A.L. (1998). Sensory ecology, receiver biases and sexual selection. *Trends in Ecology and Evoultion* **13**: 415–20.

Erickson, C.J. (1985). Mrs Harvey's parrot and some problems of socioendocrine response. In *Perspectives in Ethology*, Vol. 6, ed. P.P.G. Bateson and P.H. Klopfer. New York: Plenum, pp. 261–86.

Esch, H., Esch, I. & Kerr, W.E. (1965). Sound: an element common to communication of stingless bees and to dances of honeybees. *Science* **149**: 320–1.

Evans, S.M. (1968). *Studies in Invertebrate Behaviour*. London: Heinemann Education.

Ewert, J.P. (1980). *Neuroethology: An Introduction to the Neurophysiological Fundamentals of Behavior*. Berlin: Springer-Verlag.

Ewert, J.P. (1997). Neural correlates of key stumulus and releasing mechanism: a case study and two concepts. *Trends in Neurosciences* **20**: 332–9.

Ewert, J.P. & Traud, R. (1979). Releasing stimuli for anti-predator behaviour in the common toad *Bufo bufo* L. *Behaviour* **68**: 170–80.

Fagen, R.M. (1981). *Animal Play Behaviour*. Oxford: Oxford University Press.

Fagen, R. & Fagen, J. (2004). Juvenile survival and benefits of play behaviour in brown bears, *Ursus arctos*. *Evolutionary Ecological Research* **6**: 89–102.

Ferrari, P.F., Paukner, A., Ionica, C. & Suomi, S.J. (2009). Reciprocal face-to-face communication between rhesus macaque mothers and their newborn infants. *Current Biology* **19**(20): 1768–72.

Ferrari, P.F., Visalberghi, E., Paukner, A. *et al.* (2006). Neonatal imitation in rhesus macaques. *PLoS Biol* **3**(9):e302.

Fisher, R.A. (1958). *The Genetical Theory of Natural Selection*. Oxford: Clarendon Press.

Fitzgibbon, C.D. & Fanshawe, J.H. (1988). Stotting in Thomson's gazelles: an honest signal of condition. *Behavioral Ecology and Sociobiology* **23**: 69–74.

Fitzpatrick, M.J., Ben-Shahar, Y., Smid, H.M. *et al.* (2005). Candidate genes for behavioural ecology. *Trends in Ecology and Evolution* **20**: 96–104.

Fitzsimmons, J.T. (1972). Thirst. *Physiological Reviews* **52**: 468–561.

Fitzsimmons, J.T. & Le Magnen, J. (1969). Eating as a regulatory control of drinking. *Journal of Comparative and Physiological Psychology* **67**: 273–83.

Fleishman, L.J. (1992). The influence of the sensory systems and the environment on motion patterns in the visual displays of anoline lizards and other vertebrates. *American Naturalist* **139**: S36–S61.

Fleming, A.S. & Blass, E.M. (1994). Psychobiology of the early mother–young relationship. In *Causal Mechanisms of Behavioural Development*, ed. J.A. Hogan and J.J. Bolhuis. Cambridge: Cambridge University Press, pp. 212–41.

Follett, B.K., Mattocks, P.W. & Farner, D.S. (1974). Circadian function in the photoperiodic induction of gonadotrophin secretion in the whiote-crowned sparrow *Zonatrichia leucophrys gainbelli*. *Proceedings of the National Academy of Sciences* **71**: 1666–9.

Ford, N.B. & Low, J.R. (1984). Sex pheromone source location by garter snakes: a mechanism for detection of direction in nonvolatile trails. *Journal of Chemical Ecology* **10**: 1193–99.

Francis, R.C. (1992). Sexual lability in teleosts: developmental factors. *Quarterly Review of Biology* **67**: 1–18.

Fraser, D. (2008). *Understanding Animal Welfare. The Science in its Cultural Context*. Oxford & Chichester: Wiley Blackwell.

Fraser Darling, F. (2008). *A Herd of Red Deer*. Edinburgh: Luath Press.

Fuller, J.L. & Thompson, W.R. (1978). *Foundations of Behavior Genetics*. St Louis, MO: C.V. Mosby.

Gadgil, M. (1982). Changes with age in the strategy of social behavior. *Perspectives in Ethology* **5**: 489–502.

Galoch, Z. & Bischof, H.J. (2006). Behavioural responses to video playback by zebra finch males. *Behavioural Processes* **74**: 21–6.

Gallup, G.G., Povinelli, D.J., Suarez, S.D. *et al.* (1995). Further reflections on self-recognition in primates. *Animal Behaviour* **50**: 1533–42.

Gannon, D.P., Barros, N.B., Nowacek, D.P. *et al.* (2005). Prey detection by bottlenose dolphins *Tursiops truncatus*: an experimental test of the passive listening hypothesis. *Animal Behaviour* **69**: 709–20.

Garcia, J. & Koelling, R. (1966). Relation of cue to consequence in avoidance learning. *Psychonomic Science* **4**: 123–4.

Gerhardt, H.C. (1974). The significance of some spectral features in mating call recognition in the green treefrog *Hyla cinerea*. *Journal of Experimental Biology* **61**: 229–41.

Gibson, R.M. & Bradbury, J.W. (1985). Sexual selection in lekking sage grouse: phenotypic correlates of male mating success. *Behavioral Ecology and Sociobiology* **18**: 117–23.

Gill, F.B. & Wolf, L.L (1975). Economics of feeding territoriality in the golden-winged sunbird. *Ecology* **56**: 333–45.

Godfray, H.C.J. (1995). Evolutionary theory of parent–offspring conflict. *Nature* **376**: 133–8.

Godwin, J. (1994). Behavioural aspects of protandrous sex change in the anemonefish, *Amphiprion melanopus*, and endocrine correlates. *Animal Behaviour* **48**: 551–67.

Goldizen, A.W. (1987). Tamarins and marmosets: communal care of offspring. In *Primate Societies*, ed. B.B. Smuts, D.L. Cheney, R.M. Seyfarth, R.W. Wrangham and T.T. Struhsaker. Chicago: University of Chicago Press, pp. 34–43.

Goodall, J. (1986). *The Chimpanzees of Gombe*. Cambridge, MA: Belknap Press of Harvard University Press.

Gottlieb, G. (1971). *Development of Species Identification in Birds*. Chicago: University of Chicago Press.

Gottlieb, G. (1976). Early development of species-specific auditory perception in birds. In *Neural and Behavioral Specificity*, ed. G. Gottlieb. New York: Academic Press, pp. 237–80.

Gottlieb, G. (1983). Development of species identification in ducklings: X. Perceptual specificity in the wood duck embryo requires sib stimulation for maintenance. *Developmental Psychobiology* **16**: 323–33.

Gottlieb, G. (1993). Social induction of malleability in ducklings: sensory basis and psychological mechanism. *Animal Behaviour* **45**: 707–19.

Gottlieb, G. (1997). *Synthesizing Nature–Nurture: Prenatal Roots of Instinctive Behavior*. New York: Routledge.

Göttmark, F. & Andersson, M. (1984). Colonial breeding reduces nest predation in the common gull. *Animal Behaviour* **32**: 485–92.

Göttmark, F., Winkler, D.W. & Andersson, M. (1986). Flock-feeding on fish schools increases individual success in gulls. *Nature* **319**: 589–91.

Gould, E., Reeves, A.J., Graziano, M.S.A. & Gross, C.G. (1999). Neurogenesis in the neocortex of adult primates. *Science* **286**: 548–52.

Gould, J.L. (1975). Honey bee recruitment: the dance language controversy. *Science* **189**: 885–93.

Gould, J.L. & Gould, C.G. (1988). *The Honey Bee*. New York: Scientific American.

Gould, J.L. and Gould, C.G. (1994). *The Animal Mind*. New York: Palgrave.

Gould, J.L. & Marler, P. (1987). Learning by instinct. *Scientific American* **256**(1): 62–73.

Gould, S.J. & Lewontin, R.C. (1979). The spandrels of San Marco and the Panglossian paradigm: a critique of the adaptationist programme. *Proceedings of the Royal Society of London B* **205**: 581–98.

Gould, J.L., Dyer, F.C. & Towne, W.F. (1985). Recent progress in understanding the honey bee dance language. *Fortschriften der Zoologie* **31**: 141–61.

Grafen, A. (1984). Natural selection, kin selection and group selection. In *Behavioral Ecology: An Evolutionary Approach*, 2nd edn., ed. J.R. Krebs and N.B. Davies. Oxford: Blackwell Scientific Publications, pp. 62–84.

Grafen, A. (1990). Biological signals as handicaps. *Journal of Theoretical Biology* **144**: 517–5.

Graves, H.B., Hable, C.P. & Jenkins, T.H. (1985). Sexual selection in *Gallus*. Effects of morphology and dominance on female spatial behaviour. *Behavioural Processes* **11**: 189–97.

Griffin, D.R. (1958). *Listening in the Dark*. New Haven, CT: Yale University Press.

Griffin, D.R. (1984). *Animal Thinking*. Cambridge, MA: Harvard University Press.

Griffin, D.R. (2001). *Animal Minds: Beyond Cognition to Consciousness*. Chicago: University of Chicago Press.

Grinnell, J., Packer, C. & Pusey, A.E. (1995). Cooperation in male lions: kinship reciprocity or reciprocity? *Animal Behaviour* **60**: 523–9.

Groothuis, T.G. & Schwabl, H. (2002). Determinants of within and among clutch variation in levels of maternal hormones in black-headed gull eggs. *Functional Ecology* **16**: 281–9.

Groothuis, T.T.G. (1993). The ontogeny of social displays: form development, form fixation and change in context. *Advances in the Study of Behavior* **22**: 269–322.

Grossman, C.J (1985). Interaction between the gonadal steroids and the immune system. *Science* **227**: 257–61.

Guhl, A.M. (1968). Social inertia and social stability in chickens. *Animal Behaviour* **16**: 219–32.

Hagedorn, M. & Heiligenberg, W. (1985). Court and spark: electric signals in the courtship of mating of gymnotid fish. *Animal Behaviour* **33**: 254–65.

Halliday, T.R. (1975). An observational and experimental study of sexual behaviour in the smooth newt, *Triturus vulgaris. Animal Behaviour* **23**: 291–322.

Halliday, T.R. & Sweatman, H.P.A. (1976). To breathe or not to breathe; the newt's problem. *Animal Behaviour* **24**: 551–61.

Hamilton, W.D. (1964). The genetical evolution of social behaviour. I and II. *Journal of Theoretical Biology* **7**: 1–52.

Hamilton, W.D. (1971). Geometry for the selfish herd. *Journal of Theoretical Biology* **31**: 295–311.

Hamilton, W.D. & Zuk, M. (1984). Heritable true fitness and bright birds: a role for parasites? *Science* **218**: 384–7.

Hamilton, W.D., Axelrod, R. & Tanses, R. (1990). Sexual reproduction as an adaptation to resist parasites (a review). *Proceedings of the National Academy of Sciences USA* **87**: 3566–73.

Hampton, N.G., Bolhuis, J.J. & Horn, G. (1995). Induction and development of a filial predisposition in the chick. *Behaviour* **132**: 451–77.

Hansell, M.H. & Overhill, R. (2005). *Bird Nests and Construction Behaviour.* Cambridge: Cambridge University Press.

Harcourt, R. (1991). The development of play in the South American fur seal. *Ethology* **88**: 191–202.

Hare, B. & Tomasello, M. (2004). Chimpanzees are more skilful in competetive than in cooperative cognitive tasks. *Animal Behaviour* **68**: 571–81.

Hare, B., Call, J. & Tomasello, M. (2001). Do chimpanzees know what conspecifics know? *Animal Behaviour* **61**: 139–51.

Harlow, H.F. (1949). The formation of learning sets. *Psychological Review* **56**: 51–65.

Harlow, H.F. & Harlow, M.K. (1965). The affectional systems. In *Behaviour of Non-Human Primates*, ed. A.M. Schrier, H.F. Harlow and F. Stollnitz. New York: Academic Press, pp. 287–334.

Hartline, H.K., Wagner, H.G. & Ratliff, F. (1956). Inhibition in the eye of *Limulus. Journal of General Physiology* **39**: 651–73.

Hauser, M., Chen, M.K., Chen, F. & Chuang, E. (2003). Give unto others: genetically unrelated cotton-top tamarin monkeys preferentially give food to those who altruistically give food back. *Proceedings of the Royal Society B* **270**: 2363–70.

Healy, S.D., de Kort, S.R. & Clayton, N.S. (2005). The hippocampus, spatial memory and food hoarding: a puzzle revisited. *Trends in Ecology and Evolution* **20**:17–22.

Heinrich, B. & Smolker, R. (1998). Play in common ravens (*Corvus corax*). In *Animal Play*, ed. M. Bekoff & J.M. Byers. Cambridge: Cambridge University Press, pp. 22–44.

Hepper, P.G. (1987). The discrimination of human odour by the dog. *Perception* **17**: 549–54.

Herman, L.M. & Arbeit, W.R. (1973). Stimulus control and auditory discrimination learning sets in the bottlenose dolphin. *Journal of Experimental Animal Behaviour* **19**: 379–94.

Heyes, C.M. (1995). Self-recognition in primates: further reflections create a hall of mirrors. *Animal Behaviour* **50**: 1533–42.

Heyes, C.M. (1998). Theory of mind in nonhuman primates. *Behavioral & Brain Sciences*, **21**, 101–34.

Hinde, R.A. (1954). Factors governing the changes in strength of a partially inborn response, as shown by the mobbing behaviour of the chaffinch (*Fringilla coelebs*), II. The waning of the response. *Proceedings of the Royal Society of London B* **142**: 331–58.

Hinde, R.A. (1960). Factors governing the changes in strength of a partially inborn response, as shown by the mobbing behvaiour of the chaffinch (*Fringilla coelebs*), III. The interaction of short-term and long-term incremental and decremntal effects. *Proceedings of the Royal Society of London B* **153**: 398–420.

Hinde, R.A. (1974). *Biological Bases of Human Social Behaviour*. New York: McGraw-Hill.

Hinde, R.A. (1977). Mother–infant separation and the nature of inter-individual relationships. Experiments with rhesus monkeys. *Proceedings of the Royal Society of London B* **196**: 29–50.

Hinde, R.A. & Rowell, T.E. (1962). Communication by postures and facial expressions in the rhesus monkey (*Macaca mulatta*). *Proceedings of the Zoological Society of London* **138**: 1–21.

Hinde, R.A. & Steel, E. (1966). Integration of the reproductive behaviour of female canaries. *Symposia of the Society for Experimental Biology* **20**: 401–26.

Hinde, R.A. & Stevenson-Hinde, J. (eds.) (1973). *Constraints on Learning*. London: Academic Press.

Hitchcock, C.L. & Sherry, D.F. (1990). Long-term memory for cache sites in the black-capped chickadee. *Animal Behaviour* **40**: 701–12.

Hodos, W. & Campbell, C.B.G. (1969). Scala Naturae: why there is no theory in comparative psychology. *Psychological Review* **76**: 337–50.

Hogan, J.A. (1967). Fighting and reinforcement in Siamese fighting fish *Betta splendens*. *Journal of Comparative and Physiological Psychology* **64**: 356–9.

Hogan, J.A. (1989). The interaction of incubation and feeding in broody junglefowl hens. *Animal Behaviour* **38**: 121–38.

Hogan, J.A. & Bolhuis, J.J. (2009). The development of behavior: trends since Tinbergen (1963). In *Tinbergen's Legacy*, ed. J.J. Bolhuis and S. Verhulst. Cambridge: Cambridge University Press, pp. 82–106.

Hollard, V.D. & Delius, J.D. (1982). Rotational invariance in visual pattern recognition by pigeons and humans. *Science* **218**: 804–806.

Hölldobler, B. (1971). Communication between ants and their guests. *Scientific American* **224**(3): 86–93.

Hölldobler, B. & Wilson, E.O. (1990). *The Ants*. London: Springer-Verlag.

Hoogland, W.G. & Sherman, P.W. (1976). Advantages and disadvantages of bank swallow coloniality. *Ecological Monographs* **46**: 33–58.

Hopkins, C.D. & Bass, A.H. (1981). Temporal coding of species recognition signals in an electric fish. *Science* **212**: 85–7.

Hoppitt, W. & Laland, K.N. (2008). Social processes influencing learning in animals: a review of the evidence. *Advances in the Study of Behavior* **38**: 105–165.

Horn, G. (1990). Neural bases of recognition memory investigated through an analysis of imprinting. *Philosophical Transactions of the Royal Society of London B* **329**: 133–42.

Horn, G. (1998). Visual imprinting and the neural mechanisms of recognition memory. *Trends in Neurosciences* **21**: 300–305.

Houston, A.I. & Mcnamara, J. (1988). A framework for the functional analysis of behavior. *Behavior and Brain Science* **11**: 117–63.

Hrdy, S. (1974). Male–male competition and infanticide among langurs (*Presbytis entellus*) of Abu, Rajastan. *Folia Primatlogica* **22**: 19–58.

Hubel, D.H. & Wiesel, T.N. (1959). Receptive fields, binocular interaction and functional architecture in the cat's visual cortex. *Journal of Physiology* **48**: 574–91.

Hughes, B.O., Duncan, I.J.H. & Brown, M.F. (1989). The performance of nest-building by domestic hens: is it more important than the construction of the nest? *Animal Behaviour* **37**: 210–14.

Humphrey, N.K. (1976). The social function of intellect. In *Growing Points in Ethology*, ed. P.P.G. Bateson and R.A. Hinde. Cambridge: Cambridge University Press, pp. 303–17.

Hunt, G.R. (1996). Manufacture and use of hook tools by New Caledonian crows. *Nature*, **379**: 249–51.

Hunt, G.R. & Gray, R.D. (2003). Diversification and cumulative evolution in New Caledonian crow tool manufacture. *Proceedings of the Royal Society of London B* **270**: 867–74.

Hunt, S., Kilner, R.M., Langmore, N.E. & Bennett, A.T.D. (2003). Conspicuous, ultraviolet-rich mouth colours in begging chicks. *Proceedings of the Royal Society of London B* **270**: S25–S28.

Huntingford, F.A. (1976). The relationship between anti-predator behaviour and aggression among conspecifics in the three-spined stickleback, *Gasterosteus aculeatus*. *Animal Behaviour* **24**: 245–60.

Hurst, J. (1989). The complex network of olfactory communication in populations of wild house mice *Mus domesticus* Rutty: markings and investigation within family groups. *Animal Behaviour* **37**: 705–25.

Hutchinson, J.B. (1976). Hypothalamic mechanisms of sexual behaviour with special reference to birds. *Advances in the Study of Behavior* **6**: 159–200.

Huxley, J.S. (1914). The courtship habits of the great crested grebe *Podiceps cristatus*. *Proceedings of the Zoological Society of London* **1914**(2): 491–562.

Immelmann, K. (1972). Sexual and other long-term aspects of imprinting in birds and other species. *Advances in the Study of Behavior* **4**: 147–74.

Immelmann, K., Rove, R. Lassek, R. & Bischof, H.-J. (1991). Influence of adult courtship experience on the development of sexual preferences in zebra finch males. *Animal Behaviour* **42**: 83–90.

Ince, S.A., Slater, P.B.J. & Weismann, C. (1980). Changes with time in the songs of a population of chaffinches. *Condor* **82**: 285–90.

Islam, M.S., Roessingh, P., Simpson, S.J. & McCavery, A.R. (1994). Effects of population density experienced by parents during mating and oviposition on the phase of hatchling desert locusts. *Proceedings of the Royal Society of London B* **257**: 93–8.

Janik, V.M., Sayigh, L.S. & Wells, R.S. (2006). Signature whistle shape conveys identity information to bottlenose dolphins. *Proceedings of the National Academy of Sciences* **103**: 8293–7.

Jarvis, J.U.M. (1981). Eusociality in a mammal: cooperative breeding in naked mole rat colonies. *Science* **212**: 571–3.

Jarvis, J.U.M., Oriain, J.M., Bennett, N.C. & Sherman, P.W. (1994). Mammalian eusociality: a family affair. *Trends in Ecology and Evolution* **9**(2): 47–51.

Jerison, H.J. (1985). Animal intelligence and encephalization. *Philosphical Transactions of the Royal Society B* **308**: 21–35.

Jerison, H.J. (2001). The evolution of neural and behavioral complexity. In *Brain Evolution and Cognition*, ed. G. Roth and M.F. Wulliman. New York: Wiley, pp. 523–54.

Johnson, J.A. (1987). Dominance rank in juvenile olive baboons, *Papio anubis*: the influence of gender, size, maternal rank and orphaning. *Animal Behaviour* **35**: 1694–708.

Johnson, J.A. (1989). Supplanting by olive baboons: dominance rank difference and resource value. *Behavioral Ecology and Sociobiology* **24**: 277–83.

Johnson, M.H., Munakata, Y. & Gilmore, R.O. (eds.) (2005). *Brain Development and Cognition: A Reader*, 2nd edn. Oxford: Blackwell.

Johnson, S.D. & Steiner, K.E. (1997) Long-tongued fly pollination and evolution of floral spur length in the *Disa draconis* complex (Orchidaceae). *Evolution* **51**(1): 45–53.

Jolly, A. (1966). *Lemur Behavior*. Chicago: University of Chicago Press.

Jones, A.E. & Slater, P.J.B. (1996). The role of aggression in song tutor choice in the zebra finch: cause or effect? *Behaviour* **133**: 103–115.

Jones, G. & Holderied, M.W. (2006). Bat echolocation calls: adaptation and convergent evolution. *Proceedings of the Royal Society B* **274**: 905–12.

Joseph, R. (1979). Effects of rearing and sex on maze learning and competitive exploration in rats. *Journal of Psychology* **101**: 37–43.

Kalko, E.K.V. (1995). Insect pursuit, prey capture and echolocation in pipistrelle bats (Microchiroptera). *Animal Behaviour* **50**: 861–80.

Kalmus, H. (1955). The discrimination by the nose of the dog of individual human odours and in particular the odours of twins. *British Journal of Animal Behaviour* **3**: 25–31.

Kammer, A. & Rheuben, E. (1976). Adult motor pattern produced by moth pupae during development. *Journal of Experimental Biology* **65**: 65–84.

Kanizsa, G. (1979). *Organization in Vision: Essays on Gestalt Perception*. New York: Praegaer Publishers.

Kater, S.B. & Rowell, C.H.F. (1973). Integration of sensory and centrally programmed components in generation of cyclical feeding activity of *Helisoma trivolvis*. *Journal of Neurophysiology* **36**: 142–55.

Katz, L.C. & Shatz, C.J. (1996). Synaptic activity and the construction of cortical circuits. *Science* **274**: 1133–8.

Kempenaers, B., Verheyen, G.R., van den Broeck, M. *et al.* (1992). Extra-pair paternity results from female preference for high-quality males in the blue tit. *Nature* **37**: 494–6.

Kennedy, J.S. (1965). Coordination of successive activities in an aphid. Reciprocal effects of settling on flight. *Journal of Experimental Biology* **43**: 489–509.

Kennedy, J.S. (1992). *The New Anthropomorphism*. Cambridge: Cambridge University Press.

Keshishian, H. & Chiba, A. (1993). Neuromuscular development in *Drosophila*: insights from single neurons and single genes. *Trends in Neurosciences* **16**(7): 279–83.

Ketterson, E.D. & Nolan, V., Jr (1994). Hormones and life histories: an integrated approach. In *Behavioral Mechanisms in Evolutionary Ecology*, ed. L.A. Real, pp. 327–53. Chicago: University of Chicago Press.

Kiley-Worthington, M. (1977). *Behavioural Problems of Farm Animals*. Stocksfield, UK: Oriel Press.

King, L.E, Douglas-Hamilton, I. & Vollrath, F. (2007). African elephants run from the sound of disturbed bees. *Current Biology* **17**: R832–R833.

Kirkpatrick, M. (1982). Sexual selection and the evolution of female choice. *Evolution* **36**: 1–12.

Knudsen, E.I. & Konishi, M. (1979). Mechanisms of sound location in the barn owl *(Tyto alba)*. *Journal of Comparative Physiology* **133**: 13–21.

Kocher, T. (2004). Adaptive evolution and explosive speciation: the cichlid fish model. *Nature Reviews Genetics* **5**: 288–98.

Köhler, W. (1927). *The Mentality of Apes*, 2nd edn. London: Kegan Paul.

Koltermann, R. (1971). 24-Std.-Periodik der Langzeiterinnerung an Duft und Farbsignale bei der Honigbiene. *Zeitschrift für Vergleichende Physiologie* **75**: 49–68.

Komisaruk, B.R. (1967). Effects of local brain implants of progesterone on reproductive behavior in ring doves. *Journal of Comparative and Physiological Psychology* **64**: 219–24.

Komisaruk, B.R., Adler, N.T. & Hutchison, J. (1972). Genital sensory field: enlargement by oestrogen treatment in female rats. *Science* **178**: 1295–8.

Konishi, M. (1965). The rôle of auditory feedback on the control of vocalization in the white-crowned sparrow. *Zeitschrift für Tierpsychologie* **22**: 770–83.

Konorski, J. (1948). *Conditioned Reflexes and the Nervous System*. Cambridge: Cambridge University Press.

Krebs, J.R. & Dawkins, R. (1984). Animal signals: mind reading and manipulation. In *Behavioural Ecology: An Evolutionary Approach*, 2nd edn., ed. J.R. Krebs and N.B. Davies. Oxford: Blackwell Scientific Publications, pp. 380–402.

Krebs, J.R., Macroberts, M.H. & Cullen, J.M. (1972). Flocking and feeding in the great tit, *Parus major*: an experimental study. *Ibis* **114**: 507–30.

Kroodsma, D.E. (1982). Learning and the ontogeny of sound signals in birds. In *Acoustic Communication in Birds*, ed. D.E. Kroodsma and D.H. Miller. New York: Academic Press, pp. 1–23.

Kruczek, M. (2007). Recognition of kin in bank voles (*Clethrionomys glareolus*). *Physiology and Behavior* **90**: 483–9.

Kruijt, J.P. & Meeuwisse, G.B. (1991). Sexual preferences of male zebra finches: effects of early and adult experience. *Animal Behaviour* **42**: 91–102.

Kruuk, H. (1972). *The Spotted Hyaena*. Chicago: University of Chicago Press.

Kummer, H. (1968). Two variations in the social organization of baboons. In *Primates: Studies in Adaptance and Variability*, ed. P. Jay. New York: Holt, Rinehart & Winston, pp. 293–312.

Kummer, H., Gotz, W. & Angst, W. (1974). Triadic differentiation: an inhibitory process protecting pair bands in baboons. *Behaviour* **49**: 62–87.

Lack, D. (1943). *The Life of the Robin*. London: H.F. & G. Witherby.

Lade, B. & Thorpe, W.H. (1964). Dove songs as innately coded patterns of specific behaviour. *Nature* **202**: 66–8.

Laland, K.N. & Galef, B.G. (eds.) (2009). *The Question of Animal Culture*. Cambridge, MA: Harvard University Press.

Laland, K.N., Richardson, P.J. & Lloyd, J.E. (1993). Animal social learning: towards a new theoretical approach. *Perspectives in Ethology* **10**: 249–77.

Lall, A.B., Seligar, B.H., Biggley, W. & Lloyd, J.E. (1980). Ecology of colors of firefly bioluminscence. *Science* **210**: 560–2.

Land, M. (2008). Biological optics: circularly polarized crustacean. *Current Biology* **18**: R348–9.

Land, M.F. (1981). Optics and vision in invertebrates. In *Handbook of Sensory Physiology*, ed. H. Autrum. New York: Springer-Verlag, pp. 471–593.

Langmore, N.E., Kilner, R.M., Butchart, S.H.M., Maurer, G., Davies, N.B. *et al.* (2005). The evolution of egg rejection by cuckoo hosts in Australia and Europe. *Behavioural Ecology* **16**: 868–92.

Le Boeuf, B.J. (1974). Male–male competition and reproductive success in elephant seals. *American Zoologist* **14**: 163–76.

Lee, P. (1983). Play as a means for developing relationships. In *Primate Social Relationships*, ed. R.A. Hinde. Oxford: Blackwell, pp. 82–9.

Lee, P. (1987). Allomothering among African elephants. *Animal Behaviour* **35**: 278–91.

Legge, S. (1996). Cooperative lions escape the Prisoner's Dilemma. *Trends in Ecology and Evolution* **11**: 2–3.

Legge, S. (2000). Siblicide in the cooperatively breeding kookaburra (*Dacelo novaeguineae*). *Behavioral Ecology and Sociobiology* **48**: 293–302.

Lehrman, D.S. (1964). The reproductive behavior of ring doves. *Scientific American* **211**(5): 48–54.

Lenington, S. (1994). Of mice, men and the MHC. *Trends in Ecology and Evolution* **9**: 455–6.

Lesku, J.A., Roth, T.C., Amlaner, C.J. & Lima, S.L. (2006). A phylogenetic analysis of sleep architecture in mammals: the integration of anatomy, physiology and ecology. *The American Naturalist* **168**: 441–53.

Lettvin, J.Y., Maturana, H.R., McCulloch, W.S. & Pitts, W.H. (1959). What the frog's eye tells the frog's brain. *Proceedings of the Institute of Radio Engineers* **47**: 1940–51.

Levine, R.B. (1986). Reorganization of the insect nervous system during metamorphosis. *Trends in the Neurosciences* **9**: 315–9.

Levine, R.B. & Truman, J.W. (1985). Dendritic reorganisation of abdominal motoneurons during metamorphosis of the moth *Manduca sexta. Journal of Neuroscience* **5**: 2424–31.

Lindauer, M. (1961). *Communication Among Social Bees.* Cambridge, MA: Harvard University Press.

Lindauer, M. (1976). Evolutionary aspects of orientation and learning. In *Communication Among Social Bees*, ed. G. Baerends, C. Beer and A. Manning. Oxford: Clarendon Press, pp. 228–42.

Lloyd, J.E. (1965). Aggressive mimicry in *Photuris* firefly *femmes fatales. Science* **149**: 653–4.

Lloyd, J.E. (1975). Aggressive mimicry in *Photuris* fireflies: signal repertoires by *femmes fatales. Science* **187**: 452–3.

Lofstedt, C., Vickers, N.J., Roelofs, W.L. & Baker, T.C. (1989). Diet related courtship success in the Oriental fruit moth *Grapholita molesta* (Tortricidae). *Oikos* **55**: 1402–8.

Lorenz, K. (1941). Vergleichende Bewegungsstudien an Anatinen. *Journal of Ornithology, Supplement* **89**: 194–294. (An English translation appeared in several parts, in Vols. 57–59 of *Avicultural Magazine*)

Lorenz, K. (1952). *King Solomon's Ring.* London: Methuen.

Lorenz, K.Z. (1966). *Evolution and Modification of Behaviour.* London: Methuen.

Luria, A.R. (1975). *The Mind of a Mnemonist.* London: Penguin.

Lythgoe, J.N. (1979). *The Ecology of Vision.* Oxford: Clarendon Press.

McBride, G., Parer, I.P. & Foenander, F. (1959). The social organization and behaviour of the feral domestic fowl. *Animal Behaviour Monographs* **2**: 127–81.

McComb, K., Moss, C., Durant, S.M., Baker, L. & Sayialel, S. (2001). Matriarchs as repositories of social knowledge in African elephants. *Science* **292**: 491–4.

McComb, K., Reby, D., Baker, L., Moss, C. & Sayialel, S. (2003). Long-distance communication of acoustic cues to social identity in Africa elephants. *Animal Behaviour* **65**: 317–29.

McFarland, D.J. (1971). *Feedback Mechanisms in Animal Behaviour.* New York: Academic Press.

McFarland, D.J. (1974). Time-sharing as a behavioural phenomenon. *Advances in the Study of Animal Behaviour* **5**: 201–25.

McFarland, D.J. & Lloyd, I. (1973). Time-shared feeding and drinking. *Quarterly Journal of Experimental Psychology* **25**: 48–61.

McFarland, DJ. & Sibly, R.M. (1975). The behavioural final common path. *Philosophical Transactions of the Royal Society of London B* **270**: 265–93.

McGrew, W.C. (2004). *The Cultured Chimpanzee: Reflections on Cultural Primatology.* Cambridge: Cambridge University Press.

Mackintosh, N.J. (1965). Discrimination learning in the octopus. *Animal Behaviour* **13**(Suppl.1): 129–34.

Mackintosh, N.J. (1983). General principles of learning. In *Animal Behavior Volume 3. Genes, Development and learning*, ed. T.R. Halliday and P.J.B. Slater. Oxford: Blackwell Scientific Publications, pp. 149–77.

Mackintosh, N.J. (ed.) (1994). *Animal Learning and Cognition*, 2nd edn. New York: Academic Press.

Mackintosh, N.J., Wilson, B. & Boakes, R.A. (1985). Differences in mechanisms of intelligence among vertebrates. *Philosophical Transactions of the Royal Society of London B* **308**: 53–65.

Macphail, E. (1985). Vertebrate intelligence: the null hypothesis. *Philosophical Transactions of the Royal Society of London B* **308**: 37–51.

Macphail, E.M. (1987). The comparative psychology of intelligence. *Behavior and Brain Sciences* **10**: 645–95.

MacPhail, E.M. (1998). *The Evolution of Consciousness*. Oxford: Oxford University Press.

MacPhail, E.M. & Bolhuis, J.J. (2001). The evolution of intelligence: adaptive specialisations *versus* general process. *Biological Reviews of the Cambridge Philosophical Society (London)* **76**: 341–64.

Maier, S.F. and Schneirla, T.C. (1935). *Principles of Animal Psychology*. New York: McGraw-Hill.

Maier, S.F., Watkins, L.R. & Fleshner, M. (1994). Psychoneuroimmunology. *American Psychologist* **49**: 1004–17.

Maklakov, A.A., Bilde, T. & Lubin, Y. (2003). Vibratory courtship in a web-building spider: signalling quality or stimulating the female? *Animal Behaviour* **66**: 623–30.

Manger, P.R. & Pettigrew, J.D. (1995). Electroreception and the feeding behaviour of platypus (*Ornithorhynchus anatinus*: Monotremata: Mammalia). *Philosophical Transactions of the Royal Society of London B* **347**: 359–81.

Mann, J., Whitehead, H. & Tyack, P.L. (eds.) (2000). *Cetacean Societies: Field Studies of Dolphins and Whales*. Chicago: University of Chicago Press.

Manning, A. & McGill, T.E. (1974). Neonatal androgen and sexual behavior in female house mice. *Hormones and Behavior* **6**: 19–31.

Manning, J.C. & Goldblatt, P. (1997). The *Moegistorhynchus longirostris* (Diptera: Nemestrinidae) pollination guild: long-tubed flowers and a specialized long-proboscid fly pollination system in southern Africa. *Plant Systematics and Evolution* **206**: 51–69.

Manoli, D.S., Meissner, G.W. & Baker, B.S. (2006). Blueprints for behavior: genetic specification of neural circuitry for innate behaviors. *Trends in Neurosciences* **29**(8): 444–51.

Marchetti, K. (1993). Dark habitats and bright birds illustrate the role of environment in species divergence. *Nature* **362**: 149–52.

Marino, L. (2002). Convergence of complex cognitive abilities in cetaceans and primates. *Brain Behavior and Evolution* **59**: 21–32.

Markl, L. (1977). Adaptive radiation of mechanoreception. In *Sensory Ecology*, ed. M.A. Ali. New York: Plenum, pp. 49–69.

Marler, M. & Peters, S. (1977). Selective vocal learning in a sparrow. *Science* **198**: 519–21.

Marler, P. & Pickert, R. (1984). Species-universal microstructure in a learned birdsong: the swamp sparrow (*Melospiza georgiana*). *Animal Behaviour* **32**: 673–89.

Marler, P. & Tamura, M. (1964). Culturally transmitted patterns of vocal behavior in sparrows. *Science* **146**: 1483–6.

Marshall, D.A. & Moulton, D.G. (1981). Olfactory sensitivity to α-ionone in humans and dogs. *Chemical Senses* **6**: 53–61.

Martin, G.M. & Lett, B.T. (1985). Formation of associations of colored and flavoured food with induced sickness in five main species. *Behavioral and Neural Biology* **43**: 223–37.

Martin, P. & Bateson, P. (1985a). The influence of experimentally manipulating a component of weaning on the development of play in domestic cats. *Animal Behaviour* **33**: 511–18.

Martin, P. & Bateson, P. (1985b). The ontogeny of locomotor play behaviour in the domestic cat. *Animal Behaviour* **33**: 502–10.

Martin, P. & Caro, T.M. (1985). On the functions of play and its role in behavioral development. *Advances in the Study of Behaviour* **15**: 59–103.

Martin, R.D. (1968). Reproduction and ontogeny in tree-shrews (*Tupaia belangeri*) with reference to their general behaviour and taxonomic relationships. *Zeitschrift für Tierpsychologie* **25**: 409–95.

Mason, G.J. & Latham, N. (2004). Can't stop, won't stop: is stereotypy a reliable indicator of welfare? *Animal Welfare* **13**: S57–S69.

Mason, G.J. & Turner, M.A. (1993). Mechanisms involved in the development and control of stereotypies. In: *Perspectives in Ethology*, Vol. 10, ed. P.P.G. Bateson, P.H. Klopfer and N.S. Thompson. New York: Plenum, pp. 53–85.

Masters, W.M. & Markl, H. (1981). Vibration signal transmission in spider orb web. *Science* **213**: 363–5.

Masters, W.M. & Moffat, A.J.M. (1986). Transmission of vibration in a spider's web. In *Spiders*, ed. W.A. Shear. Stanford, CA: Stanford University Press, pp. 49–69.

Maynard Smith, J. (1964). Group selection and kin selection. *Nature* **201**: 1145–7.

Maynard Smith, J. (1978). Optimization theory in evolution. *Annual Review of Ecology and Systematics* **9**: 31–56.

Maynard Smith, J. (1982). *Evolution and the Theory of Games*. Cambridge: Cambridge University Press.

Maynard Smith, J. & Price, G.R. (1973). The logic of animal conflict. *Nature* **246**: 15–18.

Maynard Smith, J. & Riechert, S.E. (1984). A conflicting-tendency model of spider agonistic behaviour: hybrid–pure population line comparisons. *Animal Behaviour* **32**: 564.

Mech, L.D. & Boitani, L. (eds.) (2003). *Wolves: Behavior, Ecology and Conservation*. Chicago. University of Chicago Press.

Meddis, R. (1983). The evolution of sleep. In *Sleep Mechanisms and Functions*, ed. A. Mayes. London: Van Nostrand, pp. 57–106.

Menzel, R. & Erber, J. (1978). Learning and memory in bees. *Scientific American* **239**(1): 80–7.

Menzel, R., Giurfa, M., Gerber, B. & Hellstern, F. (2001). Cognition in insects: the honeybee as a case study. In *Brain Evolution and Cognition*, ed. G. Roth and M.F. Wulliman. New York: Wiley, pp. 333–66.

Menzel, R., Greggers, U. & Hammer, M. (1993). Functional organization of appetitive learning and memory in a generalist pollinator, the honey bee. In *Insect Learning: Ecological and Evolutionary Perspectives*, ed. D. Papaj and A. Lewis. London: Chapman & Hall, pp. 79–125.

Metcalfe, N.B. & Furness, R.W. (1984). Changing priorities: the effect of pre-migratory fattening on the trade-off between foraging and vigilance. *Behavioral Ecology and Sociobiology* **15**: 203–6.

Michelson, A. (1989). Ein mechanisches Modell der tanzenden der Honigbiene. *Biologie in unserer Zeit* **19**(4): 121–6.

Michelson, A., Andersen, B.B., Storm, J., Kirchner, W.H. & Lindauer, M. (1992). How honeybees perceive communication dances, studied by means of a mechanical model. *Behavioral Ecology and Sociobiology* **30**: 143–50.

Milinski, M. (1984). A predator's cost of overcoming the confusion effect of swarming prey. *Animal Behaviour* **32**: 1157–62.

Milinski, M. (2006). The major histocompatibility complex, sexual selection and mate choice. *Annual Review Ecology, Evolution and Systematics* **37**: 159–86.

Milinski, M. & Heller, R. (1978). Influence of a predator on the optimal foraging behaviour of sticklebacks (*Gasterosteus aculeatus*). *Nature* **275**: 642–4.

Miller, N.E. (1956). Effects of drugs on motivation: the value of using a variety of measures. *Annals of the New York Academy of Sciences* **65**: 318–33.

Miller, N.E. (1957). Experiments on motivation. Studies combining psychological, physiological and pharmacological techniques. *Science* **126**: 1271–8.

Miller, N.E., Sampliner, R.I. & Woodrow, P. (1957). Thirst reducing effects of water by stomach fistula versus water by mouth, measured by both a consummatory and an instrumental response. *Journal of Comparative and Physiological Psychology* **50**: 1–5.

Mills, M.G.L. & Gorman, M.L. (1987). The scent-marking behaviour of the spotted hyaena *Crocuta crocuta* in the Southern Kalahari. *Journal of Zoology, London* **212**: 483–97.

Mitani, J., Call, J., Kappeler, P., Palambit, R. & Silk, J. (2012). *The Evolution of Primate Societies*. Chicago: University of Chicago Press.

Mittermeier, R.R. & Cheney, D.L. (1987). Conservation of primates and their habitats. In *Primate Societies*, ed. B.B. Smuts, D.L. Cheney, R.M. Seyfarth, R.W. Wrangham and T.T. Struhsaker. Chicago: University of Chicago Press, pp. 477–90.

Moberg, G.P. (1985). Biological response to stress: key to assessment of animal well-being? In *Animal Stress*, ed. G, Moberg. Bethesda, MD: American Physiological Society, pp. 27–49.

Mock, D.W., Lamey, T.C., Williams, C.F. & Pelletier, A. (1987). Flexibility in the development of heron sibling aggression: an intraspecific test of the prey-size hypothesis. *Animal Behaviour* **35**: 1386–93.

Moehlman, P.D. (1979). Jackal helpers and pup survival. *Nature* **277**: 382–3.

Møller, A. (1987). Social control of deception among status signalling house sparrows. *Behavioral Ecology and Sociobiology* **20**: 307–11.

Møller, A. & Birkhead, T.R. (1994). The evolution of plumage brightness in birds is related to extrapair paternity. *Evolution* **48**: 1089–1100.

Monaghan, P., Charmantier, A., Nussey, D.H. & Ricklefs, R.E. (2008). The evolutionary ecology of senescence. *Functional Ecology* **22**: 371–8.

Money, J. & Ehrhardt, A.A. (1973). *Man and Woman, Boy and Girl*. Baltimore, MD: Johns Hopkins University Press.

Moore, B.R. (1973). The role of directed Pavlovian reactions in simple instrumental learning in the pigeon. In *Constraints on Learning*, ed. R.A. Hinde and J. Stevenson-Hinde. London: Academic Press, pp. 159–88.

Morgan, L. (1894). *An Introduction to Comparative Psychology*. London: Scott.

Morris, D. (1958). Homosexuality in the ten-spined stickleback (*Pygosteus pungitius* L.). *Behaviour* **4**: 233–61.

Moss, C. & Colbeck, M. (1992). *Echo of the Elephants*. London: BBC Books.

Moynihan, M. (1967). Comparative aspects of communication in New World primates. In *Primate Ethology*, ed. D. Morris. London: Weidenfield & Nicholson, pp. 236–66.

Munn, C.A. (1986). Birds that 'cry wolf'. *Nature* **319**: 143–5.

Munn, N.L. (1950). *Handbook of Psychological Research on the Rat*. Boston, MA: Houghton Mifflin.

Murphey, R.K. (1985). Competition and chemoaffinity in insect sensory systems. *Trends in the Neurosciences* **8**: 120–5.

Murphey, R.K. (1986). The myth of the inflexible invertebrate: competition and synaptic remodelling in the development of invertebrate nervous systems. *Journal of Neurobiology* **17**: 585–91.

Narins, P.M. & Capranica, R.R. (1976). Sexual differences in the auditory system of the tree frog, *Eleutherodactylus coqui*. *Science* **192**: 378–80.

Neel, J.V. (1999). The 'Thrifty Genotype' in 1998. *Nutrition Reviews* **57**: 2–9.

Nelson, B. (1980). *Seabirds, their Biology and Ecology.* London: Hamlyn.

Newman, E.A. & Hartline, P.H. (1982). The infrared 'vision' of snakes. *Scientific American* **246**(3): 98–107.

Nilsson, J.A. & Smith, H.G. (1988). Incubation feeding as a male tactic for early hatching. *Animal Behaviour* **36**: 641–7.

Noble, D. (2008). Genes and causation. *Philosophical Transactions of the Royal Society A* **366**: 3001–15.

Nottebohm, F. (1989). From birdsong to neurogenesis. *Scientific American* **260**(2): 56–61.

Nottebohm, F. (2005). The neural basis of birdsong. *PLoS Biology* **3**(5): e164.

Odling-Smee, F.J., Laland, K.N. & Feldman, M.W. (2003). *Niche Construction: The Neglected Process In Evolution.* Princeton, NJ: Princeton University Press.

O'Donnell, R.P., Ford, N.B., Shione, R. & Mason, R.T. (2004). Male red-sided garter snakes, *Thamnophis sirtalis*, determine female mating status from pheromone trails. *Animal Behaviour* **68**: 677–83.

Olson, G.C. & Krasne, F.B. (1981). The crayfish lateral giants are command neurons for escape behaviour. *Brain Research* **214**: 89–100.

Olsson, I.A.S. & Keeling, L.J. (2005). Why in earth? Dustbathing in jungle and domestic fowl reviewed from a Tinbergian perspective. *Applied Animal Behaviour Science* **93**: 259–82.

Oppenheim, R.W. (1974). The ontogeny of behavior in the chick embryo. *Advances in the Study of Behavior* **5**: 133–72.

Packer, C. (1977). Reciprocal altruism in *Papio anubis*. *Nature* **265**: 441–3.

Packer, C. (1986). The ecology of sociality in felids. In *Ecological Aspects of Social Evolution*, ed. D.I. Rubenstein and R.W. Wrangham. Princeton, NJ: Princeton University Press, pp. 429–51.

Partridge, B.L. & Pitcher, T.J. (1980). The sensory basis of fish schools: relative roles of lateral line and vision. *Journal of Comparative Physiology* **135**: 315–25.

Passillé, A.M.B. de, Christopherson, R. & Rushen, J. (1993). Non-nutritive sucking by the calf and postprandial secretion of insulin, CCK, and gastrin. *Physiology and Behavior* **54**: 1069–73.

Paukner, A., Suomi, S.J., Visalberghi, E. & Ferrari, P.F. (2009). Capuchin monkeys display affiliation towards humans who imitate them. *Science* **325**: 880–3.

Paul, R.C. & Walker, T.J. (1979). Arboreal singing in a burrowing cricket, *Anurogryllus arboreus*. *Journal of Comparative Physiology* **132**: 217–23.

Pavlov, I.P. (1941). *Lectures on Conditioned Reflexes.* 2 vols. New York: International Publishers.

Payne, K. & Payne, R. (1985). Large-scale changes over 19 years in songs of humpback whales in Bermuda. *Zeitschrift fur Tierpsychologie* **68**: 89–114.

Payne, R. (1971). Acoustic location of prey by barn owls (*Tyto alba*). *Journal of Experimental Biology* **54**: 535–73.

Payne, R.S. & Mcvay, S. (1971). Songs of hump-back whales. *Science* **173**: 585–97.

Payne, T.L., Birch, M.C. & Kennedy, C.E.J. (eds.) (1986). *Mechanisms in Insect Olfaction.* Oxford: Clarendon Press.

Pearce, J.M. (1996). *Animal Learning and Cognition: An Introduction*, 2nd edn. Hove, UK: Psychology Press.

Penn, D.C. & Povinelli, D.J. (2007). On the lack of evidence that non-humans possess anything remotely resembling a Theory of Mind. *Philosophical Transactions of the Royal Society B.* **362**: 731–44.

Penn, D.J. (2002). The scent of genetic compatibility: sexual selection and the major histocompatibility complex. *Ethology* **108**: 1–21.

Pepperberg, I.M. (2002). In search of King Solomon's ring: cognitive and communicative studies of grey parrots (*Psittacus erithacus*). *Brain Behavior and Evolution* **59**: 54–67.

Pepperberg, I.M. & Gordon, J.D. (2005). Number comprehension by a grey parrot (*Psittacus erithacus*), including a zero-like concept. *Journal of Comparative Psychology* **119**: 1–13.

Perry, E.A. & Stenson, G.B. (1992). Observations of nursing behaviour of hooded seals, *Cystophora cristata*. *Behaviour* **122**: 1–10.

Phillips, J.B., Borland, S.C., Freake, M.J., Brassart, J. & Kirschvink, J.L. (2002). 'Fixed-axis' magnetic orientation by an amphibian: non-shoreward-directed compass orientation, misdirected homing or positioning magnetic-based map detector in a consistent aligment relative to the magnetic field? *Journal of Experimental Biology* **205**: 3903–14.

Piersma, T., van Aelst, R., Kurk, K., Berrkhoudt, H. & Mass, L.R.M. (1998). A new pressure sensory mechanism for prey detection in birds: the use of principles of seabed dynamics. *Proceedings of the Royal Society B* **265**: 1377–83.

Pihlström, H., Fortelius, M., Hemila, S., Forsman, R. & Reuter, T. (2005). Scaling of mammalian ethmoid bones can predict olfactory organ size and performance. *Proceedings of the Royal Society B* **272**: 957–96.

Pinxthen, R., Hanook, O., Eens, M. *et al.* (1993). Extra-pair paternity and intraspecific brood parasitism in the European starlings *Sturnus vulgaris*: evidence from DNA fingerprinting. *Animal Behaviour* **45**: 795–809.

Plomin, R. & Haworth, C.M.A. (2009). Genetics of high cognitive ability. *Behavior Genetics* **39**: 347–9.

Plomin, R., DeFries, J.C., McClearn, G.E. & McGuffin, P. (2000). *Behavioral Genetics: A Primer*, 4th edn. New York: W.H. Freeman.

Plotnik, J.M., de Waal, F.B.M., & Reiss, D. (2006). Self-recognition in an Asian elephant. *Proceedings of the National Academy of Science* **103**: 17053–7.

Poole, J. (1997). *Elephants*. Grantown-on-Spey, Scotland: Colin Baxter Photography Ltd.

Poole, J.H., Payne, K., Langbauer, W.H. & Moss, C.J. (1988). The social contexts of some very low frequency calls of African elephants. *Behavioral Ecology and Sociobiology* **22**: 385–92.

Potts, W.K., Manning, C.J. & Wakeland, E.K. (1991). Mating patterns in seminatural populations of mice influenced by MHC genotype. *Nature* **352**: 619–21.

Povinelli, D.J., Nelson, K.E. & Boysen, S.T. (1992). Comprehension of role reversal in chimpanzees: evidence of empathy? *Animal Behaviour* **43**: 633–40.

Pradhan, G.R. & van Schaik, C. (2008). Infanticide-driven intersexual conflict over matings in primates and its effects on social organisation. *Behaviour* **145**: 251–75.

Premack, D. & Woodruff, G. (1978). Does the chimpanzee have a theory of mind? *Behavior and Brain Science* **1**: 515–26.

Proctor, H. (1991). Courtship behaviour in the water mite *Neumania papillator*. *Animal Behaviour* **42**: 589–98.

Provine, R.R. (1976). Eclosion and hatching in cockroach first instar larvae: a stereotyped pattern of behaviour. *Journal of Insect Physiology* **22**: 127–31.

Pulido, F., Berthold, P. & van Noordwijk, A.J. (1996). Frequency of migrants and migratory activity are genetically correlated in a bird population: evolutionary implications. *Proceedings National Academy of Sciences* **93**:14642–7.

Pusey, A. (1990). Behavioural changes at adolescence in chimpanzees. *Behaviour* **115**: 203–46.

Pusey, A. & Wolf, M. (1996). Inbreeding avoidance in animals. *Trends in Ecology and Evolution* **11**: 201–6.

Pusey, A.E. & Packer, C. (1987). The evolution of sex-biased dispersal in lions. *Behaviour* **101**: 275–310.

Queller, D.C., Strassman, J.E. & Hughes, C.R. (1993). Microsatellites and kinship. *Trends in Ecology and Evolution* **8**(8): 285–8.

Rangel, J. & Seeley, T.D. (2008). The signals initiating the mass exodus of a honeybee swarm from the nest. *Animal Behaviour* **76**: 1943–52.

Rasa, A. (1987). The dwarf mongoose: a study of behaviour and social structure in relation to ecology in a small social carnivore. *Advances in the Study of Behavior* **17**: 121–63.

Reiss, D. & Marino, L. (2001). Mirror self-recognition in the bottlenose dolphin: a case of cognitive convergence. *Proceedings of the National Academy of Sciences* **98**: 5937–42.

Rescorla R.A. (1988). Pavlovian conditioning: it's not what you think it is. *American Psychologist* **43**: 151–60.

Ridley, M. (1986). The number of males in a primate troop. *Animal Behaviour* **34**: 1848–58.

Ridley, M. (1993). *The Red Queen. Sex and the Evolution of Human Nature.* London: Penguin Books.

Riedman, M. (1990). *The Pinnipeds: Seals, Sea Lions and Walruses.* Berkeley, CA: University of California Press.

Riley, J.R., Greggers, U., Smith, A.D., Reynolds, D.R. & Menzel, R. (2005). The flight paths of honeybees recruited by the waggle dance. *Nature* **435**: 205–7.

Ristau, C.A. (ed.) (1991). *Cognitive Ethology: The Minds of Other Animals.* Hillsdale, NJ: Lawrence Erlbaum Associates.

Rizzolatti, G. & Craighero, L. (2004). The mirror-neuron system. *Annual Review of Neuroscience* **7**:169–92.

Roberts, G. (2005). Cooperation through interdependence. *Animal Behaviour* **70**: 901–6.

Rodd, F.H., Hughes, K.A., Grether, G.P. & Baril, C.T. (2002). A possible non-sexual origin of mate preference: are male guppies mimicking fruit? *Proceedings of the Royal Society of London B* **269**: 475–81.

Roeder, K.D. (1967). *Nerve Cells and Insect Behavior.* Cambridge, MA: Harvard University Press.

Roessingh, P., Simpson, S.J. & James, S. (1993). Analysis of phase-related changes in behaviour of desert locust nymphs. *Proceedings of the Royal Society of London B* **252**: 43–9.

Rodgers, C.T. & Hore, P.J. (2009). Chemical magnetoreception in birds: the radical pair mechanism. *Proccedings of the National Academy of Science* **106**: 353–60.

Rolls, B.J. & Rolls, E.T. (1982). *Thirst.* Cambridge: Cambridge University Press.

Rolls, E.T. (1994). Brain mechanisms for invariant visual recognition and learning. *Behavioural Processes* **33**: 113–38.

Rolls, E.T. (2005). *Emotion Explained.* Oxford: Oxford University Press.

Rolls, E.T. (2008). *Memory, Attention and Decision-making.* Oxford: Oxford University Press.

Rolls, E.T., Grabenhorst, L., Margot, C., da Silva, M.A.A.P. & Velazco, M.I. (2008). Selective attention to affective value alters how the brain processes olfactory stimuli. *Journal of Cognitive Neuroscience* **20**: 1816–26.

Rolls, E.T., Rolls, B.T. & Rowe, E.A. (1983). Sensory-specific satiety and motivation-specific satiety for the sight and taste of food and water in man. *Physiology and Behavior* **30**: 185–92.

Romer, A.S. (1962). *The Vertebrate Body.* Philadelphia: W.B. Saunders.

Roper, T.J. (1983). Learning as a biological phenomenon. In *Animal Behavior, Vol 3. Genes, Development and Learning*, ed. P.J.B. Slater and T.R. Halliday. Oxford: Blackwell Scientific Publications, pp. 178–212.

Ros, A.F.H., Correla, M., Wingfield, J.C. & Oliveira, R.F. (2009). Mounting an immune response correlates with decreased androgen levels in male peafowl, *Pavo cristatus*. *Journal of Ethology* **27**: 209–14.

Rosenthal, G.G. & Evans, C.S. (1998). Female preference for swords in *Xiphophorus helleri* refects a bias for large apparent size. *Proceedings of the National Academy of Science* **95**: 4431–6.

Rosenzweig, M.R. (1984). Experience, memory and the brain. *American Psychologist* **39**: 65–76.

Rossel, S. & Wehner, R. (1986). Polarization vision in bees. *Nature* **323**: 128–31.

Roth, G. (2001). The evolution of consciousness. In *Brain Evolution and Cognition*, ed. G. Roth and M.F. Wulliman. New York: Wiley, pp. 555–82.

Roth, L.M. (1948). An experimental laboratory study of the sexual behaviour of *Aëdes aegypti* (L.). *American Midland Naturalist* **40**: 265–352.

Rowland, W.J. (1989). Mate choice and the supernormality effect in female sticklebacks (*Gasterosteus aculeatus*). *Behavioral Ecology and Sociobiology* **24**: 433–8.

Rowland, W.J. (2000). Habitutation and devlopment of response specificity to a sign stimulus: male preference for female courtship posture in stickleback. *Animal Behaviour* **60**: 63–8.

Rowley, I. & Chapman, G. (1986). Cross-fostering, imprinting and learning in two sympatric species of cockatoos. *Behaviour* **96**: 1–16.

Rozin, P. (1976). The selection of food by rats, humans and other animals. *Advances in the Study of Behavior* **6**: 21–76.

Rozin, P. & Kalat, J.W. (1971). Specific hungers and poison avoidance as adaptive specializations of learning. *Psychological Review* **78**: 459–86.

Rueppell, O., Pankiw, T., Nielsen et al. (2004). The genetic architecture of the behavioral ontogeny of for aging in honeybee workers. *Genetics* **167**: 1767–79.

Russell, A.F., Youing, A.J., Spong, G., Jordan, N.R. & Clutton-Brock, T.H. (2007). Helpers increase the reproductive potential of offspring in cooperative meerkats. *Proceedings of the Royal Society B*, **274**: 513–20.

Rutter, M. (1991). A fresh look at 'maternal deprivation'. In *The Development and Integration of Behaviour: Essays in Honour of Robert Hinde*, ed. P. Bateson. New York: Cambridge University Press.

Rutter, M. (2002). Nature, nurture, and development: from evangelism through science towards policy and practice. *Child Development* **73**: 1–21.

Ryan, M.J. (1998). Sexual selelction, receiver biases and the evolution of sex differences. *Science* **281**: 1999–2003.

Ryan, M.J. & Wilczynski, W. (1991). Evolution of intra-specific variation in the advertisement call of the cricket frog (*Acris crepitans*, Hylidae). *Biological Journal of the Linnean Society* **44**: 249–71.

Ryan, M.J., Tuttle, M.D. & Rand, A.S. (1982). Bat predation and sexual advertisement in a neotropical anuran. *American Naturalist* **119**: 136–9.

Ryan, M.J., Tuttle, M.D. & Taft, L.K. (1981). The costs and benefits of frog chorusing behavior. *Behavioral Ecology and Sociobiology* **8**: 273–8.

Sachser, N. & Kaiser, S. (1996). Prenatal social stress masculinizes the females' behavior in guinea-pigs. *Physiological Behaviour* **60**: 589–94.

Sacks, O. (1986). *The Man Who Mistook his Wife for a Hat*. London: Picador.

Savage-Rumbaugh, E.S., Williams, S.L., Furuichi, T. & Kano, T. (1996). Language perceived: *Paniscus* branches out. In *Great Ape Societies*, ed. W.C. McGrew, L.F. Marchant and T. Nishida. Cambridge: Cambridge University Press, pp. 173–84.

Schaller, G. (1972). *The Serengeti Lion: a Study of Predator Prey Relations*. Chicago: University of Chicago Press.

Scheih, H., Langner, G., Tidemann, C., Coles, R.B. & Guppy, A. (1986). Electroreception and electrolocation in platypus. *Nature* **319**: 401–2.

Scheller, R.H. & Axel, R. (1984). How genes control an innate behavior. *Scientific American* **250**(3): 44–52.

Schjelderup-Ebbe, T. (1935). Social behaviour of birds. In *Handbook of Social Psychology*, ed. C. Murchison. Worcester, MA: Clark University Press, pp. 947–72.

Schleidt, W.M., Schleidt, M. & Magg, M. (1960). Störungen der Mutter-Kind-Beziehung bei Trüthuhnern durch Gehörverlust. *Behaviour* **16**: 254–60.

Schwabl, H., Mock, D.W. & Gieg, J.A. (1997). A hormonal mechanism for parental favouritism. *Nature*, **386**: 231.

Scott, J.P. & Fuller, J.L. (1965). *Genetics and the Social Behavior of the Dog*. Chicago: University of Chicago Press.

Searcy, W.A. & Brenowitz, E.A. (1988). Sexual differences in species recognition of avian song. *Nature* **332**: 152–4.

Searcy, W.A. & Marler, P. (1981). A test for responsiveness to song structure and programming in female sparrows. *Science* **213**: 926–8.

Searcy, W.A., Marler, P. & Peters, S.S. (1981). Species song discrimination in adult female song and swamp sparrows. *Animal Behaviour* **29**: 997–1003.

Seed, A.M., Emery, N.J. & Clayton, N.S. (2009). Intelligence in corvids and apes: a case of convergent evolution? *Ethology* **115**: 401–20.

Seeley, T.D. (1982). How honeybees find a home. *Scientific American* **247**: 158–68.

Seeley, T.D. (1992). The tremble dance of the honey bee: message and meanings. *Behavioral Ecology and Sociobiology* **31**(6): 375–83.

Seeley, T.D. & Visscher, P.K. (1988). Assessing the benefits of cooperation in honeybee foraging: search costs, forage quality, and competitive ability. *Trends in Ecology and Evolution* **22**: 229–37.

Seeley, T.D. & Visscher, P.K. (2004). Quorum sensing during nest-site selection by honeybee swarms. *Behavioral Ecology and Sociobiology* **56**: 594–601.

Seeley, T.D. & Visscher, P.K. (2008). Sensory coding of nest-site value in honeybee swarms. *Journal of Experimental Biology* **211**: 3691–7.

Seligman, M.E.P. & Hager, J.L. (eds.) (1972). *Biological Boundaries of Learning*. Englewood Cliffs, NJ: Prentice-Hall.

Sevenster, P. (1961). A causal analysis of a displacement activity (fanning in *Gasterosteus aculeatus* L.). *Behaviour (Suppl.)* **9**: 1–170.

Sevenster-Bol, A.C.A. (1962). On the causation of drive reduction after a consummatory act (in *Gasterosteus aculeatus* L.). *Archives neérlandaises de Zoologie* **15**: 175–236.

Seyfarth, R.M. & Cheyney, D.L. (1984). Grooming, alliances and reciprocal altruism in vervet monkeys. *Nature*, **308**: 541–3.

Seyfarth, R.M. & Cheney, D.L. (1986). Vocal development in vervet monkeys. *Animal Behaviour* **34**: 1640–58.

Seyfarth, R.M., Cheney, D.L. & Marler, P. (1980a). Vervet monkey alarm calls: evidence of predator classification and semantic communication. *Science* **210**: 801–3.

Seyfarth, R.M., Cheney, D.L. & Marler, P. (1980b). Vervet monkey alarm calls – semantic communication in a free-ranging primate. *Animal Behaviour* **28**: 1070–94.

Shapiro, D.Y. (1979). Social behavior, group structure and the control of sex reversal in hermaphroditic fish. *Advances in the Study of Behavior* **10**: 43–102.

Sharpe, L.L. (2005a). Play fighting does not affect subsequent fighting success in wild meerkats. *Animal Behaviour* **69**:1023–9.

Sharpe, L.L. (2005b). Play does not enhance social cohesion in a cooperative mammal. *Animal Behaviour* **70**: 551–8.

Sharpe, L.L. (2005c). Frequency of social play does not affect dispersal partnerships in wild meerkats. *Animal Behaviour* **70**: 559–69.

Sherman, G. & Visscher, P.K. (2002). Honeybee colonies achieve fitness through dancing. *Nature*, **419**: 920–2.

Sherrington, C.S. (1906). *The Integrative Action of the Nervous System*. New York: Scribner's.

Sherrington, C.S. (1917). Reflexes elicitable in the cat from pinna, vibrissae and jaws. *Journal of Physiology* **51**: 404–31.

Sherry, D. (1985). Food storage by birds and mammals. *Advances in the Study of Behavior* **15**: 153–88.

Sherry, D.F., Mrosovsky, N. & Hogan, J.A. (1980). Weight loss and anorexia during incubation in birds. *Journal of Comparative and Physiological Psychology* **94**: 89–98.

Shettleworth, S.J. (1983). Memory in food-hoarding birds. *Scientific American* **248**(3): 102–10.

Shettleworth, S.J. (2003). Memory and hippocampal specialization in food-storing birds: challenges for research on comparative cognition. *Brain Behavior and Evolution* **62**: 108–116.

Shettleworth, S.J. (2009). *Cognition, Evolution and Behavior*, 2nd edn. Oxford University Press. New York.

Short, R.V. & Balaban, E. (eds.) (1994). *The Differences Between the Sexes*. Cambridge: Cambridge University Press.

Silk, J.B. (2007a). The strategic dynamics of cooperation in primate groups. *Advances in the Study of Behaviour* **37**: 1–41.

Silk, J.B. (2007b). The adaptive value of sociality in mammalian groups. *Philosophical Transactions of the Royal Society B* **362**: 539–59.

Simmons, J.A. & Stein, R.A. (1980). Acoustic imaging in bat sonar: echolocating signals and the evolution of echolocation. *Journal of Comparative Physiology A* **135**: 61–84.

Simpson, S.J., McCaffery, A.R. & Hagele, B.F. (1999). A behavioural analysis of phase change in the desert locust. *Biological Reviews of the Cambridge Philosophical Society* **74**: 461–80.

Singleton, I. & van Schaik, C.P. (2002). The social organisation of a population of Sumatran orang-utans. *Folia Primatologica* **73**: 1–20.

Skinner, B.F. (1938). *The Behavior of Organisms*. New York: Appleton-Century-Crofts.

Slater, P.J.B. & Ince, S.A. (1979). Cultural evolution in chaffinch song. *Behaviour* **71**: 146–66.

Smuts, B.B., Cheney, D.L., Seyfarth, R.M., Wrangham, R.W. & Struhsaker., T.T. (eds.) (1987). *Primate Societies*. Chicago: University of Chicago Press.

Sokolowski, M.B. (2001). *Drosophila*: genetics meets behaviour. *Nature Review of Genetics* **2**: 879–90.

Spalding, D. (1873). Instinct: with original observations on young animals. *Macmillan's Magazine* **27**: 282–93. (Reprinted in *British Journal of Animal Behaviour* **2**: 1–11, 1954.)

Sparks, J. (1982). *The Discovery of Animal Behaviour*. London: Collins, BBC.

Speakman, J.R. (2008). Thrifty genes for obesity and diabetes, an attractive but flawed idea and an alternative scenario: the 'drifty gene' hypothesis. *International Journal of Obesity* **32**: 1611–7.

Spinka, M., Newberry, R.C. & Bekoff, M. (2001). Mammalian play: training for the unexpected. *Quarterly Review of Biology* **76**: 141–68.

Stander, P.E. (1992). Co-operative hunting in lions: the role of the individual. *Behavioral Ecology and Sociobiology* **29**: 445–54.

Stent, G.S. (1980). The genetic approach to developmental neurobiology. *Trends in the Neurosciences* **3**: 49–51.

Stent, G. (1981). Genetics and the development of the nervous system. *Annual Review of Neuroscience* **4**: 163–94.

Stephens, D.W. & Krebs, J.R. (1986). *Foraging Theory*. Princeton, NJ: Princeton University Press.

Stevens, M. (2005). The role of eyespots as anti-predator mechanisms, principally demonstrated in the Lepidoptera. *Biological Reviews* **80**: 573–88.

Stevens, M. & Cuthill, I.C. (2007). Hidden messages: are ultraviolet signals a special channel in avian communication? *BioScience* **57**: 501–7.

Stoehr, A.M. & Kokko, H. (2006). Sexual dimorphism in immunocompetence: what does life history theory predict? *Behavioural Ecology* **17**: 751–6.

Strier, K.B. (2006). *Primate Behavioral Ecology*, 3rd edn. Boston, MA: Allyn & Bacon.

Stringer, C. (2011). *The Origin of Our Species*. London: Allen Lane.

Stringer, C. & Andrews, P. (2005). *The Complete World of Human Evolution*. London: Thames & Hudson.

Struhsaker, T.T. & Leland, L. (1987). Colobines: infanticide by adult males. In *Primate Societies*, ed. B.B. Smuts, D.L. Cheney, R.M. Seyfarth, R.W. Wrangham and T.T. Struhsaker. Chicago: University of Chicago Press, pp. 38–97.

Sutherland, L., Holbrook, R. & De Perera Burt, T. (2009). Sensory system affects orientational strategy in a short-range spatial task in blind and eyed morphs of the fish *Astyanax fasciatus*. *Ethology* **115**: 504–15.

Swaddle, J.P., McBride, L. & Malhotra, S. (2006). Female zebra finches prefer unfamiliar males but not when watching non-interactive video. *Animal Behaviour* **72**: 161–7.

Teitelbaum, P. (1955). Sensory control of hypothalamic hyperphagia. *Journal of Comparative and Physiological Psychology* **48**: 156–63.

Teitelbaum, P. & Epstein, A.N. (1984). The lateral hypothalamic syndrome: recovery of feeding and drinking after lateral hypothalamic lesions. *Psychological Review* **69**: 74–90.

ten Cate, C., Los, L. & Schilperood, L. (1984). The influence of differences in social experience on the development of species recognition in zebra finch males. *Animal Behaviour* **32**: 852–60.

ter Hofstede, H.M., Ratcliffe, J.M. & Fullard, J.H. (2008). Nocturnal activity positively correlated with auditory sensitivity in noctuid moths. *Biology Letters* **4**: 262–265.

Thornhill, R., Gangestad, S.W., Miller, R., *et al.* (2003). Major histocompatibility complex genes, symmetry and body scent attractiveness in men and women. *Behavioural Ecology* **14**: 668–678.

Thorpe, W.H. (1961). *Bird Song*. Cambridge: Cambridge University Press.

Thorpe, W.H. (1963). *Learning and Instinct in Animals*, 2nd edn. London: Methuen.

Tiefer, L. (1978). The context and consequences of contemporary sex research: a feminist perspective. In *Sex and Behavior: Status and Prospectus*, ed. T.E. McGill, D.A. Dewsbury and B.D. Sachs. New York: Plenum, pp. 363–85.

Tinbergen, N. (1951). *The Study of Instinct*. Oxford: Oxford University Press.

Tinbergen, N. (1952). 'Derived' activities, their causation, biological significance, origin and emancipation during evolution. *Quarterly Review of Biology* **27**: 1–32.

Tinbergen, N. (1959). Comparative studies of the behaviour of gulls (Laridae): a progress report. *Behaviour* **15**: 1–70.

Tinbergen, N. (1963). On the aims and methods of ethology. *Zeitschrift für Tierpsychologie* **20**: 410–33.

Tinbergen, N. & Perdeck, A.C. (1950). On the stimulus situation releasing the begging response in the newly-hatched herring gull chick (*Larus a. argentatus* Pont). *Behaviour* **3**: 1–38.

Tinbergen, N., Broekhuysen, G.J., Feekes, F., Houghton, J.C., Kruuk, H. & Szuk, E. (1962). Egg-shell removal by the black-headed gull *Larus ridibundus* L.; a behavioural component of camouflage. *Behaviour* **19**: 74–117.

Toates, F. (1986). *Motivational Systems*. Cambridge: Cambridge University Press.

Toates, F. (1995). *Stress: Conceptual and Biological Aspects*. Chichester: John Wiley.

Tomer, R., Denes, A.S., Tessmar-Raible, K. & Arendt, D. (2010). Profiling by image registration reveals common origin of annelid mushroom bodies and vertebrate pallium. *Cell* **142**: 800–809.

Trainor, B.C. & Basolo, A.L. (2000). An evaluation of video playback using *Xiphophorus helleri*. *Animal Behaviour* **59**: 8–89.

Trainor, B.C. & Basolo, A.L. (2006). Location, location, location: stripe position effects on female sword preference. *Animal Behaviour* **71**: 35–140.

Tregenza, T. & Weddell, N. (2002). Polyandrous females avoid costs of inbreeding. *Nature* **415**: 71–3.

Trillmich, F. (1996). Parental investment in pinnipeds. *Advances in the Study of Behavior* **25**: 533–77.

Trivers, R.I. (1974). Parent–offspring conflict. *American Zoologist* **14**: 249–64.

Truman, J.W., Thorn, R.S. & Robinow, S. (1992). Programmed neuronal death in insect development. *Journal of Neurobiology* **23**: 1295–311.

Tyndale-Biscoe, C.H. (1973). *The Life of Marsupials*. London: Edward Arnold.

Vallin, A., Jakobson, S., Lind, J. & Wiklund, C. (2005). Prey survival by predator intimidation: an experimental study of peacock butterfly defence against blue tits. *Proceedings of the Royal Society B* **272**: 1203–207.

van Rhijn, J.G. (1980). Communication by agonistic displays: a discussion. *Behaviour* **74**: 284–93.

van chaik, C.P. (1983). Why are diurnal primates living in groups? *Behaviour* **87**: 120–44.

van Schaik, C.P. & van Hoof, J.A.R.A.M. (1996). Toward an understanding of the orangutan's social system. In *Great Ape Societies*, ed. W.C. McGrew, L.F. Marchant and T. Nishida. Cambridge: Cambridge University Press, pp. 3–15.

Venkatamaran, A. (1998). Male-biased sex ratios and their significance for cooperative breeding in dhole (*Cuon alpinus*) packs. *Ethology* **104**: 671–84.

Vestergaard, K. (1980). The regulation of dustbathing and other patterns in the laying hen: a Lorenzian approach. In *The Laying Hen and its Environment*, ed. R. Moss. The Hague: Martinus Nijhoff, pp. 101–20.

Vince, M.A. (1969). Embryonic communication, respiration and the synchronization of hatching. In *Bird Vocalizations*, ed. R.A. Hinde. Oxford: Oxford University Press, pp. 233–60.

Vince, M.A. (1993). Newborn lambs and their dams: the interaction that leads to suckling. *Advances in the Study of Behavior* **22**: 239–68.

Vines, G. (1981). Wolves in dog's clothing. *New Scientist* **91**: 648–52.

vom Saal, F.S. & Bronson, F. (1980). Sexual characteristics of adult females correlate with their blood testosterone levels during development in mice. *Science* **208**: 597–9.

von Frisch, K. (1967). *The Dance Language and Orientation of Bees*. Cambridge, MA: The Belknap Press of Harvard University Press.

Vowles, D.M. (1965). Maze learning and visual discrimination in the wood ant (*Formica rufa*). *British Journal of Psychology* **56**: 15–31.

Waage, J.K. (1979). Dual function of the damselfly penis: sperm removal and transfer. *Science* **203**: 916–18.

Walker, S. (1983). *Animal Thought*. London: Routledge & Kegan Paul.

Walsh, E.G. (1964). *Physiology of the Nervous System*, 2nd edn. London: Longman.

Ward, C., Bauer, E.B. & Smuts, B.B. (2008). Partner preferences and asymmetries in social play among domestic dog, *Canis lupus familiaris*, littermates. *Animal Behaviour* **76**: 1187–99.

Warner, R.R. (1987). Female choice of sites versus mates in a coral reef fish *Thalassoma bifasciatum*. *Animal Behaviour* **41**: 375–82.

Warren, J.M. (1965). Primate learning in comparative perspective. In *Behaviour of Non-human Primates*, Vol. 1, ed. A.M. Schrier, H.F. Harlow and F. Stollnitz. New York: Academic Press, pp. 249–81.

Warren, J.M. (1973). Learning in vertebrates. In *Comparative Psychology: A Modern Survey*, ed. D.A. Dewsbury and D.A. Rethlingshafer. New York: Academic Press, pp. 471–509.

Watson, J.B. (1924). *Behaviorism*. Chicago: University of Chicago Press.

Webster, J. (2005). *Animal Welfare: Limping Towards Eden*. Oxford: Blackwell Publishing.

Wedekind, C., Seebeck, T., Bettens, F. & Paephe, A.J. (1995). MHC-dependent mate preferences in humans. *Proceedings of the Royal Society of London B* **260**: 245–59.

Weilgart, L., Whitehead, H. & Payne, K. (1996). A colossal convergence. *American Scientist* **84**: 278–87.

Wells, M.J. (1962). Early learning in *Sepia*. *Symposia of the Zoological Society of London* **8**: 149–69.

Wenner, A.M. (2002). The elusive honey bee dance 'language' hypothesis. *Journal of Insect Behaviour* **15**: 859–78.

West, M.J. & King, A.P. (1988). Female visual displays affect the development of male song in the cowbird. *Nature* **334**: 244–6.

West, S.A., Griffin, A.S. & Gardner, A. (2007). Social semantics, altruism, cooperation, mutualism, strong reciprocity and group selection. *Journal of Evolutionary Biology* **20**: 415–432.

Whiten, A. & Byrne, R.W. (1997). (eds.) *Machiavellian Intelligence II. Extensions and Evaluations*. Cambridge: Cambridge University Press.

Whiten, A. (in press). Primate social learning, traditions and culture. In *The Evolution of Primate Societies*, ed. J. Mitani, J. Call, P. Kappeler, R. Palombit and J. Silk. Chicago: University of Chicago Press, pp. xxx–xxx.

Wier, A.A.S, Chappell, J. & Kacelnik, A. (2002). The shaping of hooks by New Caledonian crows. *Science* **297**: 981.

Wier, A.A.S. & Kacelnik, A. (2006). A New Caledonian crow (*Corvus moneduloides*) creatively re-designs tools by bending or unbending aluminium strips. *Animal Cognition* **9**: 317–34.

Wiersma, C.A.G. (1947). Giant nerve fibre system of the crayfish. *Journal of Neurophysiology* **10**: 23–38.

Wilcox, R.S. (1979). Sex discrimination in *Gerris remigris*: role of a surface wave signal. *Science* **206**: 1325.

Wiley, R.H. (1973a). The strut display of male sage grouse: a 'fixed' action pattern. *Behaviour* **47**: 129–52.

Wiley, R.H. (1973b). Territoriality and non-random mating in sage grouse, *Centrocercus urophasianus*. *Animal Behaviour Monographs* **6**(2): 87–129.

Wilkinson, G.S. (1984). Reciprocal food sharing in the vampire bat. *Nature* **308**: 181–4.

Wilson, E.O. (1965). Chemical communication in the social insects. *Science* **149**: 1064–71.

Wilson, E.O. (1971). *The Insect Societies*. Cambridge, Massachusetts: Belknap Press of Harvard University Press.

Wilson, E.O. (1975). *Sociobiology*. Cambridge, Massachusetts: Belknap Press of Harvard University Press.

Wilson, J. (1992). A re-assessment of the signficance of status signaling in populations of wild great tits, *Parus major*. *Animal Behaviour* **43**: 999–1009.

Wilson, M.L., Hauser, M.D. & Wrangham, R.W. (2001). Does participation in intergroup conflict depend on numerical assessment, range location, or rank for wild chimpanzees? *Animal Behaviour* **61**:1203–16.

Wiltschko, W. & Wiltschko, R. (2001). Light-dependent magnetoreception in birds: the behaviour of European Robins, *Erithecus rubecula*, under monochromatic light of various wavelengths and intensities. *Journal of Experimental Biology* **204**: 3295–302.

Wingfield, J.C., Hegner, R.E., Duffy, A.M., Jr & Ball, G.F (1990). The 'Challenge hypothesis': theoretical implications for patterns of testosterone secretion, mating systems, and breeding strategies. *American Naturalist* **136**: 829–46.

Winkler, D.W. & Sheldon, F.W. (1993). Evolution of nest construction in swallows (Hirundinidae): A molecular phylogenetic perspective. *Proceedings of the National Academy of Sciences* **90**: 5705–7.

Winslow, J.T., Hastings, N., Carter, C.S., Harbaugh, C.R., & Insel, T.R. (1993). A role for central vasopressin in pair bonding in monogamous prairie voles. *Nature* **365**: 545–548

Winn, P. (1995). The lateral hypothalamus and motivated behavior: an old syndrome reassessed and a new perspective gained. *Current Directions in Psychological Science* **4**: 1182–7.

Winn, P., Tarbuck, A. & Dunnett, S.B. (1984). Ibotenic acid lesions of the lateral hypothalamus: comparisons with the electrolytic lesion syndrome. *Neuroscience* **12**: 225–40.

Wirtshafter, D. & Davis, J.D. (1977). Set points, settling points and the control of body weight. *Physiology and Behaviour* **19**: 75–8.

Woolfenden, G.E. & Fitzpatrick, J.W. (1984). *The Florida Scrub Jay*. Princeton, NJ: Princeton University Press.

Wrangham, R.W. (1987). Evolution of social structure. In *Primate Societies*, ed. B.B. Smuts, D.L. Cheney, R.M. Seyfarth, R.W. Wrangham and T.T. Struhsaker. Chicago: University of Chicago Press, pp. 282–96.

Wyatt, T.D. (2003). *Pheromones and Animal Behaviour. Communication by Smell and Taste.* Cambridge: Cambridge University Press.

Wyatt, T.D. (2010). Pheromones and signature mixtures: defining species-wide signals and variable cues for identity in both invertebrates and vertebrates. *Journal of Comparative Physiology A.* **196**: 685–700.

Ydenberg, R. & Dill, L.M. (1986). The economics of fleeing from predators. *Advances in the Study of Behavior* **16**: 229.

Ydenberg, R. & Houston, A.I. (1986). Optimal trade-offs between competing behavioural demands in the great tit. *Animal Behaviour* **34**: 1041–50.

Zahavi, A. (1975). Mate selection: a selection for a handicap. *Journal of Theoretical Biology* **53**: 205–14.

Zahavi, A. (1991). On the definition of sexual selection, Fisher's model, and the evolution of waste and of signals in general. *Animal Behaviour* **42**: 501–3.

Zarrow, M.X., Denberg, V.H. & Anderson, C.O. (1965). Rabbit: frequency of suckling in the pup. *Science* **150**: 1835–6.

Zeki, S. (1993). *A Vision of the Brain*. Oxford: Blackwell Scientific Publications.

Zuk, M. (1994). Immunology and the evolution of behavior. In *Behavioral Mechanisms in Evolutionary Ecology*, ed. L.A. Real. Chicago: University of Chicago Press, pp. 354–68.

FIGURE CREDITS

 ## Chapter 1

1.1   Photograph by M.S. Dawkins
1.2   © iStockphoto.com/Marcus Jones
1.3   From Camhi 1984
1.4   Drawing by Nigel Mann
1.5   Modified from Groothuis 1993
1.6   After Walsh 1964
1.9   Modified from Sherrington 1906
1.10 Modified from Sherrington 1906
1.11 From Hinde 1954
1.12 Modified from McFarland 1974
1.13 Modified from Sherrington 1906

 ## Chapter 2

2.1   Drawings by Nigel Mann
2.2   Drawing by Nigel Mann
2.3   (a) © iStockphoto.com/Steve Smith; (b) © iStockphoto.com/Eric Isselée

2.4   From Morris 1952
2.5   (a) © The Trustees of the British Museum; (b) Photograph by Alan and Sandy Carey. © Photolibrary.com
2.6   Photograph by Steen Drozd Lund. © Photolibrary.com
2.7   Photograph by Karen McComb
2.8   After Tinbergen 1959
2.9   Photograph by Kristin Mosher. © Photolibrary.com
2.10   Photograph by Graeme Chapman; from Rowley and Chapman 1986
2.11   Modified from Gottlieb 1976
2.12   From den Hartog *et al.* 2007
2.13   From Dilger 1962
2.14   Photographs by H. C. Bennet-Clark
2.15   Modified from Blakemore and Cooper 1970
2.16   Drawn from photographs in Bentley and Hoy 1970
2.17   Redrawn from Levine 1986
2.18   Photograph by Richard Ling
2.19   Modified from Catchpole and Slater 2008
2.20   (a) Photograph by Mark MacEwan. © Photolibrary.com; (b) Photograph by Mark Deeble & Victoria Stone. © Photolibrary.com
2.21   (a) Modified from Tyndale-Biscoe 1973; (b) © Shutterstock.com
2.22   Photograph by Fred Bruemmer. © Photolibrary.com
2.23   Modified from Hinde 1974
2.24   (a) © Photolibrary.com; (b) © Photolibrary.com; (c) © iStockphoto.com/Eric Gevaert; (d) © iStockphoto.com/John Carnemolla; (e) © iStockphoto.com/suemack; (f) (g) and (h) © shutterstock.com; (i) Photograph by Pete Veilleux; (j) from Bekoff and Byers 1998
2.25   (a) Erbenge meinschaft Lorenz ; (b) Photo by Nina Leen/Time Life Pictures/Getty Images
2.26   Modified from Hinde 1974
2.27   Photograph by M. L. Ryder
2.28   Reproduced with permission from the International Crane Foundation
2.29   Photograph by C. ten Cate
2.30   Modified from Bateson 1983
2.32   From Ince *et al.* 1980; Thorpe 1961; i photograph © iStockphoto.com/Andrew Hewe
2.33   Photo © iStockphoto.com/Richard Rodvold; Illustration modified from Marler and Tamura 1964
2.34   Modified from Catchpole 1980
2.36   Modified from Catchpole and Slater 2008; photograph © iStockphoto.com/Anna Kravchuk

 ## Chapter 3

3.1   Modified from Tinbergen 1951
3.2   © Cordelia Molloy/Science Photo Library

Chapter 4

Chapter 5

5.1	Modified from Munn 1950
5.2	From Clark 1960a. Photograph by Dino Simeonidis. © Photolibrary.com
5.3	Used with permission from the Alan Mason Chesney Medical Archives of The Johns Hopkins Medical Institutions
5.4	Illustration by Simon Tegg
5.5	(a) © shutterstock.com; (b) © iStockphoto.com/Terry Lawrence; (c) Photo by Tim Flach/Stone/Getty Images; (d) Photograph by Brian Kenney. © Photolibrary.com
5.6	Modified from Maier and Schneirla 1935
5.7	Drawn from photographs in Moore 1973
5.8	Modified from Konorski 1948
5.9	Modified from Menzel and Erber 1978
5.10	Modified from Menzel *et al.* 1993
5.11	Drawing by Nigel Mann
5.12	Modified from Hollard and Delius 1982
5.13	From Köhler 1927
5.14	Modified from Mackintosh *et al.* 1985
5.15	Modified from Warren 1965
5.16	Redrawn from Whiten, *in press*
5.17	Photograph by W. McGrew
5.18	Photograph by Ch. Boesch
5.19	Reproduced with permission from *Nature*
5.20	Photograph from the Mary Evans Picture Library
5.21	Redrawn from Hunt 1996
5.22	Photographs by Chris Bird from Bird and Emery 2009 a and b
5.23	Photograph by William Munoz
5.24	Reproduced with permission from the Great Ape Trust
5.25	Photograph by Sue Savage-Rumbaugh
5.26	Photographs reproduced with permission from W. McGrew
5.27	Photograph by Frans De Waal
5.28	Modified from Povinelli *et al.* 1992

 ## Chapter 6

6.1	(a) Photograph by Henry Bennet-Clark
6.2	(a) From Tinbergen *et al* 1962; (b) Photo by Nina Leen/Time & Life Pictures/Getty Images
6.3	Photograph John Manning
6.4	(a) © shutterstock.com

6.5 © shutterstock.com

6.6 From the Boy's Own Paper, date unknown

6.7 Modified from Cade 1981

6.8 (a and b) modified from Berthold and Querner 1981; drawing by Nigel Mann

6.9 Modified from Barthold and Hellbig 1992

6.10 © iStockphoto.com/Anton Harder

6.11 (a) Photograph by Raymond Mendez. © Photolibrary.com; (b) drawing by Priscilla Barrett from a photograph by David Curl

6.12 Drawing by Priscilla Barrett from Wilson 1975

6.13 Modified from Russell *et al.* 2007

6.14 Photograph by Doug Mock

6.17 © iStockphoto.com/Debra Feinman

6.18 Photograph © iStockphoto.com/Gaspare Messina

6.19 Photograph by Oxford Scientific. © Photolibrary.com

6.20 (a); (b) © Anne Frijsinger and Mat Vestjens

6.21 Modified from Bschary and Schäffer 2002

6.22 (a) Dr Morley Read/Science Photo Library; (b) Photograph by Geoff Higgins. © Photolibrary.com; (c) © shutterstock.com

6.23 Drawing by Nigel Mann

6.24 Modified from Catchpole *et al.* 1984; from Catchpole and Slater 2008

6.25 Photograph by the National Oceanic and Atmospheric Administration

6.26 Modified from Searcy & Marlow 1981 and Searcy *et al.* 1981

6.27 © shutterstock.com

6.28 From Huxley 1914

6.29 (a) © iStockphoto.com/Vladimir Gorsky; (b) © iStockphoto.com/Karel Broz; (c) © iStockphoto.com/indykb

Chapter 7

7.1 Photograph by Richard Packwood. © Photolibrary.com

7.2 Photograph by Accent Alaska

7.3 Modified from Göttmark and Andersson 1984

7.4 Photograph by Ashleigh Griffin

7.5 © Todd Gustafson

7.6 Drawing by Nigel Mann

7.7 Modified from Lindauer 1961, reproduced with permission from Harvard University Press © President and Fellows of Harvard College

7.8 Photograph by Steve Garvie

7.9 Photograph by Bryan Nelson

7.10 Photograph by Mark Hamblin. © Photolibrary.com

INDEX